Recent Developments in Bayesian Econometrics and Their Applications

Stepan Mazur • Pär Österholm
Editors

Recent Developments in Bayesian Econometrics and Their Applications

Festschrift in Honour of Sune Karlsson

Springer

Editors

Stepan Mazur
School of Business
Örebro University
Örebro, Sweden

Pär Österholm
School of Business
Örebro University
Örebro, Sweden

ISBN 978-3-032-00109-2 ISBN 978-3-032-00110-8 (eBook)
https://doi.org/10.1007/978-3-032-00110-8

This Springer imprint is published by the registered company Springer Nature Switzerland AG
The registered company address is: Gewerbestrasse 11, 6330 Cham, Switzerland

To Sune

Preface

This book is a tribute to Sune Karlsson. In conjunction with his 65th birthday, we celebrate his contributions to the field of Bayesian econometrics with this Festschrift. New contributions in his main field of research—that is, time series econometrics—are presented. These consist of both methodological contributions and empirical applications in macroeconomics and finance. Overall, the book consists of 10 chapters written by experts in the field of Bayesian econometrics, each offering advanced research and insights.

It would not be right to characterize Sune solely by his academic and scientific achievements. While successful researchers naturally attract attention, much of Sune's popularity among colleagues and friends comes from qualities beyond his research. Working with Sune is truly a privilege; he combines competence, humility, friendliness, and efficiency, creating a collaborative environment that is both productive and enjoyable.

When launching a project like this, one never knows how it will work out in the end. In this case, it turned out to be quite an enjoyable project to manage. We would like to express our sincere thanks to all the authors of the chapters for their contributions and commitment to this Festschrift.

We hope this book finds a special place on Sune's shelf, serving as a testament to his contributions and the lasting impact he has had on the field of Bayesian econometrics.

Örebro, Sweden
February, 2025

Stepan Mazur
Pär Österholm

Contents

Contributors

Olha Bodnar School of Business, Örebro University, Örebro, Sweden

Taras Bodnar Department of Management and Engineering, Linköping University, Linköping, Sweden

Joshua C. C. Chan Economics Department, Purdue University, West Lafayette, IN, USA

Jamie L. Cross Melbourne Business School, University of Melbourne, Parkville, VIC, Australia

Oskar Gustafsson Department of Statistics, Stockholm University, Stockholm, Sweden

Chenghan Hou Hunan University, Hunan, China

Tamás Kiss School of Business, Örebro University, Örebro, Sweden

Robert Kohn School of Economics, University of New South Wales, Kensington, NSW, Australia

Gary Koop University of Strathclyde, Glasgow, UK
ESCoE, London, UK

Stepan Mazur School of Business, Örebro University, Örebro, Sweden

Stuart McIntyre University of Strathclyde, Glasgow, UK
ESCoE, London, UK

James Mitchell Federal Reserve Bank of Cleveland, Cleveland, OH, USA
ESCoE, London, UK

Hoang Nguyen Department of Management and Engineering, Linköping University, Linköping, Sweden

Vilhelm Niklasson Department of Mathematics, Stockholm University, Stockholm, Sweden

Oscar Oelrich Department of Statistics, Stockholm University, Stockholm, Sweden

Pär Österholm School of Business, Örebro University, Örebro, Sweden
National Institute of Economic Research, Stockholm, Sweden

Aubrey Poon University of Kent, Canterbury, UK
ESCoE, London, UK

Yaling Qi Economics Department, Purdue University, West Lafayette, IN, USA

Emily Tallman Department of Statistical Science, Duke University, Durham, NC, USA

Herman K. van Dijk Tinbergen Institute, Erasmus University Rotterdam, Rotterdam, Netherlands
Norges Bank, Oslo, Norway

Mattias Villani Department of Statistics, Stockholm University, Stockholm, Sweden

Mike West Department of Statistical Science, Duke University, Durham, NC, USA

Ping Wu University of Strathclyde, Glasgow, UK
ESCoE, London, UK

Introduction

Stepan Mazur and Pär Österholm

Abstract This chapter gives a brief introduction to the Festschrift. A short overview of Sune Karlsson's career and research is provided. In addition, the content of each chapter is presented.

Substantial progress has been made in the field of Bayesian econometrics over the last four decades. One important part of the explanation for this development can be found in the significant advances that we have seen regarding computational power as both hardware and software have improved rapidly. With computational resources becoming broadly available, researchers have been able to approach areas that need to be addressed using numerical methods—something that is often the case in Bayesian econometrics.

This book presents new contributions to this expanding field. Reflecting the research conducted by Sune throughout his career, the focus is on time series econometrics and applications thereof in macroeconomics and finance. Particular attention will be paid to Bayesian vector autoregressive (VAR) models and forecasting—two areas where Sune has made notable contributions.

Sune joined the Bayesian movement early. After finishing his undergraduate degree in statistics at Uppsala University in 1985, he began his PhD studies in economics at Purdue University the same year. It was during these studies that he started working on Bayesian VAR models—a field pioneered shortly before by Litterman (1980, 1986) and Doan et al. (1984). This resulted in a thesis on this topic in 1989 (Karlsson, 1989). Having returned to Sweden in 1989 for a position at the Stockholm School of Economics, he later published two contributions on Bayesian

S. Mazur (✉)
School of Business, Örebro University, Örebro, Sweden
e-mail: stepan.mazur@oru.se

P. Österholm
School of Business, Örebro University, Örebro, Sweden

National Institute of Economic Research, Stockholm, Sweden
e-mail: par.osterholm@oru.se

S. Mazur, P. Österholm (eds.), *Recent Developments in Bayesian Econometrics and Their Applications*, https://doi.org/10.1007/978-3-032-00110-8_1

VAR models together with his former supervisor, Rao Kadiyala. In Kadiyala and Karlsson (1993), the effect on forecasting performance of employing priors that previously had not been used in Bayesian VARs was assessed, and Kadiyala and Karlsson (1997)—a much-cited paper in the literature on Bayesian VARs—studied the implementation of different Monte Carlo methods when evaluating the posterior distribution.

Later on, Sune broadened his research agenda. He, for example, did work on panel data models (in a non-Bayesian setting). This includes Karlsson and Löthgren (2000), who studied the power and interpretation of panel unit root tests, and Andersson and Karlsson (2001), who proposed and evaluated resampling procedures for panel data models. Bayesian econometrics was still a part of his research agenda though, such as his work on forecast combination (Jacobson & Karlsson, 2004; Eklund & Karlsson, 2007) and structural breaks and lag length (Hultblad & Karlsson, 2008). In 2013, he also returned to Bayesian VARs as he wrote the chapter "Forecasting with Bayesian Vector Autoregression" for the *Handbook of Economic Forecasting*.

In 2005, Sune became professor of statistics at the School of Business at Örebro University. His interest in Bayesian econometrics has made a significant impact on the research environment there. When he first arrived, he was the only Bayesian at the School of Business. Over time, there has become an increasing number of colleagues that conduct research in Bayesian econometrics or employ such methods for their empirical analysis. A substantial share of the research output in statistics and economics now relies on Bayesian analysis. A fair bit of this research is coauthored work and Sune is frequently one of the coauthors. This is not a coincidence; speaking out of personal experience, we note that it is a privilege to work with someone who combines competence, humility, friendliness, and efficiency in an excellent way. Sune also shares his knowledge in nonacademic fora. He is a long-standing member of both Statistics Sweden's Scientific Council and Statistics Sweden's CPI Board; since 2019, he is also a member of the European Statistical Advisory Committee.

In more recent years, some of Sune's research has been more empirically orientated, aiming to answer economic questions. One area that has been given a fair amount of attention has been on whether macroeconomic relationships have been stable over time or not. This includes the Phillips curve (Karlsson & Österholm, 2020b, 2023), Okun's law (Karlsson & Österholm, 2020a), and the relation between money and inflation (Berger et al., 2023; Edvinsson et al., 2025). All of the studies on stability have been conducted in a Bayesian VAR framework with drifting parameters and stochastic volatility. An important aspect has been that of model selection, where the methods developed by Chan and Eisenstat (2018a,b) have been relied upon. But while more empirical work has dominated recently, Sune has also conducted methodological work. For example, Karlsson et al. (2023) present a Bayesian VAR with stochastic volatility where the error terms are drawn from distributions which can have both heavy tails and skewness.

Given the importance of Bayesian VARs and forecasting in shaping Sune's research agenda throughout his career, these two topics are the focus of the next chapter of this Festschrift. In this chapter, Jamie Cross and Herman van Dijk provide an overview in terms of a bibliometric analysis of papers within the Web of Science that are devoted to forecasting with Bayesian VARs. The analysis assesses the evolution of the number of publications over time, establishes the leading topics, articles and authors, and by whom and where the research is conducted. In addition to the bibliometric analysis, the authors also provide a reflection on how they anticipate that the increasingly important field of machine learning will affect that of Bayesian VARs. It can sadly be noted that this chapter is one of Herman van Dijk's last contributions to academic literature. After a long and productive career, he passed away in January 2025.

This is followed by a number of chapters using Bayesian VAR frameworks. In the chapter "Large Bayesian Tensor VARs with Stochastic Volatility", Joshua Chan and Yaling Qi work with high-dimensional Bayesian VARs. Following the work of Wang et al. (2022), they address the challenges that such models pose in terms of heavy parameterization using a tensor decomposition. Chan and Qi extend the framework though by allowing for heteroskedastic disturbances (in terms of stochastic volatility)—a feature that a growing literature has identified as an important aspect when modeling macroeconomic time series; see, for example, Cogley and Sargent (2005), Clark (2011), and Karlsson and Österholm (2023). The forecasting performance of various specifications of the Bayesian tensor VAR is assessed in an application employing 40 US macroeconomic time series. It is concluded that the Bayesian tensor VAR outperforms a more traditionally specified Bayesian VAR with a Minnesota prior.

With the goal of providing quarterly historical estimates of subregional output growth in the United Kingdom, Gary Koop, Stuart McIntyre, James Mitchell, Aubrey Poon, and Ping Wu develop a mixed-frequency Bayesian VAR in the chapter "Measuring Subregional Economic Activity: Missing Frequencies and Missing Data". The model uses national, regional, and subregional data and accordingly has a large number of equations; in addition to having to address the issue of high dimensionality, the authors also—in contrast to related research, such as Schorfheide and Song (2015)—have to take cross-sectional restrictions into account (since subregional output must add up to regional output and regional output national output). The historical data produced are finally used to study various aspects of subregional business cycle dynamics.

Bayesian VARs with stochastic volatility is again the topic in the chapter "VAR Models with Fat Tails and Dynamic Asymmetry". Here Tamás Kiss, Stepan Mazur, Hoang Nguyen, and Pär Österholm develop a model which allows for disturbances that have both fat tails and dynamic skewness. In doing this, they extend the work of Karlsson et al. (2023) who had a similar setting but where the skewness was constant over time. Using data on US industrial production, consumer prices, and economic policy uncertainty, they conduct analysis both within and out of sample. The results indicate some moderate support for time-varying skewness and that shocks to economic policy uncertainty have a dampening effect on both industrial production growth and CPI inflation.

The chapter "International Transmission of Macroeconomic Uncertainty in Small Open Economies: An Empirical Approach" concludes the contributions relying on Bayesian VARs through a study of the transmission of domestic and international sources of macroeconomic uncertainty shocks for three commonly analyzed inflation-targeting small open economies: Australia, Canada, and New Zealand. For that analysis, Jamie Cross, Chenghan Hou, and Aubrey Poon offer a VAR model with common stochastic volatility in the mean that extends the common stochastic volatility VAR model developed by Carriero et al. (2016). The results show that both domestic and foreign sources of macroeconomic uncertainty shape the economic climate in each of the small open economies.

In the chapter "Modeling Local Predictive Ability Using Power-Transformed Gaussian Processes", Oscar Oelrich and Mattias Villani present work on forecast combination. They add to the literature on this topic—introduced by Bates and Granger (1969) and recently surveyed by Wang et al. (2023)—through a contribution in which they focus on the accuracy of an expert's forecast density. More specifically, Oelrich and Villani propose a power-transformed Gaussian process for modeling local predictive ability. The proposed model demonstrates strong performance in predicting bike-sharing data (Fanaee-T & Gama, 2014), outperforming almost all comparison models. The only exception is the local version of the linear prediction pool introduced by Geweke and Amisano (2011) and further developed in Oelrich et al. (2024).

The chapter "Spectral Domain Likelihoods for Bayesian Inference in Time-Varying Parameter Models" focuses on analysis of time-varying parameter models in the frequency domain. In this chapter, Oskar Gustafsson, Mattias Villani, and Robert Kohn explore the finite-sample performance of posterior distributions obtained from local Whittle likelihoods for time-varying parameter models through several simulation experiments. This can be contrasted with the majority of related previous studies which have focused on the frequentist properties of the Whittle maximum likelihood estimator; see, for example, Sykulski et al. (2019) and Subba Rao and Yang (2021). The block Whittle likelihood in Dahlhaus (1997) and the dynamic Whittle likelihood in Tang et al. (2023) are compared. The results show that both the dynamic and block Whittle likelihoods demonstrate good performance in all experiments, with the dynamic Whittle having an edge.

The final two chapters deal with portfolio analysis in a Bayesian framework. The chapter "Bayesian Regularization of the Tangency Portfolio" explores the tangency portfolio, which allows investment in a risk-free asset in addition to the risky assets; see, for example, Karlsson et al. (2021). Here, Olha Bodnar, Taras Bodnar, and Vilhelm Niklasson extend the results of Bodnar et al. (2024, 2025), introducing regularization techniques for tangency portfolio selection and providing a more efficient method to address parameter uncertainty problems in practice. In particular, two new priors—based on the Laplace prior and the spike-and-slab prior—are developed for the tangency portfolio weights. Furthermore, under both priors, the posterior distributions of the tangency portfolio weights are derived in terms of stochastic representations.

The chapter "Predictive Decision Synthesis for Portfolios: Betting on Better Models" finalizes the book by addressing portfolio decision goals when multiple forecasting models are to be explored and combined. In this chapter, Emily Tallman and Mike West discuss and develop Bayesian dynamic modeling and predictive decision synthesis for portfolio analysis. More specifically, Tallman and West develop Bayesian predictive decision synthesis (BPDS), introduced in Tallman and West (2023), in the context of financial forecasting for portfolio decisions. The case study of BPDS, focusing on financial forecasting of international exchange rate time series and portfolio rebalancing, demonstrates its potential to improve the trade-off between risk and return. Moreover, it offers additional methodology for understanding relevant score functions and setting target scores in portfolio management.

References

Andersson, M. K. & Karlsson, S. (2001). Bootstrapping error component models. *Computational Statistics, 16*, 221–231.

Bates, J. M. & Granger, C.W. J. (1969). The combination of forecasts. *Journal of the Operations Research Society, 20*(4), 451–468.

Berger, H., Karlsson, S., & Österholm, P. (2023). A note of caution on the relation between money growth and inflation. *Scottish Journal of Political Economy, 70*, 479–496.

Bodnar, O., Bodnar, T., & Niklasson, V. (2024). Constructing Bayesian tangency portfolios under short-selling restrictions. *Finance Research Letters, 62*, 105065.

Bodnar, O., Bodnar, T., & Niklasson, V. (2025). Incorporating different sources of information for Bayesian optimal portfolio selection. *Journal of Business and Economic Statistics, 43*(2), 365–377.

Carriero, A., Clark, T. E., & Marcellino, M. (2016). Common drifting volatility in large Bayesian VARs. *Journal of Business and Economic Statistics, 34*(3), 375–390.

Chan, J. C. C. & Eisenstat, E. (2018a). Bayesian model comparison for time-varying parameter VARs with stochastic volatility. *Journal of Applied Econometrics, 33*, 509–532.

Chan, J. C. C. & Eisenstat, E., (2018b). Comparing hybrid time-varying parameter VARs. *Economics Letters, 171*, 1–5.

Clark, T. E. (2011). Real-time density forecasts from VARs with stochastic volatility. *Journal of Business and Economic Statistics, 29*, 327–341.

Cogley, T. & Sargent, T. J. (2005). Drifts and volatilities: Monetary policies and outcomes in the post WWII US. *Review of Economic Dynamics, 8*, 262–302.

Dahlhaus, R. (1997). Fitting time series models to nonstationary processes. *Annals of Statistics, 25*, 1–37.

Doan, T., Litterman, R., & Sims, C. (1984). Forecasting and conditional projection using realistic prior distributions. *Econometric Reviews, 3(1)*, 1–100.

Edvinsson, R., Karlsson, S., & Österholm, P. (2025). Does money growth predict inflation in Sweden? Evidence from vector autoregressions using four centuries of data. *Empirical Economics, 68*, 1613–1635.

Eklund, J. & Karlsson, S. (2007). Forecast combination and model averaging using predictive measures. *Econometric Reviews, 26*, 329–363.

Fanaee-T, H. & Gama, J. (2014). Event labeling combining ensemble detectors and background knowledge. *Progress in Artificial Intelligence, 2*, 113–127.

Geweke, J., & Amisano, G. (2011). Optimal prediction pools. *Journal of Econometrics, 164*(1), 130–141.

Hultblad, B. & Karlsson, S. (2008). Bayesian simultaneous determination of structural breaks and lag lengths. *Studies in Nonlinear Dynamics and Econometrics, 12*, Article 4.

Jacobson, T. & Karlsson, S. (2004). Finding good predictors for inflation: A Bayesian model averaging approach. *Journal of Forecasting, 23*, 479–496.

Kadiyala, R. K. & Karlsson, S. (1993). Forecasting with generalized Bayesian vector autoregressions. *Journal of Forecasting, 12*, 365–378.

Kadiyala, R. K. & Karlsson, S. (1997). Numerical methods for estimation and inference in Bayesian VAR-models. *Journal of Applied Econometrics, 12*, 99–132.

Karlsson, S. (1989). *Bayesian Analysis of Vector Autoregressions*. PhD Thesis, Purdue University.

Karlsson, S. (2013). Forecasting with Bayesian vector autoregression. In: G. Elliott & A. Timmermann (Eds.), *Handbook of Economic Forecasting* (Vol. 2, pp. 791–897). Elsevier.

Karlsson, S. & Löthgren, M. (2000). On the power and interpretation of panel unit root tests. *Economics Letters, 66*, 249–255.

Karlsson, S., Mazur, S., & Muhinyuza, S. (2021). Statistical inference for the tangency portfolio in high dimension. *Statistics, 55*, 532–560.

Karlsson, S., Nguyen, H., & Mazur, S. (2023). Vector autoregression models with skewness and heavy tails. *Journal of Economic Dynamics and Control, 146*, 104580.

Karlsson, S. & Österholm, P. (2020a). A hybrid time-varying parameter Bayesian VAR analysis of Okun's law in the United States. *Economics Letters, 197*, 109622.

Karlsson, S. & Österholm, P. (2020b). A note on the stability of the Swedish Phillips curve. *Empirical Economics, 59*, 2573–2612.

Karlsson, S. & Österholm, P. (2023). Is the US Phillips curve stable? Evidence from Bayesian VARs. *Scandinavian Journal of Economics, 125*, 287–314.

Litterman, R. B. (1980). *A Bayesian Procedure for Forecasting with Vector Autoregressions*. Working Paper, Massachusetts Institute of Technology.

Litterman, R. B. (1986). Forecasting with Bayesian vector autoregressions — Five years of experience. *Journal of Business and Economic Statistics, 4*, 25–38.

Oelrich, O., Villani, M., & Ankargren, S. (2024). Local prediction pools. *Journal of Forecasting, 43*, 103–117.

Schorfheide, F. & Song, D. (2015). Real-time forecasting with a mixed-frequency VAR. *Journal of Business and Economic Statistics, 33*, 366–380.

Subba Rao, S. & Yang, J. (2021). Reconciling the Gaussian and Whittle likelihood with an application to estimation in the frequency domain. *Annals of Statistics, 49*, 2774–2802.

Sykulski, A. M., Olhede, S. C., Guillaumin, A. P., Lilly, J. M., & Early, J. J. (2019). The debiased Whittle likelihood. *Biometrika, 106*, 251–266.

Tallman, E. & West, M. (2023). Bayesian predictive decision synthesis. *Journal of the Royal Statistical Society (Ser B), 86*, 340–363.

Tang, Y., Kirch, C., Lee, J. E., & Meyer, R. (2023). *Bayesian Nonparametric Spectral Analysis of Locally Stationary Processes*. arXiv preprint arXiv:2303.11561.

Wang, X., Hyndman, R. J., Li, F., & Kang, Y. (2023). Forecast combinations: An over 50-year review. *International Journal of Forecasting, 39*, 1518–1547.

Wang, D., Zheng, Y., Lian, H., & Li, G. (2022). High-dimensional vector autoregressive time series modeling via tensor decomposition. *Journal of the American Statistical Association, 117*, 1338–1356.

Forecasting with Bayesian Vector Autoregressions Revisited

Jamie L. Cross and Herman K. van Dijk

Abstract Given the excellent contributions from Sune Karlsson on forecasting with Bayesian vector autoregression (BVAR) models, we reflect here on the major contributions and developments in that literature since its inception in the 1980s. Methodologically, we conduct a bibliometric analysis of 270 research articles that are contained within the Web of Science database. Results across four key areas of the literature are presented: publication trends, leading research topics, leading articles and authors, and social structure of research. Throughout the article, we also provide some more personal insights into Karlsson's contributions and impact on the field, and to Bayesian VAR models more generally.

1 Introduction

[The] role of the VAR model has to some degree been taken over by the current crop of DSGE models, a new generation of theory-based models, which are—at times—ill at ease with the data. The role of the VAR model as the baseline, serious, model for economic forecasting is, however, unchallenged. The popularity stems in part from its relative simplicity, flexibility, and ability to fit the data but, of course, also from its success as a forecasting device (Karlsson, 2013).

In many ways the above quote from Karlsson (2013) remains true just over a decade later. The old litmus test for a new forecasting model's success was to provide superior point forecast accuracy compared to the random walk model. Nowadays, it is increasingly common to compare the model's point and density forecasts against the Bayesian vector autoregression (BVAR). Scholars with knowledge

J. L. Cross (✉)
Melbourne Business School, University of Melbourne, Parkville, VIC, Australia
e-mail: j.cross@mbs.edu

H. K. van Dijk
Tinbergen Institute, Erasmus University Rotterdam, Rotterdam, Netherlands

Norges Bank, Oslo, Norway

S. Mazur, P. Österholm (eds.), *Recent Developments in Bayesian Econometrics and Their Applications*, https://doi.org/10.1007/978-3-032-00110-8_2

of the BVAR will quickly note that it reduces to a system of random walk equations under the usual Minnesota prior developed by Litterman (1979), Litterman (1980) and popularized through research at the Federal Reserve Bank of Minneapolis (Doan et al., 1984; Litterman, 1986). However, when faced with the data, Bayes theorem allows for strong signals between variables in the data to pull the parameter estimates away from the prior toward the likelihood function. One of the reasons for the aforementioned transition is thus that the BVAR offers the benefits of having the parsimony of the random walk model and the information content of the often over parameterized VAR model. Another reason for the transition is that this combination is not only theoretically elegant, but has also proven to be extremely beneficial for empirical forecasting in economics and finance. For this reason, the use of the BVAR for forecasting purposes extends beyond scholarly investigation, into practical work by central banks, policy institutions, and industry practitioners.

In his PhD thesis (Karlsson, 1989), Karlsson proposed various methods for posterior and predictive inference with BVARs. The main technical contribution of this work was that Karlsson introduced various alternative priors to the Minnesota prior that allow for dependence between the equations of the BVAR model. This includes now commonly used prior specifications like the normal-diffuse prior (Zellner, 1971), and lesser used priors like the extended natural conjugate (ENC) prior (Morales, 1970; Drèze, 1976; Drèze & Richard, 1983). His key insight was to recognize that a VAR can be viewed as a special case of both multivariate regression models and simultaneous equation models. As is often the case in Bayesian work, the main difficulty in implementing these priors for practical analysis was that the posterior distributions become intractable. To overcome this hurdle, Karlsson showed that it is possible to use numerical methods like importance sampling (IS) (Kloek & Van Dijk, 1978) and Markov chain Monte Carlo (MCMC) (Geman & Geman, 1984; Gelfand & Smith, 1990) to evaluate the posterior moments of the BVAR and produce forecasts (see Kadiyala and Karlsson (1993) for the former IS implementation and Kadiyala and Karlsson (1997) for the latter MCMC methods). In terms of practical application, it was in forecasting exercises that BVARs showed their true power (see Edlund & Karlsson, 1993 and Kadiyala & Karlsson, 1993, 1997 for early contributions). Of course, nowadays these insights are often taken for granted, but this was truly the first time that such prior distributions and computational methods had been used in the BVAR setting.

Karlsson (2013) provides an excellent general coverage of the theoretical ideas and computational methods for posterior and predictive inference with BVARs. After introducing readers to general procedures for Bayesian forecasting and inference, he shows how these methods are used in the context of BVARs. He then covers a range of practical topics on how to estimate reduced form and structural BVAR models, account for cointegration, construct conditional forecasts, add additional features like time-varying parameters and stochastic volatility, and estimate large BVARs with many variables.

Given the excellent coverage of methods and algorithms in that chapter, we do not repeat them here. Instead, we aim to provide a broader discussion of the literature that uses BVARs for forecasting in economics and finance. To facilitate the discussion, we conduct a novel bibliometric analysis of the literature. Using this analysis, we present novel insights on publication trends, identify the leading research topics, articles and authors in this field, and detail how the developments of this literature have truly been a global effort. Throughout the analysis, we also add some more personal insights into Karlsson's contributions and impact on the field.

The contents of this chapter are organized as follows: Sect. 2 presents an overview of the bibliometric analysis tools used. Section 3 presents the results. Section 4 discusses machine learning and BVARs. Finally, Sect. 5 concludes.

2 Bibliometric Analysis

To identify the most important practical lessons and methodological developments over the past five decades in the literature, we make use of bibliometric analysis. Such analyses have become increasing popular in the natural and social sciences, but are less common in economics and finance.[1] A practical generalist guide to bibliometric analysis is provided by Donthu et al. (2021).

To create a bibliographic database, we first search the Web of Science for keywords in "All Fields" of an article across three categories:

1. *Bayesian Vector Autoregressions*: "Bayesian vector autoregression," or "Bayesian vector autoregressive," or "Bayesian VAR," or "BVAR"; AND
2. *Forecasting*: "Forecasting," or "Forecast," or "Forecasts"; AND
3. *Economics and Finance*: "Finance" OR "Economic" OR "Macroeconomic" OR "Macroeconomics" OR "Inflation" OR "GDP" OR "Exchange Rate" OR "Interest Rate" OR "Returns" OR "Stock" OR "Unemployment."

This search resulted in 270 published articles with the first paper being (Litterman, 1986). We emphasize that while earlier papers like Litterman (1980) and Doan et al. (1984) exist, they are not contained within the Web of Science database. Nor is the excellent recent contribution of Karlsson et al. (2023). A caveat of our analysis is therefore that it is limited to the articles contained in the Web of Science database at the date of our search, which was September 27, 2024.

As a next step, we exported from Web of Science the "Full Record and Cited References" for all 270 articles and loaded them into the R package "Bibliometrix." Given the aims and scope of our analysis, we conduct both a performance analysis and a scientific mapping of articles and authors in this field. The performance metric used here is "total citations," as reported by Web of Science. This metric then feeds into a network analysis to construct a scientific mapping that identifies the most influential publications in the field, and relationships among publications. Themes and topics that are of particular interest to the field are identified using a co-word analysis of articles in the network.

[1] The only bibliometric analysis that we have seen in these fields is by Baştürk et al. (2014).

3 Main Results

3.1 Increasing Trends in the Number of Publications

Figure 1 provides a timeline of the total number of articles on forecasting with BVARs in economics and finance published annually. The research began with Litterman (1986), but really started to take off after the Global Financial Crisis of 2007–2008. Since 2009, at least five articles on the topic have been published per year, and this number has doubled to at least 10 over the past decade. As of September, 2024, 18 articles have been published on the topic this year.

The maximum, minimum, mean, and median number of citations per article are 721, 0, 21, and 7, respectively. The fact that the mean citation count is greater than the median indicates that the distribution is right skewed. We find that 52 articles have more citations than the mean, with 25 articles having more than 50 citations, and only 10 articles having more than 100 citations.

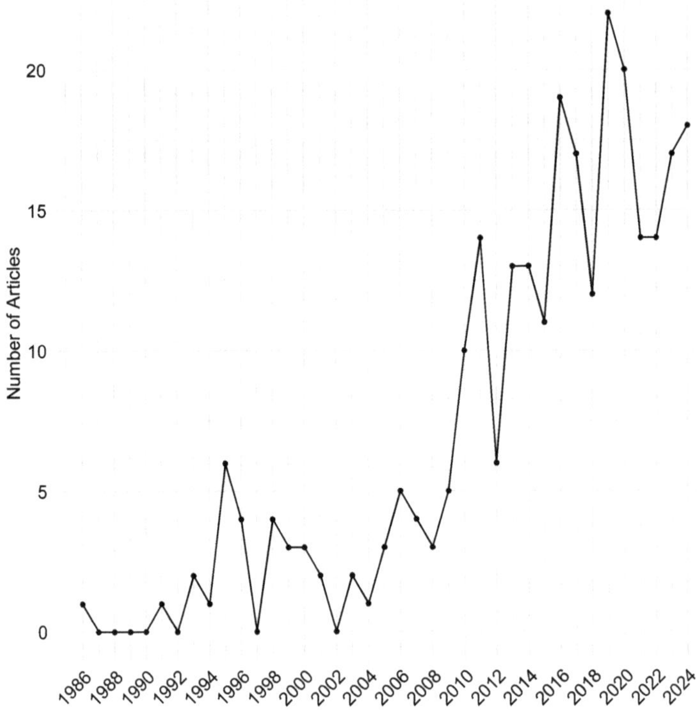

Fig. 1 Annual number of scientific articles published on forecasting with Bayesian vector autoregression from the database of 270 articles in Web of Science

Fig. 2 Word cloud of top 50 most frequent keywords by authors from the database of 270 articles in Web of Science

3.2 *Leading Topics*

Figure 2 provides an illustration of important topics in this field. It shows the top 50 most frequent keywords specified by authors in published papers on forecasting with BVARs in economics and finance contained in the Web of Science database. The relative size of the words is proportional to their frequency, with larger font size indicating more frequent usage, and vice versa.

While forecasting is the main focus, we see that the scope of the literature has expanded to include nowcasting and that the use of real-time data is also common. As mentioned earlier, an important transition in the literature relates to forecast accuracy, where there has been a shift in focus away from solely reporting point forecasts, to also quantifying uncertainty using complete predictive densities. Undoubtedly part of the reason for this change in focus was the pioneering work of policymakers at the Bank of England to communicate their uncertainty about forecasts using the so-called fan charts (see, e.g., Mitchell & Hall, 2005).[2] Another reason is the development of computationally efficient methods for sampling from the posterior of BVAR models under various prior distributions (e.g., Chan, 2020 and references therein), and the large advances in computing technologies over the past few decades (see van Dijk, 2024 for further discussion).

From an empirical perspective, we see that commonly analyzed variables are key macroeconomic indicators: inflation, unemployment, and exchange rates, and from finance, house prices and asset bubbles are commonly studied. It is also relatively common to use BVARs for the forecasting of big data. As mentioned in the introduction, one of the reasons for this is that the hyperparameters of the prior

[2] While the Bank of England was the pioneer when it comes to fan charts, such charts have also been published by other central banks (e.g., Sveriges Riksbank and Norges Bank) and the IMF.

distribution provide a tool for explicit regulation of shrinkage toward a random walk. While this idea was first proposed in Litterman (1980), and built on in Kadiyala and Karlsson (1993, 1997), among others, it was largely popularized by the work of Bańbura et al. (2010) after the financial crisis. While most of the literature focuses on datasets from the USA, we see that it is also common for scholars to examine the suitability of BVARs for forecasting in the Euro Area.

In terms of methodology, we find that the use of shrinkage priors, e.g., the Minnesota prior, and the specification of stochastic volatility and cointegrating relationships are commonly explored. Karlsson (2013) provides an excellent introduction to each of these topics. Another common area of investigation not covered in that chapter is the relative performance of BVARs and DSGE models. For that we refer readers to the handbook chapter of Del Negro and Schorfheide (2013) for a case in favor of DSGE models, and to Gürkaynak et al. (2013) for the case in favor of large-scale BVARs. In our opinion, both DSGE and BVARs provide useful tools for forecasting in macroeconomics. However for forecasting, BVARs are often a more practical choice given their relatively low computational cost, additional flexibility to have features like time-varying parameters, and generally comparable forecast accuracy.

Finally, one of the first articles to tackle the issue of model selection in VAR-based forecasting models is Edlund and Karlsson (1993). While frequentist estimation is used, they find that the VAR model successfully predicts turning points in the unemployment rate. An interesting insight from their empirical analysis is that the VAR's forecasting performance of the unemployment rate is improved by accounting for cointegration relationships between the other variables in the system, despite the fact that the unemployment rate is not included in the cointegrating relations. While cointegration methods have fallen out fashion in recent years, this suggests that it might be worth reigniting the literature by investigating the usefulness of cointegration with modern techniques.

3.3 *Leading Articles and Authors*

Figure 3 illustrates the top ten articles in the field ranked by the highest total number of citations. In this context, we have used the total number of global citations as the metric, which is defined as the number of citations each article has received across the entire Web of Science database. This is in contrast to local citations, which are limited to those provided within the 270 papers in our database. Our use of global citations is intentional, as we aim to identify the most influential articles in the broader scholarly community. We believe that this approach offers a more comprehensive measure of the overall impact and recognition these articles have achieved in the field.

We find that the seminal paper by Litterman (1986) has the most prominent position with 721 citations. This is followed by Sims and Zha (1998), who extend the focus from Bayesian point forecasting to a framework suitable for

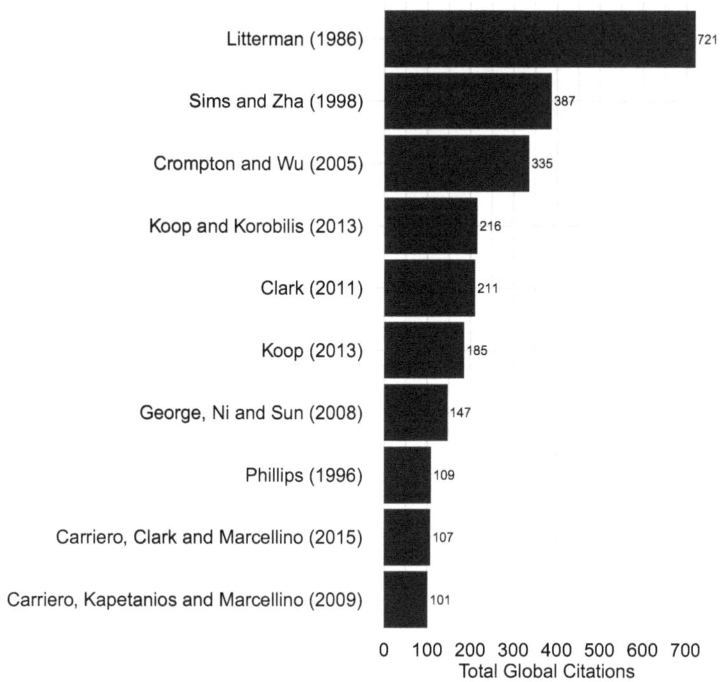

Fig. 3 Top 10 articles by total number of global citations from the database of 270 articles in Web of Science

generating probability assessments of forecasts for policy analysis using VARs. Third is Crompton and Wu (2005), who use BVAR models in an application that forecasts energy consumption in China. In fourth, Koop and Korobilis (2013) develop a novel algorithm for estimation and forecasting in large BVARs with time-varying parameters. Their underlying idea is to use "forgetting factors" to expand and contract the dimension of the model over time. In the same year, Koop (2013)—ranked sixth—demonstrated that the simple Minnesota prior forecasts well in medium and large VARs. In fifth, Clark (2011) showed that specifying stochastic volatility in the BVAR improves the accuracy of real-time forecasts in the USA. In our experience, this model also provides a hard to beat benchmark for macroeconomic forecasting over a decade later. Seventh, George et al. (2008) introduce an alternative prior to the Minnesota prior, known as the stochastic search variable selection (SSVS) prior, that allows for different degrees of regularization across the BVARs coefficients. The basic idea is to specify a two-state Gaussian mixture prior over each coefficient in which in the first distribution the variance is relatively small, and in the second distribution the variance is relatively large. Set in this manner, the SSVS prior clusters coefficients into those that require strong regularization and those that do not. Eighth, Phillips (1996) presents a general framework for Bayesian model determination and demonstrates this using BVARs.

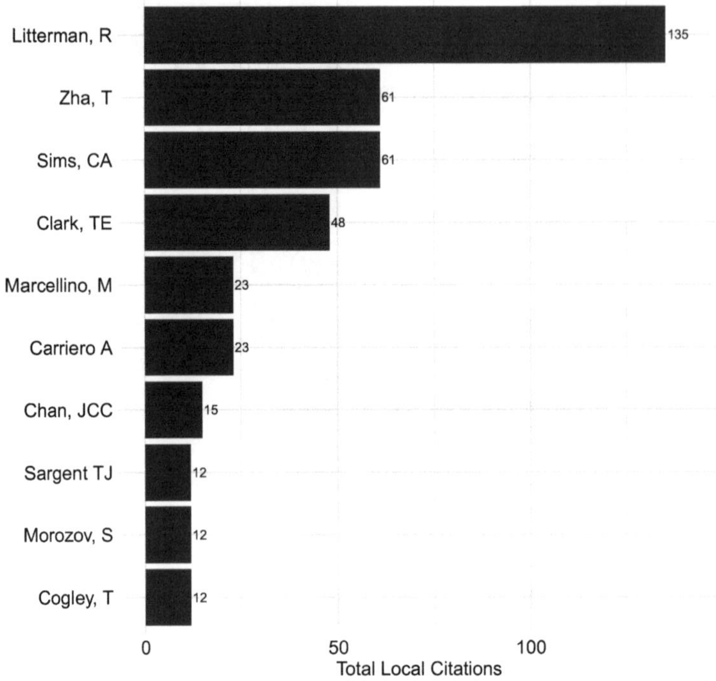

Fig. 4 Top 10 authors by the total number of local citations for articles published on forecasting with Bayesian vector autoregression in economics and finance from the database of 270 articles in Web of Science

Finally, the ninth and tenth articles are Carriero et al. (2015) and Carriero et al. (2009), respectively. The former study considers the forecast accuracy of BVARs under various specification choices, and the latter shows that large BVARs provide better forecasts of the exchange rate than a random walk across 33 currency pairs.

Next, in Fig. 4, we identify the top ten authors ranked by the total number of local citations in the database. The dataset comprises 270 articles authored by 488 individuals, with an average of 2.36 coauthors per article, highlighting the collaborative nature of the field. Additionally, 50 of these articles were single-authored, demonstrating a mix of individual and collaborative contributions to the literature. Since our focus is on assessing the impact of authors specifically within the context of this collection of works, we consider local citations to be a more appropriate measure than global citations, which were previously used for broader impact assessment.

We find that the author with the most local impact is Robert Litterman. This is perhaps unsurprising given his seminal papers in the field (Litterman, 1986). Next we see that Chris Sims and Tao Zha share second-third due to their highly cited coauthored article Sims and Zha (1998). In fourth, fifth, and sixth are Todd Clark, Andrea Carriero, and Massimiliano Marcellino, whom have coauthored many articles in the field. In seventh is Joshua Chan, who has many single-authored articles that develop efficient estimation algorithms for various types of BVARs. In eighth to tenth are Timothy Cogley, Sergei Morozov, and Thomas J. Sargent. The final three scholars are equally cited due to their influential paper on the use of Bayesian fan charts for UK inflation (Cogley et al., 2005). We also here give honorable mention to Gary Koop who collected a large number of global citations in Fig. 3.

3.4 Social Structure

An observation from Fig. 4 is that there is a common clustering of authors in different regions. In Europe, scholars like Koop, Carriero, Marcelino are major contributors to the field, while Sims, Zha, Cogley, Morozov, and Sargent are located in the USA. Chan was originally working in Australia and has also relocated to the USA. Here we investigate the spread of authors in more detail.

A detailed breakdown of the social structure of common groups of coauthors is provided in Fig. 5. It is important to note that we here exclude the 50 of the 270 articles that were single-authored, and examine the remaining 220 papers that had multiple authors.

It is seen that there exists a variety in the number of coauthors, ranging from two, up to four or five, and that the spread takes place over several continents. This emphasizes that research in the area of forecasting with BVARs has become more of a group process (as also occurred in several econometric research topics). We also see that there are six distinct clusters of coauthorship. Notably, the cluster in red is a group of researchers who have historically worked together at Örebro University with Professor Karlsson. While much of their collaborations have been in forecasting with BVARs, they have also used BVARs for analyzing economic questions. For instance, Karlsson and Österholm (2020a) present a hybrid time-varying parameter Bayesian VAR analysis of Okun's law; Karlsson and Österholm (2020b) investigate whether the relation between the corporate bond-yield spread and the real economy has changed over time; Karlsson and Österholm (2023) investigate whether the slope of the US Phillips curve has changed over time; and Edvinsson et al. (2025) use four centuries of data to address the age-old question: "Does money growth predict inflation?".

In addition to this group of Sweden-based researchers, there are other commonly seen groups of coauthorship like: Giannone and Lenza, who have also done much work with Primiceri (e.g., Giannone et al., 2015, 2019); Carriero, Clark and Marcellino who have done much work on VARs with stochastic volatility in mean (e.g., Carriero et al., 2018, 2020), and various other BVAR-related topics; Koop, who has worked on many topics in BVARs both independently and interdependently (see, e.g., Koop, 2013, 2014, and Koop et al., 2019, among others); and Gupta, who has used BVARs in a large quantity of empirical works with various coauthors (e.g.,

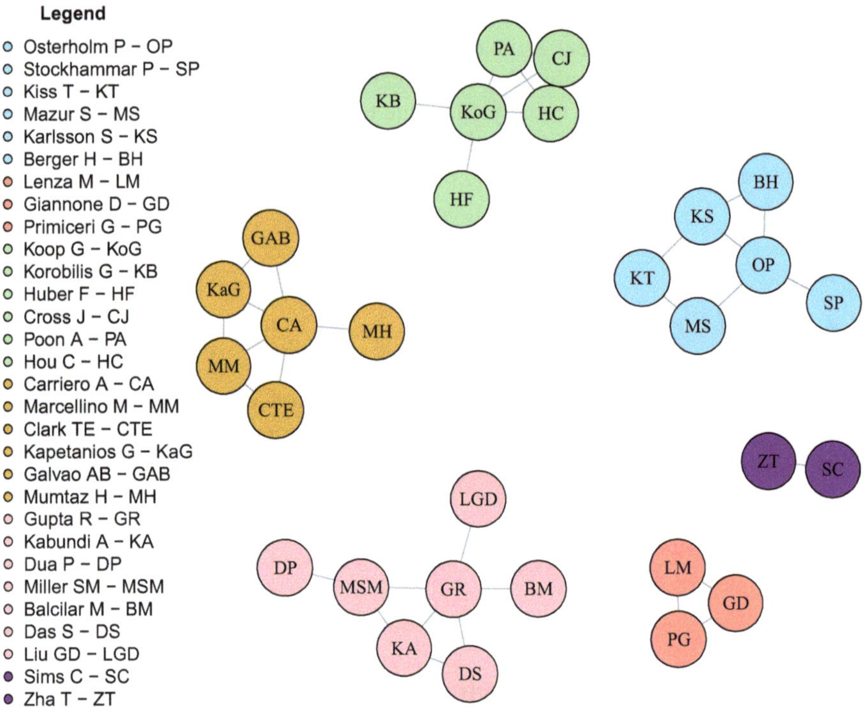

Fig. 5 Social structure of coauthorship for 30 scholars working on forecasting with Bayesian vector autoregression in economics and finance from the database of 220 coauthored articles in Web of Science

Gupta & Kabundi, 2010; Bekiros et al., 2015, and Gupta & Wohar, 2017, among others).

Finally, Fig. 6 demonstrates that research in this field has evolved into a genuinely global effort, encompassing a diverse range of countries and regions. The arrows in the figure highlight coauthorship connections between various countries, indicating a complex network of international collaboration. Contributions to the field have originated from all continents (except Antarctica), with significant coauthorship that crosses continental boundaries, showcasing the extent of global scholarly integration.

While the primary hubs of research activity are centered in the USA and UK, substantial contributions have also emerged from a variety of countries worldwide. Notably, researchers from Australia, China, Germany, Italy, Romania, South Africa, and Sweden have made significant strides, contributing valuable perspectives and insights to the ongoing discourse. This diverse participation underscores the multidisciplinary and multinational nature of the research, reflecting an increasingly interconnected academic community. By bridging geographical distances and fostering collaborative efforts across borders, the research community is advancing the field with a collective and shared global vision.

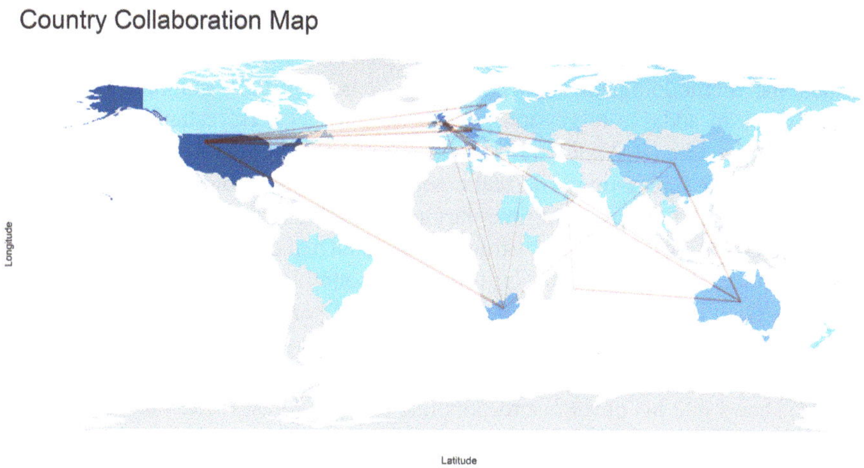

Fig. 6 Global network of scholars working on forecasting with Bayesian vector autoregression from the database of 220 coauthored articles in Web of Science

4 Machine Learning and BVARs

Recent advances in machine learning have opened new avenues for a wide range of applications, including macroeconomic forecasting. Before concluding, we offer a brief reflection on these developments and how we anticipate their intersection with BVARs. Rather than providing a bibliometric analysis, we present a curated selection of empirical and methodological studies in the field, noting that it is intended as a personal perspective rather than an exhaustive review.

While the term machine learning (and artificial intelligence more broadly) has exploded in recent years, early contributions in the literature on macroeconomic forecasting were made more than a decade ago. For instance, Leung et al. (2000) introduce a multilayered feed-forward neural network for forecasting stock indices, and Ahmed et al. (2010) provide a broader introduction of machine learning methods into economic forecasting. The latter paper in particular apply numerous machine learning models to the entire dataset of series used in the M3 time series competition data (around a thousand time series in total). The models are: Bayesian neural networks, CART regression trees, Gaussian processes, generalized regression, K-nearest neighbor regression, multilayer perceptron, neural networks, radial basis functions, and support vector regression. The results for best model based on average rank of accuracy (defined as smallest symmetric mean absolute percentage error (SMAPE)) across each series are split between the multilayer perceptron and the Gaussian process regression. More recently, Smeekes and Wijler (2018) examine the forecast performance of various shrinkage methods: factor, shrinkage, and hybrid models, on a large macroeconomic dataset; and Medeiros et al. (2021) demonstrate the benefits of the random forecast for forecasting inflation over traditional methods like the autoregressive (AR) model, unobserved components model with stochastic volatility (UC-SV), LASSO and ridge regressions, and a homoscedastic BVAR.

On the methodological side, Huber and Rossini (2022) combine the BVAR with Bayesian additive regression tree (BART) to get a Bayesian additive vector autoregressive tree (BAVART) model; Clark et al. (2023) propose the use of BART models for tail forecasting; Hauzenberger et al. (2022) show that BART can be used to model the dynamics of time-varying parameters in a BVAR model, both in terms of conditional mean and variance; and Hauzenberger et al. (2024) develop Bayesian neural networks (BNNs) for macroeconomic forecasting; they find that this new class of BNNs provides superior forecast performance over other models that are commonly used in machine learning like random forests, Bayesian additive regression trees (BART), and also the UC-SV model. Finally, Goulet Coulombe (2024) shows that his novel macroeconomic random forest (MRF) model provides competitive macroeconomics forecasts over a range of models, including ARs with a range of features: factors, time-varying parameters, self-exciting threshold dynamics, and smooth transition dynamics, and regressions with typical shrinkage methods: LASSO and ridge.

A related area of research uses regularization tools from machine learning as priors within the BVAR. Examples include the normal-gamma prior in Huber and Feldkircher (2019) and the horseshoe prior in Follett and Yu (2019). Using such priors, Gefang et al. (2023) develop variational inference methods that overcome the computational hurdle in large VARs with time-varying volatilities. Interestingly, while Cross et al. (2020) show that specifying the Minnesota prior within the BVAR-SV often produces better point and density forecasts compared to machine-learning-based global-local priors, Chan (2021) shows that combining these methods into what he calls "Minnesota-type adaptive hierarchical priors" generally outperforms both adaptive hierarchical priors and Minnesota priors.

Goulet Coulombe et al. (2022) provides evidence that the usefulness of machine learning methods for macroeconomic forecasting can be summarized by the fact that such methods are better able to capture nonlinearities than traditional econometric methods, and thus better capture periods with high macroeconomic uncertainty and financial stress. However, such methods are rarely compared to BVARs. In fact, the only paper we found that uses a BVAR when comparing machine learning forecasts is Medeiros et al. (2021). Given the known success of the BVAR-SV model for macroeconomic forecasting, in future work, it would be interesting to further explore this both in terms of out-of-sample point and density forecast performance, and also in terms of their usefulness as a tool for reliable scenario-based policy analysis. Another area for future research will be to establish reliable methods of constructing density forecasts for many of the ML models, which currently only produce point forecasts. This is an area where the merits of simulation-based Bayesian inference can be further explored.

5 Conclusions

Motivated by Sune Karlsson's work on forecasting with Bayesian vector autoregressions (BVARs), we have undertaken a bibliometric analysis to explore the development and trends within this subject. Our results reveal several key insights: First, there is a strong upward trend in the number of publications per year, indicating growing interest and advancements in this field. Second, the types of BVAR specifications and applications have expanded significantly. Finally, the research has benefited from contributions of numerous authors and collaborations across multiple continents.

In concluding, we reemphasize the pivotal influence of Sune Karlsson's early work on prior distributions and estimation methods (Kadiyala & Karlsson, 1993, 1997), which has truly been foundational to the progression and sophistication of research in BVAR forecasting. His contributions laid the groundwork for subsequent methodological innovations and set the stage for the field's development. His survey chapter (Karlsson, 2013) also provides an excellent introduction to BVARs for new scholars and practitioners entering the field and is a great refresher for those more experienced readers.

Competing Interests The authors have no conflicts of interest to declare that are relevant to the content of this chapter.

Acknowledgments The present paper should not be reported as representing the views of Norges Bank. The views expressed are those of the authors, and do not necessarily reflect those of Norges Bank.

References

Ahmed, N. K., Atiya, A. F., El Gayar, N., & El-Shishiny, H. (2010). An empirical comparison of machine learning models for time series forecasting. *Econometric Reviews, 29*(5–6), 594–621.

Bańbura, M., Giannone, D., & Reichlin, L. (2010). Large Bayesian vector autoregressions. *Journal of Applied Econometrics, 25*(1), 71–92.

Baştürk, N., Çakmaklı, C., Ceyhan, S. P., & van Dijk, H. K. (2014). On the rise of Bayesian econometrics after Cowles Foundation. *Œconomia. History, Methodology, Philosophy, 4–3*, 381–447.

Bekiros, S., Gupta, R., & Paccagnini, A. (2015). Oil price forecastability and economic uncertainty. *Economics Letters, 132*, 125–128.

Carriero, A., Kapetanios, G., & Marcellino, M. (2009). Forecasting exchange rates with a large Bayesian VAR. *International Journal of Forecasting, 25*(2), 400–417.

Carriero, A., Clark, T. E., & Marcellino, M. (2015). Bayesian VARs: Specification choices and forecast accuracy. *Journal of Applied Econometrics, 30*(1), 46–73.

Carriero, A., Clark, T. E., & Marcellino, M. (2018). Measuring uncertainty and its impact on the economy. *Review of Economics and Statistics, 100*(5), 799–815.

Carriero, A., Clark, T. E., & Marcellino, M. (2020). Assessing international commonality in macroeconomic uncertainty and its effects. *Journal of Applied Econometrics, 35*(3), 273–293.

Chan, J. C. C. (2020). Large Bayesian vector autoregressions. In *Macroeconomic forecasting in the era of big data: Theory and practice* (pp. 95–125). Springer.

Chan, J. C. C. (2021). Minnesota-type adaptive hierarchical priors for large Bayesian VARs. *International Journal of Forecasting, 37*(3), 1212–1226.
Clark, T. E. (2011). Real-time density forecasts from Bayesian vector autoregressions with stochastic volatility. *Journal of Business & Economic Statistics, 29*(3), 327–341.
Clark, T. E., Huber, F., Koop, G., Marcellino, M., & Pfarrhofer, M. (2023). Tail forecasting with multivariate Bayesian additive regression trees. *International Economic Review, 64*(3), 979–1022.
Cogley, T., Morozov, S., & Sargent, T. J. (2005). Bayesian fan charts for UK inflation: Forecasting and sources of uncertainty in an evolving monetary system. *Journal of Economic Dynamics and Control, 29*(11), 1893–1925.
Crompton, P., & Wu, Y. (2005). Energy consumption in China: Past trends and future directions. *Energy Economics, 27*(1), 195–208.
Cross, J. L., Hou, C., & Poon, A. (2020). Macroeconomic forecasting with large Bayesian VARs: Global-local priors and the illusion of sparsity. *International Journal of Forecasting, 36*(3), 899–915.
Del Negro, M., & Schorfheide, F. (2013). DSGE model-based forecasting. In *Handbook of economic forecasting* (vol. 2, pp. 57–140). Elsevier.
Doan, T., Litterman, R., & Sims, C. (1984). Forecasting and conditional projection using realistic prior distributions. *Econometric Reviews, 3*(1), 1–100.
Donthu, N., Kumar, S., Mukherjee, D., Pandey, N., & Lim, W. M. (2021). How to conduct a bibliometric analysis: An overview and guidelines. *Journal of Business Research, 133*, 285–296.
Drèze, J. H. (1976). Bayesian limited information analysis of the simultaneous equations model. *Econometrica, 44*(5), 1045–1075.
Drèze, J. H., & Richard, J.-F. (1983). Bayesian analysis of simultaneous equation systems. *Handbook of Econometrics, 1*, 517–598.
Edlund, P.-O., & Karlsson, S. (1993). Forecasting the Swedish unemployment rate: VAR vs. transfer function modelling. *International Journal of Forecasting, 9*(1), 61–76.
Edvinsson, R., Karlsson, S., & Österholm, P. (2025). Does money growth predict inflation? Evidence from vector autoregressions using four centuries of data. *Empirical Economics, 68*, 1613–1635.
Follett, L., & Yu, C. (2019). Achieving parsimony in Bayesian vector autoregressions with the horseshoe prior. *Econometrics and Statistics, 11*, 130–144.
Gürkaynak, R. S., Kısacıkoglu, B., & Rossi, B. (2013). Do DSGE models forecast more accurately out-of-sample than VAR models? In *VAR models in macroeconomics–new developments and applications: Essays in Honor of Christopher A. Sims* (pp. 27–79). Emerald Group Publishing Limited.
Gefang, D., Koop, G., & Poon, A. (2023). Forecasting using variational Bayesian inference in large vector autoregressions with hierarchical shrinkage. *International Journal of Forecasting, 39*(1), 346–363.
Gelfand, A. E., & Smith, A. F. M. (1990). Sampling-based approaches to calculating marginal densities. *Journal of the American Statistical Association, 85*(410), 398–409.
Geman, S., & Geman, D. (1984). Stochastic relaxation, Gibbs distributions, and the Bayesian restoration of images. *IEEE Transactions on Pattern Analysis and Machine Intelligence, 6*, 721–741.
George, E. I., Sun, D., & Ni, S. (2008). Bayesian stochastic search for VAR model restrictions. *Journal of Econometrics, 142*(1), 553–580.
Giannone, D., Lenza, M., & Primiceri, G. E. (2015). Prior selection for vector autoregressions. *Review of Economics and Statistics, 97*(2), 436–451.
Giannone, D., Lenza, M., & Primiceri, G. E. (2019). Priors for the long run. *Journal of the American Statistical Association, 114*(526), 565–580.
Goulet Coulombe, P. (2024). The macroeconomy as a random forest. *Journal of Applied Econometrics, 39*(3), 401–421.

Goulet Coulombe, P., Leroux, M., Stevanovic, D., & Surprenant, S. (2022). How is machine learning useful for macroeconomic forecasting? *Journal of Applied Econometrics, 37*(5), 920–964.
Gupta, R., & Kabundi, A. (2010). Forecasting macroeconomic variables in a small open economy: A comparison between small- and large-scale models. *Journal of Forecasting, 29*(1–2), 168–185.
Gupta, R., & Wohar, M. (2017). Forecasting oil and stock returns with a Qual VAR using over 150 years of data. *Energy Economics, 62*, 181–186.
Hauzenberger, N., Huber, F., Koop, G., & Mitchell, J. (2022). Bayesian modeling of TVP-VARs using regression trees. arXiv preprint arXiv:2209.11970.
Hauzenberger, N., Huber, F., Klieber, K., & Marcellino, M. (2024). Bayesian neural networks for macroeconomic analysis. *Journal of Econometrics, 249*, 105843.
Huber, F., & Feldkircher, M. (2019). Adaptive shrinkage in Bayesian vector autoregressive models. *Journal of Business & Economic Statistics, 37*(1), 27–39.
Huber, F., & Rossini, L. (2022). Inference in Bayesian additive vector autoregressive tree models. *The Annals of Applied Statistics, 16*(1), 104–123.
Kadiyala, K. R., & Karlsson, S. (1993). Forecasting with generalized Bayesian vector autoregressions. *Journal of Forecasting, 12*(3–4), 365–378.
Kadiyala, K. R., & Karlsson, S. (1997). Numerical methods for estimation and inference in Bayesian VAR-models. *Journal of Applied Econometrics, 12*(2), 99–132.
Karlsson, L. S. (1989). *Bayesian analysis of vector autoregressions*. Purdue University.
Karlsson, S. (2013). *Forecasting with Bayesian vector autoregression*. In G. Elliott, C. W. J. Granger, & A. Timmermann (Eds.) *Handbook of economic forecasting* (vol. 2, pp. 791–897). Elsevier.
Karlsson, S., & Österholm, P. (2020a). A hybrid time-varying parameter Bayesian VAR analysis of Okun's law in the United States. *Economics Letters, 197*, 109622.
Karlsson, S., & Österholm, P. (2020b). The relation between the corporate bond-yield spread and the real economy: Stable or time-varying? *Economics Letters, 186*, 108883.
Karlsson, S., & Österholm, P. (2023). Is the US Phillips curve stable? Evidence from Bayesian vector autoregressions. *The Scandinavian Journal of Economics, 125*(1), 287–314.
Karlsson, S., Mazur, S., & Nguyen, H. (2023). Vector autoregression models with skewness and heavy tails. *Journal of Economic Dynamics and Control, 146*, 104580.
Kloek, T., & van Dijk, H. K. (1978). Bayesian estimates of equation system parameters: An application of integration by Monte Carlo. *Econometrica, 46*(1), 1–19.
Koop, G. (2013). Forecasting with medium and large Bayesian VARs. *Journal of Applied Econometrics, 28*(2), 177–203.
Koop, G. (2014). Forecasting with dimension switching VARs. *International Journal of Forecasting, 30*(2), 280–290.
Koop, G., & Korobilis, D. (2013). Large time-varying parameter VARs. *Journal of Econometrics, 177*(2), 185–198.
Koop, G., Korobilis, D., & Pettenuzzo, D. (2019). Bayesian compressed vector autoregressions. *Journal of Econometrics, 210*(1), 135–154.
Leung, M. T., Daouk, H., & Chen, A.-S. (2000). Forecasting stock indices: A comparison of classification and level estimation models. *International Journal of Forecasting, 16*(2), 173–190.
Litterman, R. B. (1979). Techniques of forecasting using vector autoregressions. Working paper, Federal Reserve Bank of Minneapolis, Research Department, no. 115.
Litterman, R. B. (1980). *Bayesian procedure for forecasting with vector autoregressions*. Massachusetts Institute of Technology.
Litterman, R. B. (1986). Forecasting with Bayesian vector autoregressions—five years of experience. *Journal of Business & Economic Statistics, 4*(1), 25–38.
Medeiros, M. C., Vasconcelos, G. F. R., Veiga, Á., & Zilberman, E. (2021). Forecasting inflation in a data-rich environment: The benefits of machine learning methods. *Journal of Business & Economic Statistics, 39*(1), 98–119.

Mitchell, J., & Hall, S. G. (2005). Evaluating, comparing and combining density forecasts using the KLIC with an application to the Bank of England and NIESR 'fan' charts of inflation. *Oxford Bulletin of Economics and Statistics, 67*, 995–1033.

Morales, J.-A. (1970). Bayesian full information analysis of the simultaneous equations model. In *Bayesian full information structural analysis: with an application to the study of the belgian beef market* (pp. 13–88). Springer.

Phillips, P. C. B. (1996). Econometric model determination. *Econometrica: Journal of the Econometric Society, 64*(4), 763–812.

Sims, C. A., & Zha, T. (1998). Bayesian methods for dynamic multivariate models. *International Economic Review, 39*(4), 949–968.

Smeekes, S., & Wijler, E. (2018). Macroeconomic forecasting using penalized regression methods. *International Journal of Forecasting, 34*(3), 408–430.

van Dijk, H. K. (2024). Challenges and opportunities for twenty-first century Bayesian econometricians: A personal view. *Studies in Nonlinear Dynamics & Econometrics, 28*(2), 155–176.

Zellner, A. (1971). Bayesian and non-Bayesian analysis of the log-normal distribution and log-normal regression. *Journal of the American Statistical Association, 66*(334), 327–330.

Large Bayesian Tensor VARs with Stochastic Volatility

Joshua C. C. Chan and Yaling Qi

Abstract We consider Bayesian tensor vector autoregressions (TVARs) in which the VAR coefficients are arranged as a three-dimensional array or tensor, and this coefficient tensor is parameterized using a low-rank CP (CANDECOMP/PARAFAC) decomposition. We develop a family of TVARs using a general stochastic volatility specification, which includes a wide variety of commonly used multivariate stochastic volatility and COVID-19 outlier-augmented models. In a forecasting exercise involving 40 US quarterly variables, we show that these TVARs outperform the standard Bayesian VAR with the Minnesota prior. The results also suggest that the parsimonious common stochastic volatility model tends to forecast better than the more flexible Cholesky stochastic volatility model.

1 Introduction

Large Bayesian vector autoregressions (BVARs) are now commonly used in macroeconomic forecasting and structural analysis, following early influential papers by Bańbura et al. (2010), Carriero et al. (2009), and Koop (2013). The dominant approach to tackle the challenge of parameter proliferation in large systems is to use shrinkage priors to regularize the variations in VAR coefficients; there is now an extensive literature on various shrinkage priors designed for BVARs.[1] In contrast, other dimension reduction techniques are relatively unexplored.

[1] The most widely used shrinkage priors for BVARs are the family of Minnesota priors developed in a series of papers by Doan et al. (1984), Litterman (1986), and Kadiyala and Karlsson (1993, 1997). Recent additions to this family include Giannone et al. (2015) and Chan (2022). Another popular family is the adaptive hierarchical shrinkage priors that can be represented as scale

J. C. C. Chan (✉) · Y. Qi
Economics Department, Purdue University, West Lafayette, IN, USA
e-mail: chan196@purdue.edu; qi123@purdue.edu

S. Mazur, P. Österholm (eds.), *Recent Developments in Bayesian Econometrics and Their Applications*, https://doi.org/10.1007/978-3-032-00110-8_3

We investigate the usefulness of specifying a low-rank structure on the VAR coefficients for forecasting. More specifically, we follow the approach proposed by Wang et al. (2022) to treat the VAR coefficients collectively as a three-dimensional array or tensor: For a BVAR with n endogenous variables and p lags, we arrange the $n \times n$ coefficient matrices $\mathbf{A}_1, \ldots, \mathbf{A}_p$ in the third dimension to construct the third-order tensor $\mathcal{A} \in \mathbb{R}^{n \times n \times p}$. We then model $\mathcal{A}$ using a rank-R CP decomposition. We call these BVARs constructed via the CP decomposition tensor VARs or TVARs.

Using this tensor decomposition, the number of free parameters is reduced from $n^2 p$ to $(2n + p)R$. Since the number of free parameters under this tensor decomposition grows linearly in n, it is especially suitable for applications with a large number of variables. This approach is related to the reduced-rank VAR (see, e.g., Carriero et al., 2011), which may be viewed as a special case in which the rank of $\mathcal{A}$ is reduced along one of the three possible dimensions.

Departing from the homoskedastic framework in Wang et al. (2022), we formulate the TVARs using a general stochastic volatility specification, which can represent a wide variety of multivariate stochastic volatility and COVID-19 outlier-augmented models commonly used for BVARs. This more general setup is motivated by the increasing recognition of the importance of allowing time-varying volatility for forecasting macroeconomic and financial variables, especially after the extreme economic turbulence triggered by the COVID-19 pandemic (see, e.g., Lenza and Primiceri, 2022; Carriero et al., 2024).

We develop efficient estimation procedures for these TVARs with stochastic volatility. In particular, we propose two types of algorithms to sample the components of the CP decomposition of the VAR coefficient tensor $\mathcal{A}$: We can either sample each block of the components jointly as a matrix or sample each column of the component matrix separately. The latter is inspired by the equation-by-equation estimation approach designed for BVARs proposed in Carriero et al. (2019) and Carriero et al. (2022), which can drastically reduce the computational burden when the number of endogenous variables n is very large.

We illustrate the methodology using a forecasting exercise that involves 40 US quarterly variables, such as GDP, industrial production, labor market variables, and a variety of inflation and interest rates. We consider various TVARs with different stochastic volatility specifications and compare them to a standard BVAR with the Minnesota prior in an out-of-sample forecasting exercise. The results show that TVARs clearly outperform the standard BVAR, highlighting the usefulness of the low-rank specification for the VAR coefficient tensor. We also find that models with some form of time-varying volatility substantially forecast better

mixtures of normals. Examples include the Bayesian Lasso (Park & Casella, 2008; Korobilis & Pettenuzzo, 2019), the normal-gamma prior (Griffin & Brown, 2010; Huber & Feldkircher, 2019), the horseshoe prior (Carvalho et al., 2010; Follett & Yu, 2019), and the Dirichlet-Laplace prior (Bhattacharya et al., 2015; Kastner and Huber, 2020). Naturally, one can also combine these adaptive hierarchical priors with the Minnesota priors, as proposed in Chan (2021). Hauzenberger et al. (2024) provide an excellent recent review on the state-of-the-art shrinkage priors developed for macroeconomic forecasting using BVARs.

than their homoskedastic counterparts. Interestingly, the parsimonious common stochastic volatility model of Carriero et al. (2016) tends to outperform the more flexible stochastic volatility model of Cogley and Sargent (2005), suggesting that it is foremost important to capture the strong comovements in the macroeconomic volatilities.

Our chapter is closely related to the recent work by Luo and Griffin (2023), who also consider a BVAR with a tensor decomposition. While they focus on the Cholesky stochastic volatility model of Cogley and Sargent (2005), we consider a more general setup that can represent a wide range of multivariate stochastic volatility models. Our chapter is also related to the emerging literature on modeling multidimensional tensors, where the coefficient tensors are typically parameterized using CP or Tucker decompositions; see, for example, Billio et al. (2023) and Wang et al. (2024).

The rest of this chapter is organized as follows. Section 2 first introduces a general framework of BVARs with a generic time-varying error covariance matrix and discusses how it can be used to represent a variety of popular stochastic volatility and COVID-19 outlier-augmented models. We then outline the proposed approach of arranging the VAR coefficient matrices as a third-order tensor in Sect. 3. Next, we introduce the efficient sampling algorithms in Sect. 4. Section 5 considers a recursive out-of-sample forecasting application that involves 40 US macroeconomic variables. Lastly, Sect. 6 concludes and briefly discusses some future research directions.

2 Large Bayesian VARs with Stochastic Volatility

Let $\mathbf{y}_t$ be an $n \times 1$ vector of endogenous variables at time t for $t = 1, \ldots, T$. Consider the following VAR(p):

$$\mathbf{y}_t = \mathbf{A}_1\mathbf{y}_{t-1} + \cdots + \mathbf{A}_p\mathbf{y}_{t-p} + \mathbf{u}_t,$$

where $\mathbf{A}_1, \ldots, \mathbf{A}_p$ are $n \times n$ coefficient matrices. We omit an intercept term for ease of exposition; an intercept or any exogenous variables can be added to the model with minor modifications. Let $\mathbf{A} = (\mathbf{A}_1, \ldots, \mathbf{A}_p)'$ be the $np \times n$ matrix of VAR coefficients, and let $\mathbf{x}_t = (\mathbf{y}'_{t-1}, \ldots, \mathbf{y}'_{t-p})'$ denote a vector of lag variables of dimension $np \times 1$. Then, stacking the observations over $t = 1, \ldots, T$, we can rewrite the VAR more succinctly as

$$\mathbf{Y} = \mathbf{X}\mathbf{A} + \mathbf{U}, \tag{1}$$

where the matrices $\mathbf{Y}$, $\mathbf{X}$, and $\mathbf{U}$ are, respectively, $T \times n$, $T \times np$, and $T \times n$. In a standard homoskedastic VAR, the reduced-form errors $\mathbf{u}_1, \ldots, \mathbf{u}_T$ are assumed to be independent and identically distributed (iid) as $\mathcal{N}(\mathbf{0}_n, \mathbf{\Sigma})$, where $\mathbf{0}_n$ is an $n \times 1$ vector of zeros and $\mathbf{\Sigma}$ is an $n \times n$ covariance matrix. However, it is increasingly recognized that some form of time-varying volatility is needed in modeling typical macroeconomic time series. Early influential papers such as Cogley and Sargent (2005), Primiceri (2005), and Sims and Zha (2006) have highlighted the secular variations in volatility. There is now a large empirical literature that demonstrates the importance of allowing time-varying volatility in improving model fit and forecasting performance in the context of Bayesian VARs; examples include Clark (2011), Koop & Korobilis (2013), D'Agostino et al. (2013), Clark and Ravazzolo (2015), Cross and Poon (2016), and Chan and Eisenstat (2018).

Below we outline a few stochastic volatility models suitable for large BVARs. In particular, the innovations of the VAR in (1) are now distributed as

$$\mathbf{u}_t \sim \mathcal{N}(\mathbf{0}_n, \mathbf{\Sigma}_t), \tag{2}$$

where $\mathbf{\Sigma}_t$ is a generic time-varying covariance matrix.

One of the first stochastic volatility models designed for large BVARs is the common stochastic volatility proposed in Carriero et al. (2016). Their model is motivated by the empirical observation that the estimated time-varying error variances of many macroeconomic variables have broadly similar low-frequency movements. A parsimonious way to model these comovements is to introduce a time-varying latent factor to scale the error covariance matrix via

$$\mathbf{\Sigma}_t = \mathrm{e}^{h_t}\mathbf{\Omega}, \tag{3}$$

where $\mathbf{\Omega}$ is a time-invariant covariance matrix. The log-volatility h_t is modeled using a zero-mean stationary AR(1) process:

$$h_t = \phi h_{t-1} + u_t^h, \quad u_t^h \sim \mathcal{N}(0, \sigma_h^2), \tag{4}$$

for $t = 2, \ldots, T$, where $|\phi| < 1$ and the process is initialized as $h_1 \sim \mathcal{N}(0, \sigma_h^2/(1-\phi^2))$. The common stochastic volatility model specified in (3)–(4) can be extended to incorporate other useful features. For example, Chan (2020a) introduces a general framework that can accommodate heavy-tailed, heteroskedastic, and serially dependent innovations.

Another widely used stochastic volatility specification for BVARs is the Cholesky stochastic volatility model—based on the modified Cholesky decomposition of the covariance matrix—developed in Cogley and Sargent (2005). More specifically, consider the decomposition of $\mathbf{\Sigma}_t$ via

$$\mathbf{\Sigma}_t = \mathbf{B}_0^{-1}\mathbf{D}_t(\mathbf{B}_0^{-1})', \tag{5}$$

where $\mathbf{B}_0$ is an $n \times n$ lower triangular matrix with ones on the diagonal and $\mathbf{D}_t = \mathrm{diag}(\mathrm{e}^{h_{1,t}}, \ldots, \mathrm{e}^{h_{n,t}})$. Each element of the vector $\mathbf{h}_t = (h_{1,t}, \ldots, h_{n,t})'$ is modeled via an independent autoregressive process:

$$h_{i,t} = \mu_i + \phi_i(h_{i,t-1} - \mu_i) + u_{i,t}^h, \quad u_{i,t}^h \sim \mathcal{N}(0, \sigma_i^2) \tag{6}$$

for $t = 2, \ldots, T$, where $h_{i,1}$ is initialized as $h_{i,1} \sim \mathcal{N}(\mu_i, \sigma_i^2/(1-\phi_i^2))$ for $i = 1, \ldots, n$.

This Cholesky stochastic volatility specification is more flexible than the common stochastic volatility model in (3), since the former contains n stochastic volatility processes and can accommodate more complex covolatility patterns. This flexibility, however, comes at a cost of higher model complexity. Whether this stochastic volatility specification forecasts better than alternatives in large systems is an empirical question.

Another difference between the Cholesky stochastic volatility and the common stochastic volatility is that the latter is order-invariant—i.e., parameter estimates are invariant to reordering the endogenous variables in $\mathbf{y}_t$—whereas the former is not. One practical implication is that forecasts from the Cholesky stochastic volatility model could differ substantially across different variable orderings, as documented in Arias et al. (2023) using a similar model of Primiceri (2005).

The reason why the Cholesky stochastic volatility is not order-invariant is partly due to the use of the lower triangular parameterization of $\mathbf{B}_0$ in (5). Motivated by this simple observation, Chan et al. (2024) extend the model by relaxing this lower triangular assumption and instead specify $\mathbf{B}_0$ to be any nondegenerate square matrix. They prove that the model is order-invariant. Moreover, based on the results in Bertsche and Braun (2022), $\mathbf{B}_0$ is also identified up to permutations and sign switches.

The extreme movements in many macroeconomic variables at the onset of the COVID-19 pandemic have motivated much recent work on modeling outliers in macroeconomic time series. An example is the outlier-augmented stochastic volatility model proposed by Carriero et al. (2024), which builds on the Cholesky stochastic volatility model and the discrete mixture representation for the innovations introduced in Stock and Watson (2016). This outlier-augmented model can also be represented using (2).

There are many other multivariate stochastic volatility models for $\mathbf{\Sigma}_t$. Clark and Mertens (2023) and Chan (2024) provide two recent reviews on BVARs with a wide range of stochastic volatility and outlier-augmented specifications.

3 Tensor Decomposition of VAR Coefficients

For high-dimensional settings, there are two key challenges related to the proliferation of the VAR coefficients. First, the number of VAR coefficients increases quadratically in n. In many large-scale applications, there are far more VAR coefficients than the number of observations, which makes it necessary to regularize these VAR coefficients. Second, due to the proliferation of VAR coefficients, sampling them tends to be very computationally intensive.

These two related challenges are typically tackled separately in the literature. For instance, Bayesian shrinkage priors are widely used to regularize the VAR coefficients. These include the family of Minnesota priors (Doan et al., 1984; Litterman, 1986; Kadiyala & Karlsson, 1993, 1997; Giannone et al., 2015; Chan, 2022) and various adaptive hierarchical shrinkage priors (Huber & Feldkircher, 2019; Korobilis & Pettenuzzo, 2019; Kastner and Huber, 2020; Chan, 2021). The computational challenge is addressed by developing efficient MCMC or variational methods to sample the large number of VAR coefficients (Carriero et al., 2019, 2022; Gefang et al., 2023; Bernardi et al., 2024). We instead take an alternative approach that tackles these two challenges simultaneously by imposing a low-rank structure on the VAR coefficients.

One possibility is the reduced-rank VAR (see, e.g., Carriero et al., 2011), in which the matrix $\mathbf{A}' = (\mathbf{A}_1, \ldots, \mathbf{A}_p)$ is assumed to have a reduced rank $R < n$. That is, the dimension of the column space of the VAR coefficient matrices $\mathbf{A}_1, \ldots, \mathbf{A}_p$ has rank R. Alternatively, one could impose a low-rank structure on $(\mathbf{A}'_1, \ldots, \mathbf{A}'_p)$ or $(\text{vec}(\mathbf{A}_1), \ldots, \text{vec}(\mathbf{A}_p))$, whose ranks correspond to the dimensions of the row space and vectorized matrix space of the VAR coefficient matrices, respectively. Each of these options essentially reduces the dimension along one of the three different directions.

This motivates us to follow the approach in Wang et al. (2022) to treat the VAR coefficients collectively as a three-dimensional array or tensor. That is, we arrange the $n \times n$ coefficient matrices $\mathbf{A}_1, \ldots, \mathbf{A}_p$ in the third dimension to form the third-order tensor $\mathcal{A} \in \mathbb{R}^{n \times n \times p}$. With this tensor representation, we then use a rank-R CP decomposition to construct $\mathcal{A}$:

$$\mathcal{A} = \sum_{r=1}^{R} \boldsymbol{\theta}_1^{(r)} \circ \boldsymbol{\theta}_2^{(r)} \circ \boldsymbol{\theta}_3^{(r)}, \tag{7}$$

where $\boldsymbol{\theta}_1^{(r)}, \boldsymbol{\theta}_2^{(r)} \in \mathbb{R}^n$ and $\boldsymbol{\theta}_3^{(r)} \in \mathbb{R}^p$. Each component $\boldsymbol{\theta}_1^{(r)} \circ \boldsymbol{\theta}_2^{(r)} \circ \boldsymbol{\theta}_3^{(r)}$ is a rank-1 third-order tensor whose (i, j, k) element is the product of the i-th, j-th, and k-th elements of $\boldsymbol{\theta}_1^{(r)}, \boldsymbol{\theta}_2^{(r)}$, and $\boldsymbol{\theta}_3^{(r)}$, respectively. Using this tensor decomposition with a small R, the number of parameters is reduced from $n^2 p$ to $(2n + p)R$. We refer the readers to Kolda and Bader (2009) for a general introduction to tensors and their operations.[2]

In practice, one typically sets a small value for R relative to n. This implies a low-rank approximation to the coefficient tensor $\mathcal{A}$, which might not be appropriate for some applications. For example, if the VAR coefficients $\mathbf{A}_1, \ldots, \mathbf{A}_p$ are expected to be close to diagonal (e.g., the time series do not exhibit dynamic interdependence), R needs to be at least as large as n to provide a good approximation. Another example is when different series have different seasonality patterns with seasonal

[2] The marginals $\boldsymbol{\theta}_1^{(r)}, \boldsymbol{\theta}_2^{(r)}$, and $\boldsymbol{\theta}_3^{(r)}$ are not separately identifiable without further restrictions. Since here we are mainly interested in obtaining forecasts, which are identified, we do not impose any identification restrictions.

periods s_i, so that some autoregressive coefficients with multiples of lag s_i are significantly different from 0, whereas others are not.

Instead of the CP decomposition in (7), one could consider the more general Tucker decomposition that allows different ranks across different dimensions:

$$\mathcal{A} = \sum_{r_1=1}^{R_1} \sum_{r_2=1}^{R_2} \sum_{r_3=1}^{R_3} g_{r_1 r_2 r_3} \boldsymbol{\theta}_1^{(r_1)} \circ \boldsymbol{\theta}_2^{(r_2)} \circ \boldsymbol{\theta}_3^{(r_3)}, \tag{8}$$

where $\boldsymbol{\theta}_1^{(r_1)}, \boldsymbol{\theta}_2^{(r_2)} \in \mathbb{R}^n$, $r_1 = 1, \ldots, R_1, r_2 = 1, \ldots, R_2$, $\boldsymbol{\theta}_3^{(r_3)} \in \mathbb{R}^p$, $r_3 = 1, \ldots, R_3$, and $\mathcal{G} = (g_{r_1 r_2 r_3}) \in \mathbb{R}^{n \times n \times p}$ is called the core tensor. In particular, the Tucker decomposition allows a more flexible modeling of the lag structure of the VAR coefficient matrices by using different ranks R_1, R_2, and R_3. The trade-off for this more general decomposition is the more complex sampling scheme required to sample the marginals $\boldsymbol{\theta}_1^{(r_1)}, \boldsymbol{\theta}_2^{(r_2)}$, and $\boldsymbol{\theta}_3^{(r_3)}$. We leave this possibility to future research.

4 Bayesian Estimation

In this section we describe the priors and outline the posterior simulator. In particular, we focus on the sampling of the components $\boldsymbol{\theta}_1^{(r)}, \boldsymbol{\theta}_2^{(r)}$, and $\boldsymbol{\theta}_3^{(r)}, r = 1, \ldots, R$. To that end, let $\boldsymbol{\Theta}_1 = (\boldsymbol{\theta}_1^{(1)}, \ldots, \boldsymbol{\theta}_1^{(R)})$, and similarly define $\boldsymbol{\Theta}_2$ and $\boldsymbol{\Theta}_3$. The dimensions of $\boldsymbol{\Theta}_1, \boldsymbol{\Theta}_2$, and $\boldsymbol{\Theta}_3$ are, respectively, $n \times R$, $n \times R$, and $p \times R$. For later reference, stack $\boldsymbol{\theta}_1 = \text{vec}(\boldsymbol{\Theta}_1)$, $\boldsymbol{\theta}_2 = \text{vec}(\boldsymbol{\Theta}_2')$, and $\boldsymbol{\theta}_3 = \text{vec}(\boldsymbol{\Theta}_3')$. For reasons that will become transparent later, note that $\boldsymbol{\theta}_1$ is constructed by stacking the columns of $\boldsymbol{\Theta}_1$, whereas $\boldsymbol{\theta}_2$ and $\boldsymbol{\theta}_3$ are formed by stacking the rows of $\boldsymbol{\Theta}_2$ and $\boldsymbol{\Theta}_3$, respectively.

Next, consider the following independent Gaussian priors on $\boldsymbol{\theta}_1, \boldsymbol{\theta}_2$ and $\boldsymbol{\theta}_3$:

$$\boldsymbol{\theta}_j \sim \mathcal{N}(\boldsymbol{\theta}_{j,0}, \mathbf{V}_{\boldsymbol{\theta}_j}), \quad j = 1, 2, 3. \tag{9}$$

Since the first two dimensions relate to the endogenous variables and the third to the lag length, it might be useful to consider different prior mean vectors and covariance matrices for $\boldsymbol{\theta}_1, \boldsymbol{\theta}_2$, and $\boldsymbol{\theta}_3$. For example, to implement the usual Minnesota-type prior idea of shrinking longer lags stronger to zero, one could shrink the elements at the end of $\boldsymbol{\theta}_3^{(r)}$ more than elements earlier in this vector. Finally, we consider here simple Gaussian priors, but any adaptive hierarchical shrinkage priors that have a conditionally Gaussian representation—such as the normal-gamma prior, the horseshoe prior, or the Dirichlet-Laplace prior—can be used.

In what follows, we derive the full conditional posterior distributions of $(\boldsymbol{\theta}_1 | \mathbf{Y}, \boldsymbol{\theta}_2, \boldsymbol{\theta}_3, \boldsymbol{\Sigma})$, $(\boldsymbol{\theta}_2 | \mathbf{Y}, \boldsymbol{\theta}_1, \boldsymbol{\theta}_3, \boldsymbol{\Sigma})$, and $(\boldsymbol{\theta}_3 | \mathbf{Y}, \boldsymbol{\theta}_1, \boldsymbol{\theta}_2, \boldsymbol{\Sigma})$, where $\boldsymbol{\Sigma} = \text{diag}\,(\boldsymbol{\Sigma}_1, \ldots, \boldsymbol{\Sigma}_T)$.

4.1 Sampling of $\boldsymbol{\theta}_1$

For sampling $\boldsymbol{\theta}_1$, first let

$$\begin{aligned}\boldsymbol{\Theta}_{-1} &= (\boldsymbol{\theta}_3^{(1)} \otimes \boldsymbol{\theta}_2^{(1)}, \ldots, \boldsymbol{\theta}_3^{(R)} \otimes \boldsymbol{\theta}_2^{(R)}),\\ \boldsymbol{\Theta}_{-2} &= (\boldsymbol{\theta}_3^{(1)} \otimes \boldsymbol{\theta}_1^{(1)}, \ldots, \boldsymbol{\theta}_3^{(R)} \otimes \boldsymbol{\theta}_1^{(R)}),\\ \boldsymbol{\Theta}_{-3} &= (\boldsymbol{\theta}_2^{(1)} \otimes \boldsymbol{\theta}_1^{(1)}, \ldots, \boldsymbol{\theta}_2^{(R)} \otimes \boldsymbol{\theta}_1^{(R)}).\end{aligned}$$

Then, the mode-1 matricization of $\mathcal{A}$ can be written as (see, e.g., Kolda and Bader, 2009)

$$\mathcal{A}_{(1)} = \mathbf{A}' = \boldsymbol{\Theta}_1\boldsymbol{\Theta}_{-1}'. \tag{10}$$

Furthermore, let $\mathcal{X} \in \mathbb{R}^{T\times n\times p}$ denote the third-order tensor constructed by stacking the $n \times p$ matrices $(\mathbf{y}_{t-1}, \ldots, \mathbf{y}_{t-p}), t = 1, \ldots, T$, along the first dimension so that $\text{vec}(\mathcal{X}_{t,:,:}) = \mathbf{x}_t$. It is easy to verify that its mode-1 matricization is the $T \times np$ matrix $\mathbf{X}$, i.e., $\mathcal{X}_{(1)} = \mathbf{X}$. Hence, combining (1) and (10), we have

$$\mathbf{Y} = \mathcal{X}_{(1)}\boldsymbol{\Theta}_{-1}\boldsymbol{\Theta}_1' + \mathbf{U}. \tag{11}$$

It follows that

$$\text{vec}(\mathbf{Y}') = (\mathcal{X}_{(1)}\boldsymbol{\Theta}_{-1} \otimes \mathbf{I}_n)\text{vec}(\boldsymbol{\Theta}_1) + \text{vec}(\mathbf{U}'),$$

where $\text{vec}(\mathbf{U}') \sim \mathcal{N}(\mathbf{0}_{Tn}, \boldsymbol{\Sigma})$ with $\boldsymbol{\Sigma} = \text{diag}\,(\boldsymbol{\Sigma}_1, \ldots, \boldsymbol{\Sigma}_T)$. Given the Gaussian prior on $\boldsymbol{\theta}_1 = \text{vec}(\boldsymbol{\Theta}_1)$ specified in (9), the full conditional posterior of $\boldsymbol{\theta}_1$ is given by

$$(\boldsymbol{\theta}_1 | \mathbf{Y}, \boldsymbol{\theta}_2, \boldsymbol{\theta}_3, \boldsymbol{\Sigma}) \sim \mathcal{N}\left(\widehat{\boldsymbol{\theta}}_1, \mathbf{K}_{\boldsymbol{\theta}_1}^{-1}\right),$$

where

$$\begin{aligned}\mathbf{K}_{\boldsymbol{\theta}_1} &= \mathbf{V}_{\boldsymbol{\theta}_1}^{-1} + (\boldsymbol{\Theta}_{-1}'\mathcal{X}_{(1)}' \otimes \mathbf{I}_n)\boldsymbol{\Sigma}^{-1}(\mathcal{X}_{(1)}\boldsymbol{\Theta}_{-1} \otimes \mathbf{I}_n)\\ \widehat{\boldsymbol{\theta}}_1 &= \mathbf{K}_{\boldsymbol{\theta}_1}^{-1}\left(\mathbf{V}_{\boldsymbol{\theta}_1}^{-1}\boldsymbol{\theta}_{1,0} + (\boldsymbol{\Theta}_{-1}'\mathcal{X}_{(1)}' \otimes \mathbf{I}_n)\boldsymbol{\Sigma}^{-1}\text{vec}(\mathbf{Y}')\right).\end{aligned}$$

Since $\boldsymbol{\theta}_1$ is of length nR, sampling $\boldsymbol{\theta}_1$ generally involves $O(n^3R^3)$ elementary operations. When both n and R are large, this sampling step could be computationally intensive. An alternative is to sample each $\boldsymbol{\theta}_1^{(r)}$ separately, $r = 1, \ldots, R$. For a fixed

r, this can be done by substituting

$$\boldsymbol{\Theta}_1\boldsymbol{\Theta}_{-1}' = \sum_{s=1}^{R} \boldsymbol{\theta}_1^{(s)}\boldsymbol{\theta}_{-1}^{(s)\prime},$$

where $\boldsymbol{\theta}_{-1}^{(s)} = \boldsymbol{\theta}_3^{(s)} \otimes \boldsymbol{\theta}_2^{(s)}$, into the transpose of (11) to obtain

$$\mathbf{Y}' = \sum_{s=1}^{R} \boldsymbol{\theta}_1^{(s)}\boldsymbol{\theta}_{-1}^{(s)\prime}\mathcal{X}_{(1)}' + \mathbf{U}'.$$

Next, let $\mathbf{Y}_{1r} = \mathbf{Y}' - \sum_{s\neq r} \boldsymbol{\theta}_1^{(s)}\boldsymbol{\theta}_{-1}^{(s)\prime}\mathcal{X}_{(1)}'$, and vectorizing the above equation, we have

$$\text{vec}(\mathbf{Y}_{1r}) = (\mathcal{X}_{(1)}\boldsymbol{\theta}_{-1}^{(r)} \otimes \mathbf{I}_n)\boldsymbol{\theta}_1^{(r)} + \text{vec}(\mathbf{U}').$$

If the marginal prior for $\boldsymbol{\theta}_1^{(r)}$ is

$$\boldsymbol{\theta}_1^{(r)} \sim \mathcal{N}\left(\boldsymbol{\theta}_{1,0}^{(r)}, \mathbf{V}_{\boldsymbol{\theta}_1^{(r)}}\right),$$

the posterior distribution $(\boldsymbol{\theta}_1^{(r)} | \mathbf{Y}, \{\boldsymbol{\theta}_1^{(s)}\}_{s\neq r}, \boldsymbol{\theta}_2, \boldsymbol{\theta}_3, \boldsymbol{\Sigma})$ has the form

$$(\boldsymbol{\theta}_1^{(r)} | \mathbf{Y}, \{\boldsymbol{\theta}_1^{(s)}\}_{s\neq r}, \boldsymbol{\theta}_2, \boldsymbol{\theta}_3, \boldsymbol{\Sigma}) \sim \mathcal{N}\left(\widehat{\boldsymbol{\theta}}_1^{(r)}, \mathbf{K}_{\boldsymbol{\theta}_1^{(r)}}^{-1}\right),$$

where

$$\mathbf{K}_{\boldsymbol{\theta}_1^{(r)}} = (\boldsymbol{\theta}_{-1}^{(r)\prime}\mathcal{X}_{(1)}' \otimes \mathbf{I}_n)\boldsymbol{\Sigma}^{-1}(\mathcal{X}_{(1)}\boldsymbol{\theta}_{-1}^{(r)} \otimes \mathbf{I}_n)$$

$$\widehat{\boldsymbol{\theta}}_1^{(r)} = \mathbf{K}_{\boldsymbol{\theta}_1^{(r)}}^{-1}\left(\mathbf{V}_{\boldsymbol{\theta}_1^{(r)}}^{-1}\boldsymbol{\theta}_{1,0}^{(r)} + (\boldsymbol{\theta}_{-1}^{(r)\prime}\mathcal{X}_{(1)}' \otimes \mathbf{I}_n)\boldsymbol{\Sigma}^{-1}\text{vec}(\mathbf{Y}_{1r})\right).$$

The drawback of sampling each $\boldsymbol{\theta}_1^{(r)}$ separately is that this tends to increase the autocorrelation of the constructed Markov chain. But this approach is substantially faster and remains computationally feasible even when both n and R are large.

4.2 Sampling of $\boldsymbol{\theta}_2$ and $\boldsymbol{\theta}_3$

Next, we derive the conditional distribution of $(\boldsymbol{\theta}_2 | \mathbf{Y}, \boldsymbol{\theta}_1, \boldsymbol{\theta}_3, \boldsymbol{\Sigma})$ and show that it is Gaussian. To start, we aim to write the VAR in (1) as a linear regression in $\boldsymbol{\theta}_2 = \text{vec}(\boldsymbol{\Theta}_2')$.

We first introduce some notations. Let $\mathbf{e}_i^p$ be the i-th column of $\mathbf{I}_p$. Let $\mathbf{P}$ denote the $np \times np$ commutation matrix so that $\mathbf{P}'\text{vec}(\mathbf{Z}) = \text{vec}(\mathbf{Z}')$ for any $n \times p$ matrix $\mathbf{Z}$. Explicitly, $\mathbf{P}$ can be constructed by setting the (k, l) element to be 1, i.e., $\mathbf{P}_{k,l} = 1$, if there exist i and j such that $k = (i-1)n + j$ and $l = (j-1)p + i$; otherwise, set $\mathbf{P}_{k,l} = 0$.

Proposition 1 *The VAR in* (1) *can be written as*

$$\text{vec}(\mathbf{Y}') = \sum_{i=1}^{p} (\mathbf{P}_{i2}\mathcal{X}'_{(2)} \otimes \mathbf{P}_{i1}\boldsymbol{\Theta}_{-2})\boldsymbol{\theta}_2 + \text{vec}(\mathbf{U}'), \tag{12}$$

where $\mathbf{P}_{i1} = (\mathbf{I}_n \otimes (\mathbf{e}_i^p)')\mathbf{P}'$, $\mathbf{P}_{i2} = (\mathbf{e}_i^p)' \otimes \mathbf{I}_T$ *and* $\mathcal{X}_{(2)}$ *is the mode-2 matricization of* $\mathcal{X}$.

The proof of this proposition is given in Appendix A. Now, given the representation in (12) and the Gaussian prior on $\boldsymbol{\theta}_2$ specified in (9), by standard linear regression results, one can verify that the full conditional posterior of $\boldsymbol{\theta}_2$ is given by

$$(\boldsymbol{\theta}_2 | \mathbf{Y}, \boldsymbol{\theta}_1, \boldsymbol{\theta}_3, \boldsymbol{\Sigma}) \sim \mathcal{N}\left(\widehat{\boldsymbol{\theta}}_2, \mathbf{K}_{\boldsymbol{\theta}_2}^{-1}\right),$$

where

$$\mathbf{K}_{\boldsymbol{\theta}_2} = \mathbf{V}_{\boldsymbol{\theta}_2}^{-1} + \sum_{i=1}^{p}\sum_{j=1}^{p} (\mathcal{X}_{(2)}\mathbf{P}'_{i2} \otimes \boldsymbol{\Theta}'_{-2}\mathbf{P}'_{i1})\boldsymbol{\Sigma}^{-1}(\mathbf{P}_{j2}\mathcal{X}'_{(2)} \otimes \mathbf{P}_{j1}\boldsymbol{\Theta}_{-2}),$$

$$\widehat{\boldsymbol{\theta}}_2 = \mathbf{K}_{\boldsymbol{\theta}_2}^{-1}\left(\mathbf{V}_{\boldsymbol{\theta}_2}^{-1}\boldsymbol{\theta}_{2,0} + \sum_{i=1}^{p} (\mathcal{X}_{(2)}\mathbf{P}'_{i2} \otimes \boldsymbol{\Theta}'_{-2}\mathbf{P}'_{i1})\boldsymbol{\Sigma}^{-1}\text{vec}(\mathbf{Y}')\right).$$

Therefore, one can sample $\boldsymbol{\theta}_2$ in one block. When n and R are large, it might only be feasible to sample each $\boldsymbol{\theta}_2^{(r)}$ at a time for $r = 1, \ldots, R$. We provide the details of this alternative approach in Appendix C.

Likewise, the conditional distribution of $(\boldsymbol{\theta}_3 | \mathbf{Y}, \boldsymbol{\theta}_1, \boldsymbol{\theta}_2, \boldsymbol{\Sigma})$ can be shown to be Gaussian. More specifically, we first write the VAR in (1) as a linear regression in $\boldsymbol{\theta}_3 = \text{vec}(\boldsymbol{\Theta}'_3)$. To that end, let $\mathbf{e}_i^n$ denote the i-th column of $\mathbf{I}_n$, let $\mathcal{X}_{(3)}$ represent the mode-3 matricization of $\mathcal{X}$, and define $\mathbf{Q}$ to be the $n^2 \times n^2$ commutation matrix so that $\mathbf{Q}'\text{vec}(\mathbf{Z}) = \text{vec}(\mathbf{Z}')$ for any $n \times n$ matrix $\mathbf{Z}$. Explicitly, $\mathbf{Q}$ can be constructed by setting the $\mathbf{Q}_{k,l} = 1$ if there exist i and j such that $k = (i-1)n + j$ and $l = (j-1)n + i$; otherwise, set $\mathbf{Q}_{k,l} = 0$.

Proposition 2 *The VAR in* (1) *can be represented as*

$$\text{vec}(\mathbf{Y}') = \sum_{i=1}^{n} (\mathbf{Q}_{i2}\mathcal{X}'_{(3)} \otimes \mathbf{Q}_{i1}\boldsymbol{\Theta}_{-3})\boldsymbol{\theta}_3 + \text{vec}(\mathbf{U}'), \tag{13}$$

where $\mathbf{Q}_{i1} = (\mathbf{I}_n \otimes (\mathbf{e}_i^n)')\mathbf{Q}'$ *and* $\mathbf{Q}_{i2} = (\mathbf{e}_i^n)' \otimes \mathbf{I}_T$.

Given the representation in (13), it is easy to verify that

$$(\boldsymbol{\theta}_3|\mathbf{Y}, \boldsymbol{\theta}_1, \boldsymbol{\theta}_2, \boldsymbol{\Sigma}) \sim \mathcal{N}\left(\widehat{\boldsymbol{\theta}}_3, \mathbf{K}_{\boldsymbol{\theta}_3}^{-1}\right),$$

where

$$\mathbf{K}_{\boldsymbol{\theta}_3} = \mathbf{V}_{\boldsymbol{\theta}_3}^{-1} + \sum_{i=1}^{n}\sum_{j=1}^{n}(\mathcal{X}_{(3)}\mathbf{Q}_{i2}' \otimes \boldsymbol{\Theta}_{-3}'\mathbf{Q}_{i1}')\boldsymbol{\Sigma}^{-1}(\mathbf{Q}_{j2}\mathcal{X}_{(3)}' \otimes \mathbf{Q}_{j1}\boldsymbol{\Theta}_{-3})$$

$$\widehat{\boldsymbol{\theta}}_3 = \mathbf{K}_{\boldsymbol{\theta}_3}^{-1}\left(\mathbf{V}_{\boldsymbol{\theta}_3}^{-1}\boldsymbol{\theta}_{3,0} + \sum_{i=1}^{n}(\mathcal{X}_{(3)}\mathbf{Q}_{i2}' \otimes \boldsymbol{\Theta}_{-3}'\mathbf{Q}_{i1}')\boldsymbol{\Sigma}^{-1}\text{vec}(\mathbf{Y}')\right).$$

Alternatively, one can sample each $\boldsymbol{\theta}_3^{(r)}$ at a time for $r = 1, \ldots, R$ when n and R are large as before.

Finally, sampling the time-varying error covariance matrices $\boldsymbol{\Sigma}_1, \ldots, \boldsymbol{\Sigma}_T$ naturally depends on the stochastic volatility specification used. For a wide variety of stochastic volatility specifications commonly employed in applied work, such as the common stochastic volatility and Cholesky stochastic volatility discussed earlier, efficient algorithms are available to sample the latent variables and model parameters. We refer the readers to Chan (2023a) for more details.

5 An Empirical Application

We conduct an out-of-sample forecasting exercise to evaluate the performance of the proposed tensor VARs compared to a standard benchmark. More specifically, we construct a dataset of 40 quarterly macroeconomic and financial variables from the FRED-QD database (McCracken & Ng, 2020). The sample spans from 1969Q1 to 2024Q1. The data include key macroeconomic variables such as GDP, inflation rates, labor market variables, and various interest rates. We refer the readers to Appendix B for the detailed description of the time series and their transformations.

We consider three tensor VARs: a homoskedastic tensor VAR (TVAR), TVARs with the common volatility (TVAR-CSV), and the Cholesky stochastic volatility (TVAR-SV). For now, we set the rank of the VAR coefficient tensor $\mathcal{A}$ to be $R = 1$ for all TVARs. In addition, the priors on the elements of $\boldsymbol{\theta}_j$, $j = 1, 2, 3$, are assumed to be iid $\mathcal{N}(0, 1)$. As a benchmark, we also include a standard Bayesian VAR (BVAR) with the Minnesota prior, implemented as the natural conjugate prior. In particular, the overall shrinkage hyperparameter is set to be 0.2^2, a value commonly used in empirical work (Carriero et al., 2015; Chan, 2020b). The evaluation period of the forecasting exercise begins in 2010Q1 and ends in 2024Q1. The lag length is set to be $p = 4$ for all models.

Table 1 Joint density forecast performance of the proposed tensor VARs relative to a standard Bayesian VAR with the Minnesota prior

	BVAR	TVAR	TVAR-CSV	TVAR-SV
One-quarter-ahead	−8.01	−3.30	−1.31	−2.32
Four-quarter-ahead	−8.53	−3.30	−2.21	−2.66

Table 2 Relative root mean squared forecast errors of the proposed tensor VARs compared to a standard Bayesian VAR with the Minnesota prior. The symbol * denotes that the Diebold-Mariano test for equal forecast accuracy is rejected at 5% significance level

	One-quarter-ahead			Four-quarter-ahead		
	TVAR	TVAR-CSV	TVAR-SV	TVAR	TVAR-CSV	TVAR-SV
RPI	0.93	0.93	0.93	0.99*	0.99*	0.98*
GDP	0.65	0.64	0.78	0.82	0.82	0.85
Unemployment	0.68	0.68	0.72	0.86	0.86	0.86
CPI	1.03	1.02	1.14	0.89	0.89	0.93
Fed funds rate	0.83	0.83	1.46	0.60	0.60	0.65
10-year T-bond	0.92	0.92	0.95	0.82	0.82	0.85

To assess the performance of jointly forecasting all $n = 40$ time series, we calculate the average log predictive likelihoods for each model over one- and four-quarter-ahead forecast horizons. The results are presented in Table 1. Higher values of log predictive likelihoods signify better forecast performance.

A few observations can be drawn from these forecasting results. Firstly, comparing the two homoskedastic models, TVAR and BVAR, it is clear that the former substantially outperforms the latter, suggesting that the specification of a low-rank structure on the VAR coefficient tensor $\mathcal{A}$ is more appropriate. Secondly, allowing some form of time-varying volatility clearly improves forecast performance. For example, extending a homoskedastic TVAR to a version with the common stochastic volatility increases the average log predictive likelihood from -8 to about -3 for one-step-ahead density forecasts. This finding is in line with the large body of empirical evidence that demonstrates the importance of allowing time-varying volatility in macroeconomic forecasting (Clark, 2011; D'Agostino et al., 2013; Chan, 2023a). Finally, among the two TVARs with stochastic volatility, the version with the parsimonious common stochastic volatility outperforms the one with the more flexible Cholesky stochastic volatility, indicating strong comovements in the macroeconomic volatilities.

Next, we look at the point forecast performance of the TVARs for individual time series. In particular, Table 2 reports the relative root mean squared forecast errors (RMSFEs) of a few key macroeconomic variables over the evaluation period compared to those of the benchmark BVAR. That is, for each variable and each version of TVAR, we divide its RMSFE by that of the benchmark. Values less than 1 indicate better forecast performance than the benchmark.

The results show that the TVARs tend to outperform the benchmark BVAR for both one- and four-quarter-ahead horizons. For example, the RMSFE of the

Table 3 One-quarter-ahead joint density forecast performance of the proposed tensor VARs for $R = 1, 3, 5, 10$

	TVAR	TVAR-CSV	TVAR-SV
$R = 1$	−3.30	−1.31	−2.32
$R = 3$	−3.27	−1.39	−2.36
$R = 5$	−3.40	−1.26	−2.48
$R = 10$	−3.53	−1.29	−2.53

homoskedastic TVAR for forecasting GDP is only 65% of that of the benchmark. Interestingly, allowing time-varying volatility does not appear to improve point forecasts. In fact, the more flexible TVAR-SV often provides inferior point forecasts compared to the homoskedastic TVAR (though the performance of the more parsimonious TVAR-CSV is virtually identical to TVAR).

So far we have set R, the rank of the VAR coefficient tensor $\mathcal{A}$, to be 1 for all TVARs. Naturally, one might wonder how this choice affects the density and point forecast performance. To investigate the impact of the choice of R, we present in Table 3 the one-quarter-ahead average log predictive likelihoods of the three TVARs for $R = 1, 3, 5, 10$.

It is interesting to note that the choice of R does not seem to have a large impact on the joint density forecast performance across the three TVARs. In particular, it does not affect the relative ranking of the three models; the two TVARs with time-varying volatility outperform the homoskedastic version regardless of the rank R. Overall, low-rank TVARs tend to work, as well as, if not better than, TVARs with $R = 10$.

Table 4 reports the one-quarter-ahead RMSFEs of the three TVARs for forecasting a few key macroeconomic variables. To facilitate comparison, the results for each TVAR are relative to the corresponding TVAR with $R = 1$. With the exception of CPI inflation, increasing the rank R does not appear to substantially improve point forecast performance.

Overall, these forecasting results demonstrate the benefits of the proposed approach of specifying a low-rank structure on the VAR coefficient tensor. In addition, our results highlight the importance of accommodating time-varying volatility in forecasting macroeconomic variables. Notably, in our forecasting exercise, the more parsimonious TVAR-CSV tends to forecast better than the more flexible TVAR-SV.

Finally, to assess the sampling efficiency of the posterior samplers, we report in Fig. 1 the inefficiency factors of the posterior draws from each TVAR with $R = 1, 3, 5, 10$. More specifically, we visualize the results using the boxplot, where the middle line of the box denotes the median, while the lower and upper lines represent, respectively, the 25 and 75 percentiles. The whiskers extend to the maximum and minimum. For example, the boxplot associated with TVAR-SV ($R = 1$) indicates that about 75% of the parameters and latent variables have inefficiency factors less than 13, and the largest inefficiency factor is about 50. These results show that the proposed posterior samplers are adequately efficient even in high-dimensional settings.

Table 4 One-quarter-ahead RMSFEs of TVARs with $R = 3, 5, 10$ relative to the corresponding TVARs with $R = 1$

	$R = 3$			$R = 5$			$R = 10$		
	TVAR	TVAR-CSV	TVAR-SV	TVAR	TVAR-CSV	TVAR-SV	TVAR	TVAR-CSV	TVAR-SV
RPI	1.01	1.00	1.01	1.01	1.00	1.02	1.01	1.00	1.03
GDP	1.04	1.02	0.99	1.08	1.05	1.05	1.12	1.07	1.12
Unemployment	1.02	1.01	1.00	1.02	1.01	1.05	1.05	1.01	1.15
CPI inflation	0.97	0.95	1.06	0.96	0.95	0.90	0.89	0.90	0.94
Fed funds rate	1.08	1.11	1.08	1.24	1.21	1.05	1.20	1.18	0.99
10-year T-bond	1.00	1.00	1.22	0.99	1.00	1.17	0.98	0.99	1.16

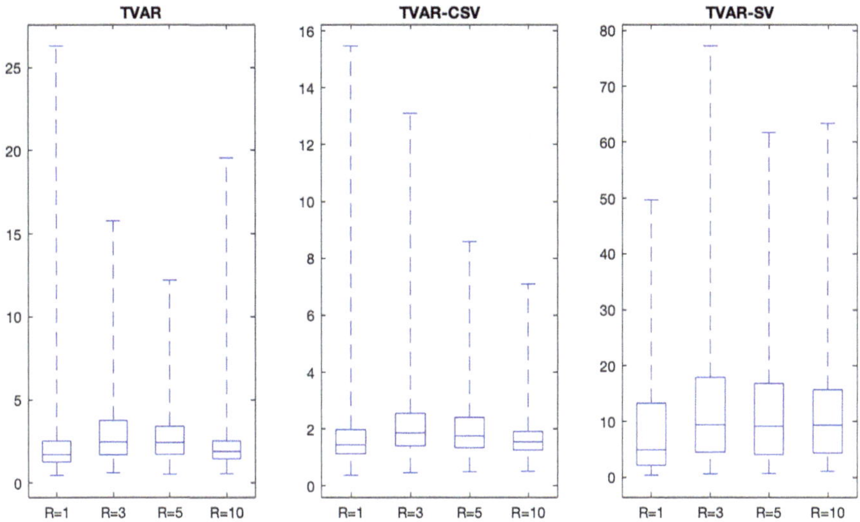

Fig. 1 Boxplots of the inefficiency factors of the posterior draws from the TVARs

6 Concluding Remarks and Future Research

We have developed Bayesian tensor VARs in which the VAR coefficients are arranged as a third-order tensor and parameterized using a rank-R CP decomposition. We then introduced efficient sampling algorithms to simulate the components of the tensor decomposition. Through a forecasting exercise, we showed that these TVARs outperform the standard BVAR with the Minnesota prior.

For future research, it would be useful to extend these TVARs to allow for time-varying coefficients in the mean equations. This tensor framework is especially suitable for developing time variation in the VAR coefficients, as the number of free parameters grows only linear in n. Of course, for large systems one might need additional shrinkage. In those cases, one can consider either the static shrinkage approach in Chan (2023b) or the dynamic shrinkage approaches proposed in Koop and Korobilis (2018) and Huber et al. (2019).

Appendix 1: Proofs of Propositions

In this appendix we provide a proof of Proposition 1. The proof of Proposition 2 is very similar and is thus omitted. We first introduce some useful notations. Let $\mathbf{e}_i^m$ denote the i-th column of $\mathbf{I}_m$; if there is no ambiguity about the dimension, we simply write $\mathbf{e}_i$. For a generic third-order tensor $\mathcal{Z}$, let $\mathcal{Z}_{(k)}$ denote its mode-k matricization for $k = 1, 2, 3$. We use the notation $\mathcal{Z}_{i_1,i_2,i_3}$ to represent the (i_1, i_2, i_3) tensor element of $\mathcal{Z}$ and $\mathcal{Z}_{(k),i,j}$ to denote the (i, j) element of $\mathcal{Z}_{(k)}$.

As an example, consider the third-order tensor $\mathcal{X} \in \mathbb{R}^{T\times n\times p}$. Its mode-2 matricization $\mathcal{X}_{(2)}$ is an $n \times Tp$ matrix, and the (i_1, i_2, i_3) tensor element $\mathcal{X}_{i_1,i_2,i_3}$ maps to the (i_2, j) matrix element $\mathcal{X}_{(2),i_2,j}$, where $j = (i_3-1)T+i_1$. Conversely, the (i, j) matrix element maps to the (i_1, i, i_3) tensor element, where $i_1 = \text{mod}(j, T)$ is the remainder after division of j by T, and $i_3 = \lceil j/T \rceil$ is the least integer greater than or equal to j/T. It can be easily verified that $j = (i_3 - 1)T + i_1$.

Naturally, any matrix multiplication involving matrices constructed by mode-k matricization can be written in terms of the elements in the original tensors. For example, consider the (i, j) element of the matrix $\mathbf{E} = \mathcal{X}'_{(2)}\mathcal{A}_{(2)}$. Let $i_3 = \lceil i/T \rceil$ and $i_1 = \text{mod}(i, T)$ so that $i = (i_3 - 1)T + i_1$. Similarly, obtain integers j_1 and j_3 so that $j = (j_3 - 1)n + j_1$. Then, we can write $\mathbf{E}_{i,j}$ in terms of elements in $\mathcal{X}$ and $\mathcal{A}$:

$$\begin{aligned}\mathbf{E}_{i,j} &= \sum_{k=1}^{n} \mathcal{X}_{(2),k,i}\mathcal{A}_{(2),k,j} \\ &= \sum_{k=1}^{n} \mathcal{X}_{(2),k,(i_3-1)T+i_1}\mathcal{A}_{(2),k,(j_3-1)T+j_1} = \sum_{k=1}^{n} \mathcal{X}_{i_1,k,i_3}\mathcal{A}_{j_1,k,j_3}.\end{aligned} \tag{14}$$

Next, we introduce a useful lemma.

Lemma 1 *The VAR in* (1) *can be written as*

$$\mathbf{Y} = \sum_{i=1}^{p} (\mathbf{e}'_i \otimes \mathbf{I}_T)\mathbf{E}\mathbf{P}(\mathbf{I}_n \otimes \mathbf{e}_i) + \mathbf{U}, \tag{15}$$

where $\mathbf{e}_i \equiv \mathbf{e}_i^p$ *is the* i*-th column of* $\mathbf{I}_p$, $\mathbf{E} = \mathcal{X}'_{(2)}\mathcal{A}_{(2)}$, *and* $\mathbf{P}$ *is the* $np \times np$ *commutation matrix in which* $\mathbf{P}_{k,l} = 1$ *if there exist integers* i *and* j *such that* $k = (i - 1)n + j$ *and* $l = (j - 1)p + i$*; otherwise,* $\mathbf{P}_{k,l} = 0$.

Proof of Lemma 1 First, note that from

$$\mathbf{Y} = \mathcal{X}_{(1)}\mathcal{A}'_{(1)} + \mathbf{U},$$

the element $\mathbf{Y}_{t,l}$ can be expressed as

$$\mathbf{Y}_{t,l} = \sum_{i=1}^{np} \mathcal{X}_{(1),t,i}\mathcal{A}_{(1),l,i} + \mathbf{U}_{t,l} = \sum_{i_1=1}^{n}\sum_{i_3=1}^{p} \mathcal{X}_{t,i_1,i_3}\mathcal{A}_{l,i_1,i_3} + \mathbf{U}_{t,l},$$

where $i = (i_3 - 1)n + i_1$. Now, it suffices to show that the (t, l) element of the sum on the right-hand side of equation (15) has the same expression and is thus equal to $\mathbf{Y}_{t,l}$.

We start by describing the typical element of the $Tp \times np$ matrix $\mathbf{EP}$. For any (i, j), obtain the pairs of integers (i_1, i_3) and (j_1, j_3) such that $i = (i_3 - 1)T + i_1$

and $j = (j_3 - 1)p + j_1$. Then, by direct computation, we have

$$\begin{aligned}
(\mathbf{EP})_{i,j} =& (\mathbf{EP})_{(i_3-1)T+i_1,(j_3-1)p+j_1} \\
=& \sum_{s=1}^{np} \mathbf{E}_{(i_3-1)T+i_1,s} \mathbf{P}_{s,(j_3-1)p+j_1} \\
=& \sum_{k=1}^{p} \sum_{l=1}^{n} \mathbf{E}_{(i_3-1)T+i_1,(k-1)n+l} \mathbf{P}_{(k-1)n+l,(j_3-1)p+j_1} \\
=& \sum_{k=1}^{p} \sum_{l=1}^{n} \mathbf{E}_{(i_3-1)T+i_1,(k-1)n+l} 1(k = j_1, l = j_3) \\
=& \mathbf{E}_{(i_3-1)T+i_1,(j_1-1)n+j_3}.
\end{aligned}$$

Then, the (t, l) element of the sum on the right-hand side of Eq. (15) is given by

$$\begin{aligned}
& \sum_{k=1}^{p} \sum_{i=1}^{Tp} \sum_{j=1}^{np} (\mathbf{e}_k' \otimes \mathbf{I}_T)_{t,i} (\mathbf{EP})_{i,j} (\mathbf{I}_n \otimes \mathbf{e}_k)_{j,l} + \mathbf{U}_{t,l} \\
&= \sum_{k=1}^{p} \sum_{i_1=1}^{T} \sum_{i_3=1}^{p} \sum_{j_3=1}^{n} \sum_{j_1=1}^{p} (\mathbf{e}_k' \otimes \mathbf{I}_T)_{t,(i_3-1)T+i_1} \mathbf{E}_{(i_3-1)T+i_1,(j_1-1)n+j_3} \\
&\quad \times (\mathbf{I}_n \otimes \mathbf{e}_k)_{(j_3-1)p+j_1,l} + \mathbf{U}_{t,l} \\
&= \sum_{k=1}^{p} \sum_{i_1=1}^{T} \sum_{i_3=1}^{p} \sum_{j_3=1}^{n} \sum_{j_1=1}^{p} (\mathbf{e}_{k,i_3}' \mathbf{I}_{T,t,i_1}) \mathbf{E}_{(i_3-1)T+i_1,(j_1-1)n+j_3} (\mathbf{I}_{n,j_3,l} \mathbf{e}_{k,j_1}) + \mathbf{U}_{t,l} \\
&= \sum_{k=1}^{p} \sum_{i_1=1}^{T} \sum_{i_3=1}^{p} \sum_{j_3=1}^{n} \sum_{j_1=1}^{p} 1(k = i_3, t = i_1) \mathbf{E}_{(i_3-1)T+i_1,(j_1-1)n+j_3} \\
&\quad \times 1(j_3 = l, k = j_1) + \mathbf{U}_{t,l} \\
&= \sum_{k=1}^{p} \mathbf{E}_{(k-1)T+t,(k-1)n+l} + \mathbf{U}_{t,l} \\
&= \sum_{k=1}^{p} \sum_{s=1}^{n} \mathcal{X}_{t,s,k} \mathcal{A}_{l,s,k} + \mathbf{U}_{t,l},
\end{aligned}$$

where the last equality holds because of (14). Hence, we have shown that the (t, l) element of the sum on the right-hand side is $\mathbf{Y}_{t,l}$, thus completing the proof. □

Proof of Proposition 1 We begin by taking the transpose of both sides of Eq. (15) in Lemma 1:

$$\mathbf{Y}' = \sum_{i=1}^{p} (\mathbf{I}_n \otimes \mathbf{e}_i')\mathbf{P}'\mathbf{E}'(\mathbf{e}_i \otimes \mathbf{I}_T) + \mathbf{U}'.$$

Next, vectorizing both sides of the above equation, we have

$$\begin{aligned}
\text{vec}(\mathbf{Y}') &= \sum_{i=1}^{p} \left((\mathbf{e}_i' \otimes \mathbf{I}_T) \otimes (\mathbf{I}_n \otimes \mathbf{e}_i')\mathbf{P}'\right) \text{vec}(\mathbf{E}') + \text{vec}(\mathbf{U}') \\
&= \sum_{i=1}^{p} (\mathbf{P}_{i2} \otimes \mathbf{P}_{i1})\text{vec}(\mathbf{E}') + \text{vec}(\mathbf{U}') \\
&= \sum_{i=1}^{p} (\mathbf{P}_{i2} \otimes \mathbf{P}_{i1})\text{vec}(\mathcal{A}_{(2)}'\mathcal{X}_{(2)}) + \text{vec}(\mathbf{U}') \\
&= \sum_{i=1}^{p} (\mathbf{P}_{i2} \otimes \mathbf{P}_{i1})\text{vec}(\mathbf{\Theta}_{-2}\mathbf{\Theta}_2'\mathcal{X}_{(2)}) + \text{vec}(\mathbf{U}') \qquad (16) \\
&= \sum_{i=1}^{p} (\mathbf{P}_{i2} \otimes \mathbf{P}_{i1})(\mathcal{X}_{(2)}' \otimes \mathbf{\Theta}_{-2})\text{vec}(\mathbf{\Theta}_2') + \text{vec}(\mathbf{U}') \\
&= \sum_{i=1}^{p} (\mathbf{P}_{i2}\mathcal{X}_{(2)}' \otimes \mathbf{P}_{i1}\mathbf{\Theta}_{-2})\boldsymbol{\theta}_2 + \text{vec}(\mathbf{U}'),
\end{aligned}$$

where $\mathbf{P}_{i1} = (\mathbf{I}_n \otimes \mathbf{e}_i')\mathbf{P}'$ and $\mathbf{P}_{i2} = \mathbf{e}_i' \otimes \mathbf{I}_T$, $\boldsymbol{\theta}_2 = \text{vec}(\mathbf{\Theta}_2')$. The fourth equality holds because $\mathcal{A}_{(2)} = \mathbf{\Theta}_2\mathbf{\Theta}_{-2}'$. □

Appendix 2: Data

The quarterly dataset is sourced from the FRED-QD database (McCracken & Ng, 2020) maintained by the Federal Reserve Bank of St. Louis. The sample spans from 1969Q1 to 2024Q1. Table 5 lists the variables and describes how they are transformed. After the appropriate transformations, they are further standardized to have zero mean and unit variance to ensure comparability.

Table 5 A list of variables and their transformations. Tcode: 1: no transformation; 2: Δx_t; 5: $\Delta \log(x_t)$; and 6: $\Delta^2 \log(x_t)$

Name	Description	T-code
RPI	Real Personal Income	5
INDPRO	IP Index	5
GDP	Real Gross Domestic Product	5
GDPDEFL	GDP deflator	6
DPCERA3M086SBEA	Real PCE	5
CMRMTSPLx	Real M & T Sales	5
HWI	Help-Wanted Index for US	2
HWIURATIO	Help Wanted to Unemployed ratio	2
CLF16OV	Civilian Labor Force	5
UNRATE	Civilian Unemployment Rate	2
PAYEMS	All Employees: Total nonfarm	5
CES0600000007	Hours: Goods-Producing	5
CPIAUCSL	CPI: All Items	6
FEDFUNDS	Effective Federal Funds Rate	2
TB3MS	3-Month T-bill	2
TB6MS	6-Month T-bill	2
GS1	1-Year T-bond	2
GS5	5-Year T-bond	2
GS10	10-Year T-bond	2
AAA	Aaa Corporate Bond Yield	2
BAA	Baa Corporate Bond Yield	2
M1SL	M1 Money Stock	5
M2SL	M2 Money Stock	5
BUSLOANS	Commercial and Industrial Loans	5
NONREVSL	Total Nonrevolving Credit	5
INVEST	Securities in Bank Credit	5
S&P 500	S&P 500	5
S&P div yield	S&P Dividend yield	2
S&P PE ratio	S&P Price/Earnings ratio	5
EXSZUSx	Switzerland/U.S. FX Rate	5
EXJPUSx	Japan/U.S. FX Rate	5
EXUSUKx	U.S./U.K. FX Rate	5
EXCAUSx	Canada/U.S. FX Rate	5
UEMPMEAN	Average Duration of Unemployment	5
AWHMAN	Hours: Manufacturing	1
ISRATIOx	Inventories to Sales Ratio	2
REALLN	Real Estate Loans	5
PPICMM	PPI: Commodities	6
PCEPI	PCE: Chain-Type Price Index	6
FPI	Fixed Private Investment	5

Appendix 3: Additional Estimation Details

In this appendix we provide technical details on the alternative approach of sampling each $\boldsymbol{\theta}_2^{(r)}$ at a time for $r = 1, \ldots, R$. To that end, we derive the conditional distribution of $\boldsymbol{\theta}_2^{(r)}$ given $\{\boldsymbol{\theta}_2^{(s)}\}_{s \neq r}$ and other model parameters.

We assume the following the marginal prior for $\boldsymbol{\theta}_2^{(r)}$:

$$\boldsymbol{\theta}_2^{(r)} \sim \mathcal{N}\left(\boldsymbol{\theta}_{2,0}^{(r)}, \mathbf{V}_{\boldsymbol{\theta}_2^{(r)}}\right).$$

Next, it follows from Eq. (16) in the proof of Proposition 1 that

$$\begin{aligned}\text{vec}(\mathbf{Y}') &= \sum_{i=1}^{p} (\mathbf{P}_{i2} \otimes \mathbf{P}_{i1}) \text{vec}(\boldsymbol{\Theta}_{-2} \boldsymbol{\Theta}_2' \mathcal{X}_{(2)}) + \text{vec}(\mathbf{U}') \\ &= \sum_{i=1}^{p} \text{vec}(\mathbf{P}_{i1} \boldsymbol{\Theta}_{-2} \boldsymbol{\Theta}_2' \mathcal{X}_{(2)} \mathbf{P}_{i2}') + \text{vec}(\mathbf{U}').\end{aligned}$$

Noting that $\boldsymbol{\Theta}_2 \boldsymbol{\Theta}'_{-2} = \sum_{s=1}^{R} \boldsymbol{\theta}_2^{(s)} \boldsymbol{\theta}_{-2}^{(s)\prime}$, where $\boldsymbol{\theta}_{-2}^{(s)} = \boldsymbol{\theta}_3^{(s)} \otimes \boldsymbol{\theta}_1^{(s)}$, one can express $\mathbf{Y}'$ as

$$\begin{aligned}\mathbf{Y}' &= \sum_{i=1}^{p} \mathbf{P}_{i1} \boldsymbol{\Theta}_{-2} \boldsymbol{\Theta}_2' \mathcal{X}_{(2)} \mathbf{P}_{i2}' + \mathbf{U}' \\ &= \sum_{i=1}^{p} \sum_{s=1}^{R} \mathbf{P}_{i1} \boldsymbol{\theta}_{-2}^{(s)} \boldsymbol{\theta}_2^{(s)\prime} \mathcal{X}_{(2)} \mathbf{P}_{i2}' + \mathbf{U}'.\end{aligned}$$

Fix r, and let $\mathbf{Y}_{2r} = \mathbf{Y}' - \sum_{i=1}^{p} \sum_{s \neq r} \mathbf{P}_{i1} \boldsymbol{\theta}_{-2}^{(s)} \boldsymbol{\theta}_2^{(s)\prime} \mathcal{X}_{(2)} \mathbf{P}_{i2}'$. Vectorize the above equation to get

$$\text{vec}(\mathbf{Y}_{2r}) = \sum_{i=1}^{p} (\mathbf{P}_{i2} \mathcal{X}_{(2)}' \otimes \mathbf{P}_{i1} \boldsymbol{\theta}_{-2}^{(r)}) \boldsymbol{\theta}_2^{(r)} + \text{vec}(\mathbf{U}').$$

Therefore, the posterior conditional distribution $(\boldsymbol{\theta}_2^{(r)} | \mathbf{Y}, \boldsymbol{\theta}_1, \{\boldsymbol{\theta}_2^{(s)}\}_{s \neq r}, \boldsymbol{\theta}_3, \boldsymbol{\Sigma})$ can be expressed as

$$(\boldsymbol{\theta}_2^{(r)} | \mathbf{Y}, \boldsymbol{\theta}_1, \{\boldsymbol{\theta}_2^{(s)}\}_{s \neq r}, \boldsymbol{\theta}_3, \boldsymbol{\Sigma}) \sim \mathcal{N}\left(\widehat{\boldsymbol{\theta}}_2^{(r)}, \mathbf{K}_{\boldsymbol{\theta}_2^{(r)}}^{-1}\right),$$

where

$$\mathbf{K}_{\boldsymbol{\theta}_2^{(r)}} = \mathbf{V}_{\boldsymbol{\theta}_2^{(r)}}^{-1} + \sum_{i=1}^{p}\sum_{j=1}^{p}(\mathcal{X}_{(2)}\mathbf{P}_{i2}' \otimes \boldsymbol{\theta}_{-2}^{(r)\prime}\mathbf{P}_{i1}')\boldsymbol{\Sigma}^{-1}(\mathbf{P}_{j2}\mathcal{X}_{(2)}' \otimes \mathbf{P}_{j1}\boldsymbol{\theta}_{-2}^{(r)})$$

$$\widehat{\boldsymbol{\theta}}_2^{(r)} = \mathbf{K}_{\boldsymbol{\theta}_2^{(r)}}^{-1}\left(\mathbf{V}_{\boldsymbol{\theta}_2^{(r)}}^{-1}\boldsymbol{\theta}_{2,0}^{(r)} + \sum_{i=1}^{p}(\mathcal{X}_{(2)}\mathbf{P}_{i2}' \otimes \boldsymbol{\theta}_{-2}^{(r)\prime}\mathbf{P}_{i1}')\boldsymbol{\Sigma}^{-1}\mathrm{vec}(\mathbf{Y}_{2r})\right).$$

References

Arias, J. E., Rubio-Ramirez, J. F., & Shin, M. (2023). Macroeconomic forecasting and variable ordering in multivariate stochastic volatility models. *Journal of Econometrics, 235*(2), 1054–1086.

Bańbura, M., Giannone, D., & Reichlin, L. (2010). Large Bayesian vector auto regressions. *Journal of Applied Econometrics, 25*(1), 71–92.

Bernardi, M., Bianchi, D., & Bianco, N. (2024). Variational inference for large Bayesian vector autoregressions. *Journal of Business and Economic Statistics, 42*, 1066–1082.

Bertsche, D., & Braun, R. (2022). Identification of structural vector autoregressions by stochastic volatility. *Journal of Business & Economic Statistics, 40*(1), 328–341.

Bhattacharya, A., Pati, D., Pillai, N. S., & Dunson, D. B. (2015). Dirichlet–Laplace priors for optimal shrinkage. *Journal of the American Statistical Association, 110*(512), 1479–1490.

Billio, M., Casarin, R., Iacopini, M., & Kaufmann, S. (2023). Bayesian dynamic tensor regression. *Journal of Business & Economic Statistics, 41*(2), 429–439.

Carriero, A., Chan, J. C. C., Clark, T. E., & Marcellino, M. G. (2022). Corrigendum to "Large Bayesian vector autoregressions with stochastic volatility and non-conjugate priors". *Journal of Econometrics, 227*(2), 506–512.

Carriero, A., Clark, T. E., Marcellino, M., & Mertens, E. (2024). Addressing COVID-19 outliers in BVARs with stochastic volatility. *The Review of Economics and Statistics, 106*(5), 1403–1417.

Carriero, A., Clark, T. E., & Marcellino, M. G. (2015). Bayesian VARs: Specification choices and forecast accuracy. *Journal of Applied Econometrics, 30*(1), 46–73.

Carriero, A., Clark, T. E., & Marcellino, M. G. (2016). Common drifting volatility in large Bayesian VARs. *Journal of Business and Economic Statistics, 34*(3), 375–390.

Carriero, A., Clark, T. E., & Marcellino, M. G. (2019). Large Bayesian vector autoregressions with stochastic volatility and non-conjugate priors. *Journal of Econometrics, 212*(1), 137–154.

Carriero, A., Kapetanios, G., & Marcellino, M. (2009). Forecasting exchange rates with a large Bayesian VAR. *International Journal of Forecasting, 25*(2), 400–417.

Carriero, A., Kapetanios, G., & Marcellino, M. (2011). Forecasting large datasets with Bayesian reduced rank multivariate models. *Journal of Applied Econometrics, 26*(5), 735–761.

Carvalho, C. M., Polson, N. G., & Scott, J. G. (2010). The horseshoe estimator for sparse signals. *Biometrika, 97*(2), 465–480.

Chan, J. C. C. (2020a). Large Bayesian VARs: A flexible Kronecker error covariance structure. *Journal of Business and Economic Statistics, 38*(1), 68–79.

Chan, J. C. C. (2020b). Large Bayesian Vector Autoregressions. In P. Fuleky (Ed.) *Macroeconomic forecasting in the era of big data* (pp. 95–125). Springer.

Chan, J. C. C. (2021). Minnesota-type adaptive hierarchical priors for large Bayesian VARs. *International Journal of Forecasting, 37*(3), 1212–1226.

Chan, J. C. C. (2022). Asymmetric conjugate priors for large Bayesian VARs. *Quantitative Economics, 13*(3), 1145–1169.

Chan, J. C. C. (2023a). Comparing stochastic volatility specifications for large Bayesian VARs. *Journal of Econometrics, 235*(2), 1419–1446.

Chan, J. C. C. (2023b). Large hybrid time-varying parameter VARs. *Journal of Business and Economic Statistics, 41*(3), 890–905.

Chan, J. C. C. (2024). BVARs and stochastic volatility. In M. P. Clements, & A. B. Galvao (Eds.), *Handbook of research methods and applications in macroeconomic forecasting*. Edward Elgar.

Chan, J. C. C., & Eisenstat, E. (2018). Bayesian model comparison for time-varying parameter VARs with stochastic volatility. *Journal of Applied Econometrics, 33*(4), 509–532.

Chan, J. C. C., Koop, G., & Yu, X. (2024). Large order-invariant Bayesian VARs with stochastic volatility. *Journal of Business and Economic Statistics, 42*(2), 825–837.

Clark, T. E. (2011). Real-time density forecasts from Bayesian vector autoregressions with stochastic volatility. *Journal of Business and Economic Statistics, 29*(3), 327–341.

Clark, T. E., & Mertens, E. (2023). Stochastic volatility in Bayesian vector autoregressions. In *Oxford research encyclopedia of economics and finance*. Oxford University Press.

Clark, T. E., & Ravazzolo, F. (2015). Macroeconomic forecasting performance under alternative specifications of time-varying volatility. *Journal of Applied Econometrics, 30*(4), 551–575.

Cogley, T., & Sargent, T. J. (2005). Drifts and volatilities: Monetary policies and outcomes in the post WWII US. *Review of Economic Dynamics, 8*(2), 262–302.

Cross, J., & Poon, A. (2016). Forecasting structural change and fat-tailed events in Australian macroeconomic variables. *Economic Modelling, 58*, 34–51.

D'Agostino, A., Gambetti, L., & Giannone, D. (2013). Macroeconomic forecasting and structural change. *Journal of Applied Econometrics, 28*, 82–101.

Doan, T., Litterman, R., & Sims, C. (1984). Forecasting and conditional projection using realistic prior distributions. *Econometric Reviews, 3*(1), 1–100.

Follett, L., & Yu, C. (2019). Achieving parsimony in Bayesian VARs with the horseshoe prior. *Econometrics and Statistics, 11*, 130–144.

Gefang, D., Koop, G., & Poon, A. (2023). Variational Bayesian inference in large vector autoregressions with hierarchical shrinkage. *International Journal of Forecasting, 39*(1), 346–363.

Giannone, D., Lenza, M., & Primiceri, G. E. (2015). Prior selection for vector autoregressions. *Review of Economics and Statistics, 97*(2), 436–451.

Griffin, J., & Brown, P. (2010). Inference with normal-gamma prior distributions in regression problems. *Bayesian Analysis, 5*(1), 171–188.

Hauzenberger, N., Huber, F., & Koop, G. (2024). Macroeconomic forecasting using BVARs. In M. P. Clements, & A. B. Galvao (Eds.), *Handbook of research methods and applications in macroeconomic forecasting*. Edward Elgar.

Huber, F., & Feldkircher, M. (2019). Adaptive shrinkage in Bayesian vector autoregressive models. *Journal of Business and Economic Statistics, 37*(1), 27–39.

Huber, F., Koop, G., & Onorante, L. (2019). Inducing sparsity and shrinkage in time-varying parameter models. arXiv preprint arXiv:1905.10787.

Kadiyala, R. K., & Karlsson, S. (1993). Forecasting with generalized Bayesian vector autoregressions. *Journal of Forecasting, 12*(3–4), 365–378.

Kadiyala, R. K., & Karlsson, S. (1997). Numerical methods for estimation and inference in Bayesian VAR-models. *Journal of Applied Econometrics, 12*(2), 99–132.

Kastner, G., & Huber, F. (2020). Sparse Bayesian vector autoregressions in huge dimensions. *Journal of Forecasting, 39*(7), 1142–1165.

Kolda, T. G., & Bader, B. W. (2009). Tensor decompositions and applications. *SIAM Review, 51*(3), 455–500.

Koop, G. (2013). Forecasting with medium and large Bayesian VARs. *Journal of Applied Econometrics, 28*(2), 177–203.

Koop, G., & Korobilis, D. (2013). Large time-varying parameter VARs. *Journal of Econometrics, 177*(2), 185–198.

Koop, G., & Korobilis, D. (2018). Variational Bayes inference in high-dimensional time-varying parameter models. Available at SSRN 3246472.

Korobilis, D., & Pettenuzzo, D. (2019). Adaptive hierarchical priors for high-dimensional vector autoregressions. *Journal of Econometrics, 212*, 241–271.

Lenza, M., & Primiceri, G. E. (2022). How to estimate a VAR after march 2020. *Journal of Applied Econometrics, 37*(4), 688–699.

Litterman, R. (1986). Forecasting with Bayesian vector autoregressions—five years of experience. *Journal of Business and Economic Statistics, 4*, 25–38.

Luo, Y., & Griffin, J. E. (2023). Bayesian inference of vector autoregressions with tensor decompositions. arXiv preprint arXiv:2211.01727.

McCracken, M. W., & Ng, S. (2020). FRED-QD: A quarterly database for macroeconomic research. *Federal Reserve Bank of St. Louis Working Paper*, 2020-005.

Park, T., & Casella, G. (2008). The Bayesian Lasso. *Journal of the American Statistical Association, 103*(482), 681–686.

Primiceri, G. E. (2005). Time varying structural vector autoregressions and monetary policy. *Review of Economic Studies, 72*(3), 821–852.

Sims, C. A., & Zha, T. (2006). Were there regime switches in U.S. monetary policy? *American Economic Review, 96*(1), 54–81.

Stock, J. H., & Watson, M. W. (2016). Core inflation and trend inflation. *Review of Economics and Statistics, 98*(4), 770–784.

Wang, D., Zheng, Y., & Li, G. (2024). High-dimensional low-rank tensor autoregressive time series modeling. *Journal of Econometrics, 238*(1), 105544.

Wang, D., Zheng, Y., Lian, H., & Li, G. (2022). High-dimensional vector autoregressive time series modeling via tensor decomposition. *Journal of the American Statistical Association, 117*(539), 1338–1356.

Measuring Subregional Economic Activity: Missing Frequencies and Missing Data

Gary Koop, Stuart McIntyre, James Mitchell, Aubrey Poon, and Ping Wu

Abstract Bayesian mixed-frequency vector autoregressions (MF-VARs) are commonly used to produce timely and high-frequency estimates of low-frequency variables. A typical application uses quarterly data on output, for a given country, and monthly indicator data to produce monthly estimates of national output. But, when working at subnational levels, data limitations preclude the use of standard MF-VARs. The frequency mismatch is more complicated, key variables can have missing data, and release delays can be substantial. In this chapter, we develop a novel MF-VAR that addresses all these issues and use it to produce historical estimates of subregional output growth in the UK. The model combines information in the annual subregional data (when available) with data from the UK regions and the UK as a whole. The model is estimated using variational Bayesian methods with shrinkage priors, reflecting the "big data" setup. We use our model to produce a new database of quarterly estimates of subregional GVA growth back to the

---

We thank the editors and an anonymous referee for helpful comments. This research has been funded by the Office of National Statistics as part of the research programme of the Economic Statistics Centre of Excellence (ESCoE). The views expressed herein are those of the authors and not necessarily those of the Federal Reserve Bank of Cleveland or the Federal Reserve System.

G. Koop · S. McIntyre · P. Wu
University of Strathclyde, Glasgow, UK

ESCoE, London, UK
e-mail: gary.koop@strath.ac.uk; s.mcintyre@strath.ac.uk; ping.wu@strath.ac.uk

J. Mitchell (✉)
Federal Reserve Bank of Cleveland, Cleveland, OH, USA

ESCoE, London, UK
e-mail: james.mitchell@clev.frb.org

A. Poon
University of Kent, Canterbury, UK

ESCoE, London, UK
e-mail: a.poon@kent.ac.uk

S. Mazur, P. Österholm (eds.), *Recent Developments in Bayesian Econometrics and Their Applications*, https://doi.org/10.1007/978-3-032-00110-8_4

1960s, that importantly, because the MF-VAR imposes temporal and cross-sectional restrictions, is consistent with those official data that do exist. We illustrate the use of these new estimates by showing how they can be used to characterize the considerable heterogeneity in subregional business cycle dynamics in the UK and contribute to our understanding of regional economic resilience.

1 Introduction

Mixed-frequency vector autoregressions (MF-VARs) are a popular tool for producing timely and high-frequency estimates of macroeconomic variables such as GDP growth, given that official estimates of GDP tend to be made available only at a lower frequency and are published with a lag. Most such applications of MF-VARs to temporal disaggregation are at the national level, e.g., see Schorfheide and Song (2015). But, in many countries, there is also a keen interest in issues of regional or subregional resilience and inequality, making it desirable to produce similar estimates at a finer level of geographical granularity. However, particularly at more spatially granular (subregional) levels, efforts to produce subregional historical estimates of GDP have been stymied by the limitations of the data. At the national level, the higher-frequency "indicator" data used to help inform on the (unobserved) monthly path of GDP growth are not "missing," have decent historical coverage, and are released fairly quickly. For instance, Schorfheide and Song (2015) produce estimates of monthly GDP in the USA based on quarterly GDP data and several monthly national indicators. The release delay for quarterly GDP itself is short (i.e., the so-called advance estimate of US GDP is released by the Bureau of Economic Analysis roughly a month after the end of each quarter). National data has similar properties in other countries.

At the regional and subregional levels,[1] data availability is more restricted. In this chapter, our methods are designed around the specific characteristics of the data available in the UK. But many countries face similar data challenges at the subnational level, and our methods could be used more widely, with minor alterations. In the UK, regional and subregional data are often lower frequency, and at times even completely unavailable. In particular, subregional gross value-added (GVA) data are missing before 1998.[2] Aggregating up one level, the regional data are "better," in the sense that they are available in some form (in nominal terms pre-1998, and in real terms thereafter) back to 1970. But their frequency changes over time: they were annual through 2011, and then switched to quarterly from 2012. Quarterly output data for Scotland and Northern Ireland are available over a slightly longer time period than the other ITL1 regions of the UK. Given

[1] When we refer to regional (subregional) data, we mean at the International Territorial Level (ITL) 1 (ITL2) level. When we refer to output, we mean Gross Value Added (GVA).

[2] Real GVA and real GDP are closely related concepts. GVA is in basic prices, while GDP is in market prices (i.e., GVA plus taxes (less subsidies) on products equals GDP).

our focus on subregional analysis, we treat the regional data as starting when the regional short-term indicators (RSTI) data produced by the ONS in the UK, described below, become available from 2012. The goal of this chapter is then to incorporate subregional, regional, and national data on output growth, each with their own properties, into an MF-VAR so as to produce higher-frequency (quarterly) historical estimates of subregional output growth.

The MF-VAR we construct differs from the conventional national setup for temporal disaggregation, as seen in Schorfheide and Song (2015), in several ways. First, the MF-VAR is much larger. Even if we only use GVA growth data, with one nation, 12 regions, and 41 subregions, our MF-VAR will involve 54 equations. Adding additional indicator variables, either at the national or regional level, will further increase the dimension of the model. Second, it involves two kinds of *cross-sectional restriction*. One is based on the fact that output for any region must be the sum of outputs for its subregions. The other is that national GVA is the sum of the regions. Third, it involves a very large number of unobserved missing values to fill in. This arises since quarterly output data are never in fact observed at the subregional level and only observed subsequently to 2012 for most regions. Fourth, the subregional data are completely missing, even at the annual frequency, prior to 1998. Finally, the frequency of the regional data changes over time. The MF-VAR that we develop accommodates all of these messy data features.

We use our MF-VAR to produce a time series of quarterly subregional output growth data back to the 1960s. We illustrate the use of these new data by characterizing subregional business cycle dynamics in the UK. Such an analysis is not possible with extant lower-frequency and incomplete subregional data. We find that subregions of the UK differ significantly in terms of the timing of their switches between recessionary and expansionary regimes. Our results are thus consistent with those in Owyang et al. (2005), who find that state-level business cycles in the USA are also not all alike. We also show a growing tendency, since the UK recession of 2008, for subregions of the UK to enter "local" recessions, suggesting that business cycle dynamics in the UK have become more heterogeneous in recent years and increasingly out of sync with national phases. We end by using our new database to measure (sub-)regional "economic resilience" across several dimensions commonly considered in economic geography.

2 National, Regional, and Subregional Data

We use "official" data from the Office for National Statistics (ONS) on GVA at the subregional, regional, and national (UK) levels. Three characteristics of these data are relevant: (1) the number of variables, (2) the sample span, and (3) their frequency:

- **UK GVA Data:** At the national level, the real GVA sample begins in 1966, and is at the quarterly frequency.

- **Regional GVA Data:** There are 12 ITL1 regions in the UK. The sample begins with annual regional GVA data from 1966 through 1998. These data are available only in nominal terms so, in the absence of regional inflation data, we deflate by the UK deflator.[3] The sample then continues with real-term annual regional GVA data from 1998. Then, from 2012 to the present, the regional short-term indicators (RSTIs) database is available.[4] This database provides quarterly real GVA for the ITL1 regions. They were first published in September 2019.

  How can these regional GVA data be used to improve subregional (ITL2) output estimates? From 1966 to 1997, the regional data will have information useful for producing historical estimates of ITL2 GVA, since ITL2 data are not available for this period. Accordingly, we impose a cross-sectional restriction to ensure the model-interpolated ITL2 data, in a given region, sum to the observed ITL1 data. From 1998 to 2011, the ITL1 data do not provide any information we can exploit, since they are at the same frequency as the ITL2 data. From 2012 onward, we can exploit the fact that the ITL1 data are published at a higher frequency than ITL2 data.
- **Subregional GVA Data:** There are 41 ITL2 subregions. Real GVA data for these subregions are available on an annual basis from 1998, and align with the regional data described above over this period.

  Regional and subregional GVA data for the UK are released with much longer and more variable release delays than equivalent national data. Moreover, there were a series of disruptions to the publication schedule for these data following the onset of the COVID-19 pandemic and, as a result, our sample period ends in 2019.
- **UK macroeconomic variables:** In addition, as in Koop et al. (2020b), we incorporate four additional UK-wide quarterly macroeconomic indicators within our model, to help capture intra-year economic dynamics. These are the consumer price index, the US Dollar exchange rate with the pound sterling, the Bank of England's Bank rate, and the Brent oil price.

3 Econometric Methods

In this section, we develop our MF-VAR which combines the subregional, regional, and UK data into one model so as to produce estimates of quarterly subregional GVA growth. We emphasize that because official subregional GVA data are not available

[3] Note that as set out in the Appendix to Koop et al. (2020a), the definitions of the UK ITL1 geographies have changed over the period since 1966. For example, the South East region was divided into a South East England region and London part way through the sample. We adopt the same approach as that in Koop et al. (2020a) to address these historical geography changes at the ITL1 level. ITL2 geographies have remained constant over the period for which we observe their growth rates.

[4] https://www.ons.gov.uk/economy/grossdomesticproductgdp/datasets/quarterlycountryandregionalgdp

at any frequency prior to 1998, and so over this period our model relies exclusively on the annual regional and quarterly UK output data, the historical estimates of subregional output that our model produces prior to 1998 are likely of lower quality than our post-1998 quarterly estimates. This is because, from 1998, the subregional quarterly estimates of output produced by our MF-VAR can condition on observed official estimates of annual subregional output, as well as quarterly regional and UK data.

3.1 Notation and Key Data Features

We begin by describing some variable definitions, relationships, and defining our notational conventions. All changes and growth rates referred to below are exact (not log-differenced):

- $t = 1, ..., T$ runs at the *quarterly* frequency.
- $r1 = 1, ..., R1$ denotes the $R1$ regions in the UK.
- $r2 = 1, ..., R2$ denotes the $R2$ subregions in the UK.
- Superscripts $UK, 1, 2$ will denote UK, regional (ITL1), and subregional (ITL2) variables.
- y_t^{UK} is UK quarterly GVA growth. Data are available throughout the sample.
- $y_t^{1,r1}$ is quarterly GVA growth for region $r1$. Data are available from 2012 onward.
- $y_t^{2,r2}$ is quarterly GVA growth for subregion $r2$. Data are never available.
- $y_t^{UK,A}$ is UK annual GVA growth. Data are available throughout the sample.
- $y_t^{1,r1,A}$ is annual GVA growth for region $r1$. Data are available through the sample, but only for quarter 4.
- $y_t^{2,r2,A}$ is annual GVA growth for subregion $r2$. Data are available from 1998 onward, but only for quarter 4.

These variables are stacked into vectors, such that: $y_t^j = \left(y_t^{j,1}, ..., y_t^{j,Rj}\right)'$ is the vector of ITLj regional quarterly growth rates, for $j = 1, 2$. $y_t^{j,A}$ are the analogous ITL2 annual growth rates. $y_t = (y_t^{UK}, y_t^1, y_t^2)'$ is the vector containing all the quarterly growth rates. $y_t^A = (y_t^{UK,A}, y_t^{1,A}, y_t^{2,A})'$ is the analogous vector containing all the annual growth rates.

Quarterly variables are constrained to add up to annual quantities via the following vector of inter-temporal restrictions, as popularized by Mariano and Murasawa (2003)

$$y_t^A = \frac{1}{4}y_t + \frac{1}{2}y_{t-1} + \frac{3}{4}y_{t-2} + y_{t-3} + \frac{3}{4}y_{t-4} + \frac{1}{2}y_{t-5} + \frac{1}{4}y_{t-6}. \tag{1}$$

Koop et al. (2020b) discuss why this is the appropriate temporal disaggregation restriction to use when modeling output (GVA) in exact growth rates.

In addition, we exploit the restrictions that UK quantities are the sum of regional quantities and, in turn, regional quantities are the sum of subregional quantities. These so-called cross-sectional restrictions take the form:

$$y_t^{UK} = \sum_{r1=1}^{R1} w_t^{1,r1} y_t^{1,r1} + \eta_{t,UK}, \tag{2}$$

where $w_t^{1,r1}$ is the share of region $r1$ in UK GVA. We allow for an error in this restriction and assume $\eta_{t,UK} \sim N(0, \sigma_{UK}^2)$. The main reason to include an error is to acknowledge the fact that GVA from the regions does not exactly add up to UK GVA, due to the absence of the output from the UK Continental Shelf (UKCS) in any of the regional data. But there are other reasons for including an error. For instance, the weights in the restriction should vary over time at the quarterly frequency, but our data only allow us to calculate them at the annual frequency. Furthermore, data revisions can cause small inconsistencies between UK and regional data.

Next, we have the cross-sectional restrictions that arise from the fact that subregional output adds up to regional output. These restrictions take the form:

$$y_t^{1,r1} = \sum_{r2 \in r1} w_t^{2,r2} y_t^{2,r2} + \eta_{t,1,r1} \tag{3}$$

for $r1 = 1, ..., R1$, where $r2 \in r1$ denotes all the subregions that lie in region $r1$ and $w_t^{2,r2}$ is the share of subregion $r2$ in the GVA of the region it lies within. The errors are assumed to follow: $\eta_{t,r1} \sim N(0, \sigma_{1,r1}^2)$. However, we expect these errors to be smaller than the error in (2), since the issue relating to the UKCS is not relevant at the subregional level. But we still do not expect this subregional constraint to hold exactly, as the weights used in (3) are fixed at their observed 1998 values pre-1998 and, thereafter, are updated annually rather than quarterly. This approximation, forced on us by data limitations, again motivates inclusion of an error, $\eta_{t,1,r1}$, in (3). We choose priors for $\sigma_{1,r1}^2$ for $r1 = 1, ..., R1$ to reflect our belief that these errors should be smaller than σ_{UK}^2.

Finally, we have the cross-sectional restriction that arises from the fact that subregional data add up to UK data:

$$y_t^{UK} = \sum_{r2=1}^{R2} w_t^{2,r2} y_t^{2,r2} + \eta_{t,UK2} \tag{4}$$

for $r2 = 1, ..., R2$, where the errors $\eta_{t,UK2} \sim N(0, \sigma_{2,r2}^2)$. Note that, at first sight, it may seem redundant to incorporate both cross-sectional restrictions, since (3) implies that (4) must hold. But y_t^{UK} is released more quickly than $y_t^{1,r1}$ and, for much of the sample, is available at a higher frequency. These facts could also be exploited to help produce timely and high-frequency nowcasts of subregional data. Intuitively, when regional data are available, our model exploits them as the best source of information to update our subregional estimates. But, when they are not available, this model would fall back to exploiting information in the UK data. In other words, (4) would be useful in periods where UK data are available, but regional data have not yet been released.

3.2 *Bayesian Estimation of the Mixed-Frequency VAR*

The MF-VAR is a state-space model where the measurement equations are given by the inter-temporal and cross-sectional restrictions of the preceding subsection. The state equations of the MF-VAR are based on a VAR with y_t being the vector of dependent variables.

The MF-VAR treats some of the elements of y_t as observed and some as unobserved. The unobserved values are treated as states in a state-space model. But it is worth emphasizing that y_t contains all the quarterly growth rates for the UK, the regions, and the subregions and thus comprises $1+R1+R2 = 54$ variables, which is quite large. And, when we use data back to 1966, the subregional data at both annual and quarterly frequencies are unobserved until 1998. These facts motivate our use of Bayesian methods. These allow us to incorporate a shrinkage prior to deal with the fact that a VAR with p lags requires many coefficients to be estimated. We use the adaptive Lasso prior. Bayesian inference is implemented using the variational Bayes algorithm of Gefang et al. (2020). The remainder of this subsection provides details.

In the large VAR literature, it is increasingly common to work with a transformation of the VAR which has a diagonal error covariance matrix and, thus, allows for each equation to be estimated separately. This is done for computational reasons since working with the reduced-form VAR requires matrix manipulations involving the potentially enormous posterior covariance matrix of the VAR coefficients. To explain the transformation of the VAR, note that the error variance-covariance matrix of the reduced-form VAR, Σ, can be decomposed as $\Sigma = LDL'$, where L is lower triangular with ones on the diagonal and D is a diagonal matrix. If we multiply both sides of the VAR by L^{-1}, then the resulting transformed VAR will have a diagonal error covariance matrix, thus allowing for equation-by-equation estimation.

This transformation leads to a VAR of the following form:

$$y_t = X_t\beta + W_t\alpha + \epsilon_t, \tag{5}$$

where X_t contains lags of the dependent variables along with the quarterly UK macroeconomic indicators. W_t contains the appropriate contemporaneous elements of y_t implied by the transformation using the Cholesky decomposition of Σ. We can rewrite (5) as a series of independent equations, with the ith equation being:

$$y_{i,t} = \tilde{\mathbf{x}}_{i,t}\theta_i + \epsilon_{i,t}, \epsilon_{i,t} \sim N(0, \sigma_i^2), \tag{6}$$

where $\tilde{\mathbf{x}}_{i,t}$ is a row vector with k_i elements and θ_i is a vector containing the elements of β and α pertaining to the ith equation.

Following Gefang et al. (2023), we specify the prior distributions for the parameters in each equation:

$$\theta_i \sim N(0, \mathbf{V}_i), \tag{7}$$

$$\sigma_i^{-2} \sim G(\underline{\nu}, \underline{s}), \tag{8}$$

where G denotes the Gamma distribution.

The adaptive Lasso is a hierarchical prior which treats $\mathbf{V}_i$ as a matrix of unknown parameters with their own prior of the form:

$$\mathbf{V}_i = \text{diag}(\tau_{i,1}, \ldots, \tau_{i,k_i}). \tag{9}$$

Note that this allows for the different equations to have different prior shrinkage. The adaptive Lasso assumes

$$\tau_{i,j} \sim Expon(\tfrac{\lambda_{i,j}}{2}), \text{ for } j = 1, \ldots, k_i, \tag{10}$$

where $Expon$ is the exponential distribution, with

$$\lambda_{i,j} \sim G(\underline{a}_0, \underline{b}_0), \tag{11}$$

and k_i is the number of right-hand-side variables in equation i.

For estimation, we use the computationally efficient variational Bayes algorithm for VARs with an adaptive Lasso prior developed in Gefang et al. (2023) and extended to MF-VAR models in Gefang et al. (2020). Building on the recent work of Chan et al. (2023), to achieve further computational efficiencies, we then extend the Gefang et al. (2020) framework by utilizing a precision-based rather than Kalman filter-based method to estimate the missing quarterly values. Specifically, we implement the formulation of equation (9) of Chan et al. (2023), the joint conditional distribution of the missing data given the observed data, to approximate the missing latent states. We follow Gefang et al. (2023) and make relatively non-informative choices for the prior hyperparameters, $(\underline{a}_0, \underline{b}_0, \underline{\nu}, \underline{s})$. Specifically, we set $\underline{a}_0 = \underline{b}_0 = 0.001$, $\underline{\nu} = 5$, and $\underline{s} = 0.04$.

4 Historical Estimates of GVA Growth at the Subregional Level

Having estimated our MF-VAR, we first look at its historical estimates of regional and subregional output growth to reassure that they satisfy the necessary temporal and cross-sectional aggregation restrictions. Then, we turn to illustrating the use of the new quarterly subregional historical database for business cycle analysis and to inform on regional economic resilience.

4.1 Estimating Unobserved Regional Quarterly GVA

We compare our model-based estimates of quarterly regional and subregional GVA growth to the official, in general lower frequency, data published by the ONS. We focus on the posterior median estimates from our MF-VAR. We proceed by undertaking two sets of comparisons. First, we compare regional growth estimates to the aggregation of our subregional growth estimates. Given that a stochastic aggregation constraint is used to link these two estimates in the MF-VAR, we do not expect them to be exactly the same, but we expect them to be close. Second, we compare our subregional estimates to the published/observed annual data.

To undertake the first comparison, Figs. 1 and 2 present three estimates for each of the 12 (ITL1) regions. The first estimate shown, in red, is the time series of MF-VAR-based quarterly regional growth estimates (plotted at an annualized rate). The second, in black, is the published ONS data, which prior to 2012 are annual (hence the same growth rate is plotted for each quarter of a given year), and thereafter they are quarterly. Note that once a year (in Q4), prior to 2012, the red and black lines intersect. This shows that the inter-temporal restriction, (1), is working. Note also that, for Scotland and Northern Ireland, the post-2012 data are from the Scottish and Northern Ireland governments, respectively. The third series, in blue, is the weighted aggregation of the growth rates at the subregional level.

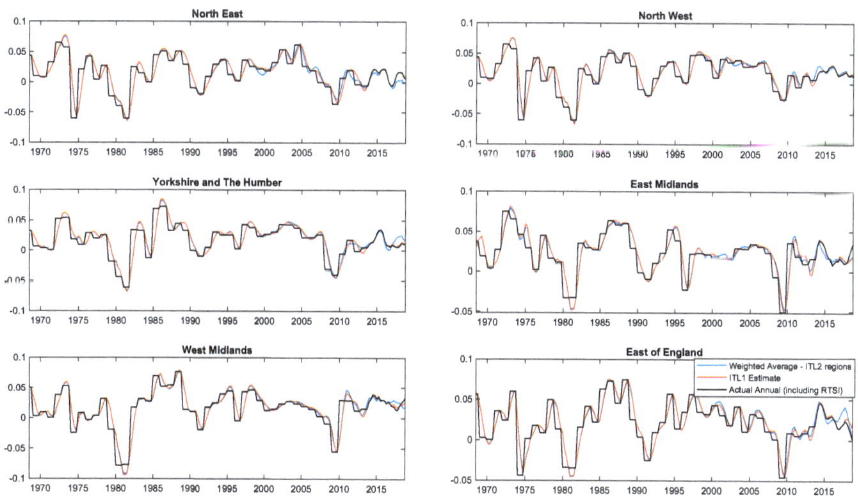

Fig. 1 ITL1 quarterly GVA growth estimates (annualized) and the cross-sectional (weighted) average of the ITL2 estimates from the MF-VAR, alongside the "official" annual estimates

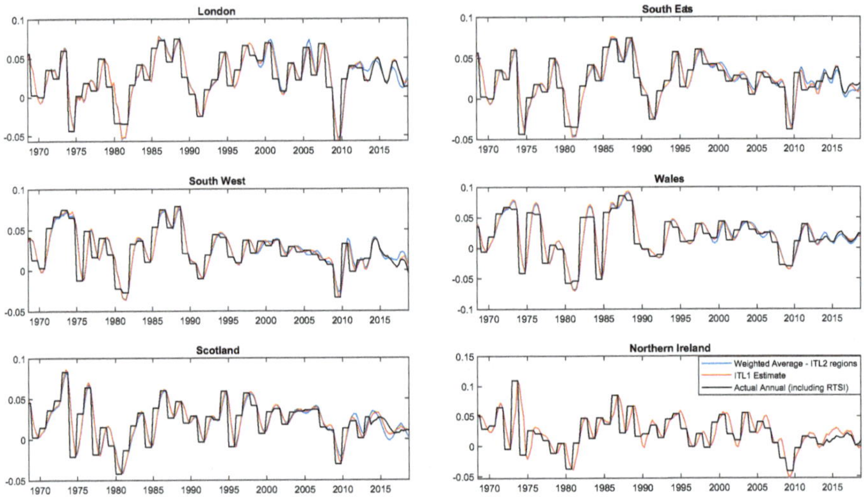

Fig. 2 ITL1 quarterly GVA growth estimates (annualized) and the cross-sectional (weighted) average of the ITL2 estimates from the MF-VAR, alongside the "official" annual estimates

We draw out a few features of these three sets of estimates. The first is that, over much of our sample period, the three series align well. This suggests that the model, and in particular the various constraints within it, is holding. After the introduction of the quarterly regional data from 2012 onward, the estimates align a little less well but still broadly track developments in the regional economies. There are a couple of reasons why the alignment is weaker in this period.

The first helps us understand the divergence between our estimates and the quarterly regional estimates in the 2012–2013 period, specifically that our model calculates annual growth rates using the inter-temporal restriction, (1), which utilizes seven lags of the quarterly regional growth rates. Until 2013Q4, we do not observe all seven lags of regional growth as "official" data. Rather, up until this point in time, the inter-temporal constraint involves a mixture of estimated states from our model and the new ("official") RSTI data. For example, when calculating annualized quarterly growth for 2012Q1 (the red line), our model uses just one-quarter of "hard" RSTI data and 6 quarters' worth of interpolated data for the unobserved quarterly regional growth rates.

The second reason is that, post-2012, the model reconciles three different datasets: the RSTI data from the ONS, the Scottish GDP data produced by the Scottish Government, and the quarterly Northern Ireland Composite Economic Index produced by NISRA. Each of these data sources is constructed in a different way. This explains why in comparing our model-based estimates, which constrain to published UK-wide data, to these published series we should not expect perfect alignment. Moreover, even though the ONS produces the annual regional data that we use in our model, as well as the quarterly regional GDP data that we use for the English regions and Wales, these data themselves do not temporally align exactly.

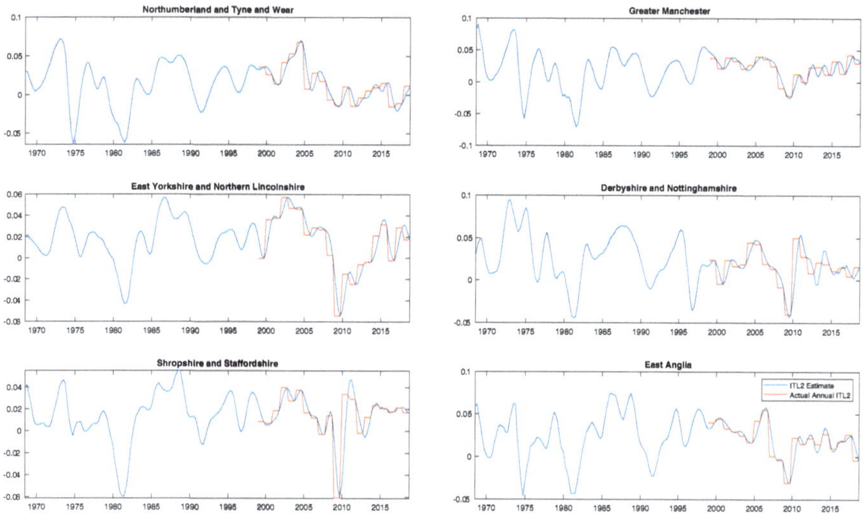

Fig. 3 ITL2 quarterly GVA growth estimates (annualized) from the MF-VAR for 6 (of 41) subregions, alongside the "official" annual estimates

Of course, in principle, they should (over matching sample periods). But release delays and data revisions complicate alignment.

Turning to the second comparison, Figs. 3 and 4 plot the MF-VAR-based subregional growth estimates against the annual estimates from the ONS, for the period from 1998 for which the latter are available. For space reasons, we focus on showing estimates for a selected subset of 12 of the 41 subregions of the UK. Again Figs. 3 and 4 show that once a year, in Q4, the MF-VAR-based estimates intersect, as they should, with the official annual estimates available from 1998.

4.2 *Heterogeneous Business Cycle Phases Across UK Subregions*

In this section, we illustrate how the new quarterly subregional data can be used to understand, at a higher temporal frequency than possible previously, the time profiles of recession and recovery in the subregions of the UK. We compare these with the five main recessions the UK as a whole has experienced since 1966. Our sample excludes, as discussed earlier, data covering the COVID-19 induced recession.

To identify the turning points that separate national and subregional business cycle expansions from contractions, we apply the popular nonparametric business cycle dating algorithm of Harding and Pagan (2002) to each of our subregional estimates (as produced by the MF-VAR) and to the UK data themselves, having

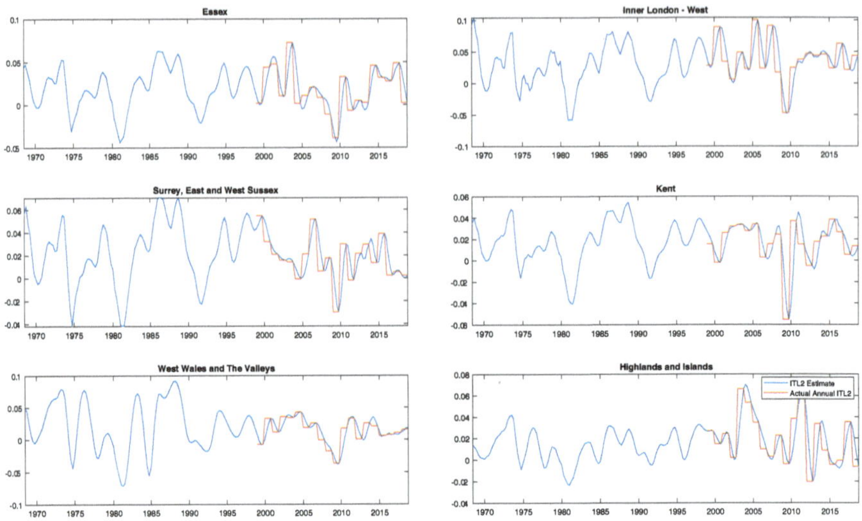

Fig. 4 ITL2 quarterly GVA growth estimates (annualized) from the MF-VAR for 6 (of 41) subregions, alongside the "official" annual estimates

transformed the growth estimates back into levels. We use as an input into the Harding-Pagan algorithm the median historical estimates of subregional GVA from our model. The Harding-Pagan dating algorithm seeks to formalize aspects of how the NBER dates business cycles in the USA, and has been found to match their turning points better than commonly used rules of thumb that characterize a recession as, for example, at least two consecutive quarters of negative growth.

The Harding-Pagan algorithm identifies five main recessions for the UK as a whole. These start in 1973Q3, 1974Q4, 1979Q3, 1990Q3, and 2008Q2. As seen in Fig. 5, our new database informs that these UK downturns do tend to be accompanied, as we might expect, with downturns also occurring in many of the regions and subregions that comprise the UK. However, we also see that many regions and subregions were in recession when the UK was not. Additionally, the recessions of the 1970s were less pervasive than the later recessions, in the sense that even when the UK as a whole entered recession, fewer regions and subregions shared this same fate in the 1970s. The tendency for subregions of the UK to experience a recession when the UK did not was weaker in the Great Moderation period, between the UK recessions of 1990 and 2008. Interestingly, consistent with lacklustre growth for the UK as a whole, since the global financial crisis there appears to be an increased tendency for UK subregions to experience local recessions, with the share of ITL2 regions in a local recession often jumping sharply. This evidences a growing tendency for regional cycles to be more volatile and often decouple from the path of the UK as a whole.

To evidence further the apparent rising incidence of local recessions, that is subregional recessions that happen without a UK recession, in Fig. 6, we zoom

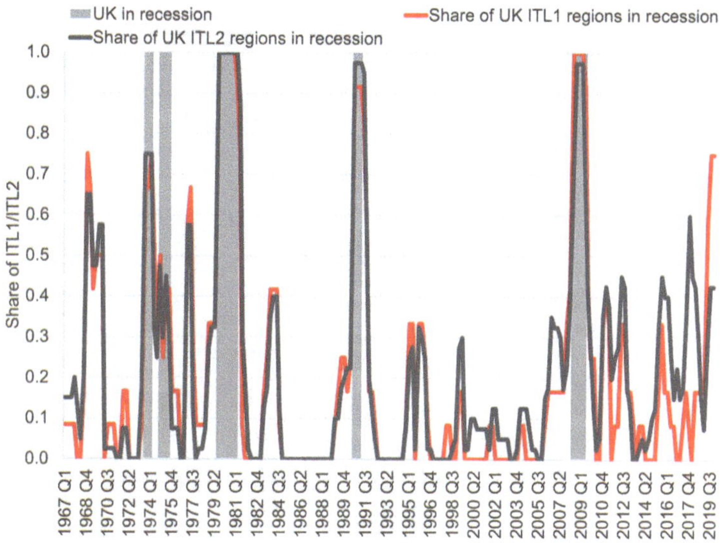

Fig. 5 UK recession bands (vertical gray bars) alongside the share (proportion) of regions (ITL1) and subregions (ITL2) in (local) recession

in to the period around the UK recession of 2008. Figure 6 plots the quarter in which each subregion, region, as well as the UK as a whole, entered and exited the 2008 UK recession associated with the global financial crisis. This figure shows considerable variation as to the frequency and timing of recessions across subregions. While the UK as a whole has been in an expansionary phase since 2009Q3, all 41 subregions have experienced at least one recession since then. Many of these regional contractions were short-lived and intra-year, and would be missed without access to our new higher-frequency database.

4.3 *Further Evidence on Subregional Business Cycle Heterogeneity*

Our new database, plus the business cycle classification of the previous subsection, let us take a deeper look at the link between national (UK) and regional/subregional business cycles, acknowledging that different regions/subregions may experience recessions at different scales, i.e., of differing magnitudes or depths. We compare the sensitivity of (sub-)regional economic growth to national economic growth with a measure of the relative (to the national economy) "gap" between a (sub-)region's average growth rates in expansionary and contractionary phases of the business cycle. This enables us to identify (sub-)regions which are more/less

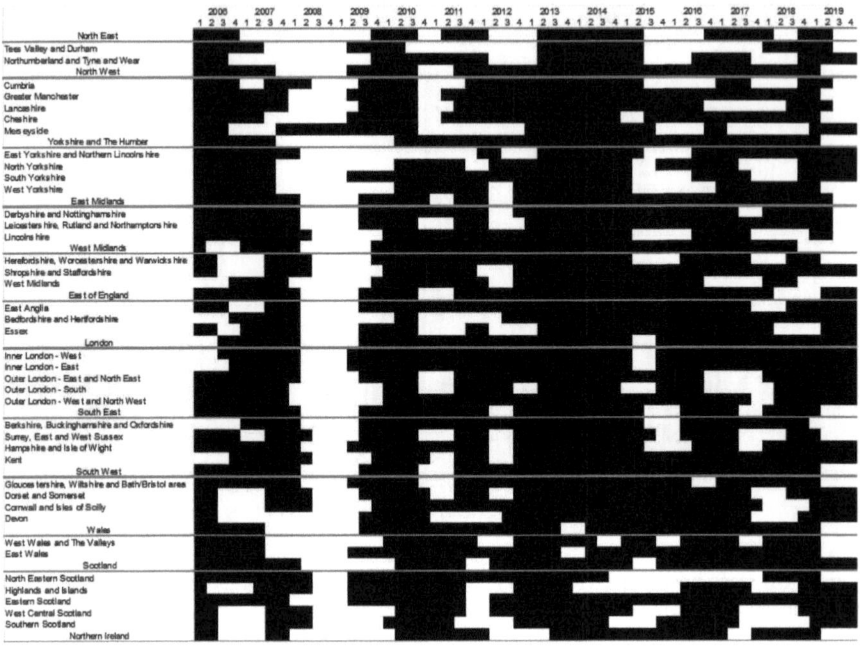

Fig. 6 Subregion recessions (in white) and expansions (in black) by quarter from 2006

aligned to the national economy in terms of both overall growth and their business cycle properties.

We measure the sensitivity of regional output to movements in national output by the estimated β coefficient in the following CAPM-style regression model. This relates, at the quarterly frequency, regional or subregional growth (as estimated by the MF-VAR) to UK growth:

$$y_t^r = \alpha + \beta y_t^{UK} + \epsilon_t. \tag{12}$$

β thus measures the volatility or sensitivity of each region's or subregion's growth to the UK (the "market") as a whole.[5] When $\beta > 1$, a region is relatively sensitive to movements in UK economic activity. When $\beta < 1$, the region is relatively insensitive.

In Fig. 7, for each subregion, we plot OLS estimates of its β against, on the Y-axis, the difference in its average growth rate in expansionary and contractionary phases *relative* to the equivalent calculation for the UK as a whole. A value of unity on the Y-axis therefore indicates that a subregion experienced the same gap between

[5] We treat y_t^{UK} as weakly exogenous in (12), given that most regions/subregions (with the possible exception of London) are small relative to the size of the UK economy.

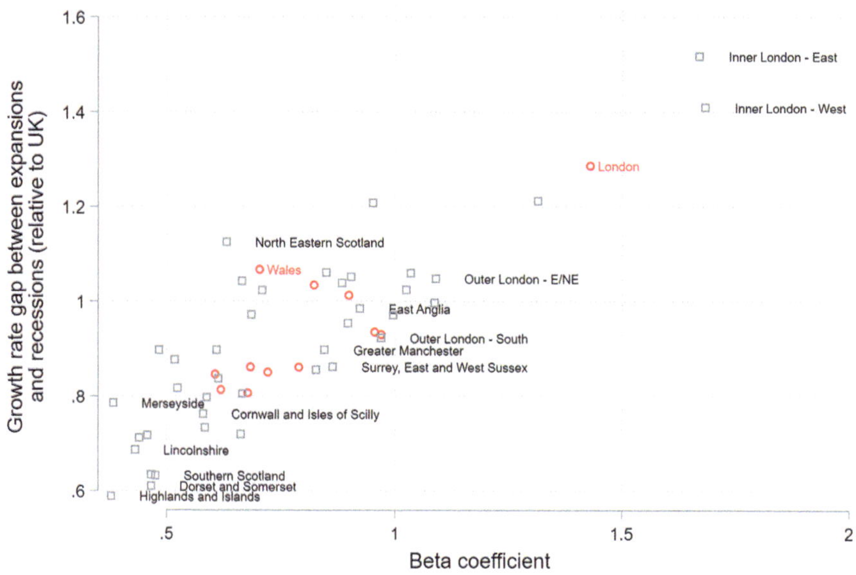

Fig. 7 Relative depth of (sub-)regional business cycles (subregional average growth rate in expansions minus recessions) versus estimates of β (the sensitivity of each subregion to the UK, as estimated via (12))

its average growth rates in expansionary and contractionary phases of the business cycle as the UK as a whole.[6]

Figure 7 illustrates that some subregions of the UK, most notably Inner London, have much larger differences between their expansionary and contractionary growth rates and, in turn, are much more sensitive to UK GDP movements than others. Given the importance of the aggregate London region (shown in red) to the UK economy, this is perhaps unsurprising, but these results make the point that London as a region is not homogeneous, with Inner and Outer London having rather distinct business cycle features. Subregions of the UK which are more rural tend to be found in the bottom-left corner of Fig. 7; they are both less sensitive to movements in UK GDP and experience much smaller differences between their growth rates in expansionary and contractionary phases of the business cycle.

Notable too is the North Eastern Scotland region, which over our sample period was heavily engaged with oil and gas activity in the North Sea. It experiences larger differences in its expansionary versus contractionary growth rates than the overall UK economy and has less sensitivity to UK GDP movements. This is not that surprising, given that its economy is largely driven by global factors. Comparing to the aggregate Scotland region, we again see the advantages of considering business cycle features at a more disaggregated spatial scale, given the evident heterogeneity

[6] ITL1 regions are shown in red, while ITL2 regions are shown in black.

within regional economies. Continuing this theme, the next section demonstrates the insights that emerge from using our new data to consider different dimensions of regional and subregional economic resilience.

4.4 Economic Resilience

This section extends the analysis of how our new regional database sheds light on business cycle heterogeneities by assessing regional economic resilience. We follow the existing literature in economic geography by distinguishing between *resistance, renewal, recovery, and reorientation*, which Martin (2012) defines as characterizing the dimensions of regional economic resilience. Improvements in each of these dimensions reflect how a regional economy can become more resilient. This relates to the wider literature on regional economic resilience in the UK (Simmie & Martin, 2010; Martin, 2012; Fingleton et al., 2012; Martin & Sunley, 2015; Sensier et al., 2024). We focus on the first three of these measures of resilience, as the fourth—*reorientation*—explores changes in a region's economic structure. This would be best explored by disaggregation of UK output not just into regions but sectors too.[7]

The first dimension of regional economic resilience is *resistance*, defined by (Martin, 2012, p. 11) as: "... the vulnerability or sensitivity of a regional economy to disturbances and disruptions, such as recessions." Intuitively, it measures the extent to which a region is able to "resist" the effects of a negative economic shock. We measure this as the peak-to-trough change (in percentage points) in the level of output (but this could be equally applied to other metrics such as employment), reflecting that more resilient regions should see smaller drops in output in the face of an economic shock. The second dimension of resilience is *recovery*, which we measure as the number of quarters it takes a given subregion to regain its precrisis (peak) level of economic activity. It reflects the expectation that more resilient regions should rebound more quickly from an economic shock. Thirdly, we measure *renewal* as the number of quarters it takes a given subregion to regain its precrisis growth rate. This fits with Martin (2012) defining *renewal* as the "extent to which [the] regional economy renews its growth path: resumption of prerecession path or hysteretic shift to a new growth path" (Martin, 2012, p. 12). The precrisis growth rate is calculated as the average annual growth rate for each region (or subregion) from the beginning of the sample (1966 Q4) until the onset of each downturn. If it is the case that there has been a significant degree of damage done by the recessionary period to the core "economic base" of a region, we might see this reflected in the region's growth rate experiencing a downward hysteretic shift.

In Fig. 8, we present a heatmap of these three measures of economic resilience across the 1979 and 2008 UK recessions. Subregions are organized by country

[7] Modeling such data would require estimation of an even larger version of our model, with even more missing data. We leave development of such a model to future research.

	1979			2008		
	Recovery Quarters	Renewal Quarters	Resistance %	Recovery Quarters	Renewal Quarters	Resistance %
Tees Valley and Durham	29	12	11	29	23	4
Northumberland and Tyne and Wear	26	11	10	16	8	5
Cumbria	33	12	7	6	6	2
Greater Manchester	28	11	11	16	15	4
Lancashire	27	11	13	24	7	8
Cheshire	24	11	13	15	8	6
Merseyside	28	12	9	10	21	2
East Yorkshire and Northern Lincolnshire	26	13	8	48	25	11
North Yorkshire	25	13	10	25	10	7
South Yorkshire	24	11	9	25	24	7
West Yorkshire	25	11	15	27	26	6
Derbyshire and Nottinghamshire	18	11	7	10	8	6
Leicestershire, Rutland and Northamptonshire	22	11	10	22	7	8
Lincolnshire	23	13	6	19	9	6
Herefordshire, Worcestershire and Warwickshire	25	11	17	11	8	6
Shropshire and Staffordshire	28	14	11	20	8	7
West Midlands	28	12	18	21	8	8
East Anglia	22	13	8	15	9	5
Bedfordshire and Hertfordshire	22	13	9	24	20	9
Essex	23	12	8	24	7	6
Inner London - West	23	12	10	16	11	8
Inner London - East	23	12	10	11	8	6
Outer London - East and North East	24	10	11	21	7	11
Outer London - South	22	10	9	44	8	10
Outer London - West and North West	19	10	8	18	10	9
Berkshire, Buckinghamshire and Oxfordshire	22	13	11	13	9	5
Surrey, East and West Sussex	21	13	8	9	8	4
Hampshire and Isle of Wight	22	13	6	9	7	4
Kent	24	14	7	21	7	7
Gloucestershire, Wiltshire and Bath/Bristol area	15	12	7	9	7	5
Dorset and Somerset	17	13	4	15	9	3
Cornwall and Isles of Scilly	23	13	7	20	9	6
Devon	22	11	6	9	7	4
West Wales and The Valleys	26	11	12	13	9	5
East Wales	25	11	11	12	9	4
North Eastern Scotland	26	24	9	6	9	2
Highlands and Islands	24	13	4	5	6	1
Eastern Scotland	24	13	8	18	18	3
West Central Scotland	24	13	6	19	11	5
Southern Scotland	24	13	5	21	8	5

Fig. 8 Comparison of *recovery*, *renewal*, and *resistance* of recessionary contractions in the UK in 1979 and 2008. The shading intensity reflects each cell's value, with the highest and lowest in each column shown in dark red and blue, respectively, and the median in white

(England, Wales, and Scotland), and within England the subregions are further (broadly) organized from North to South. What emerges is a clear sense of the geographic (subregional) pattern of these two national downturns being rather different. In particular, across the three measures of resilience, the 1979 recession hit subregions in the South of England much less than those in the North. The 2008 recession, though, had a significant impact in the North but also in London according to the resistance measure of resilience, while the South of England was again much less affected in general, London aside. As expected, looking across the three measures of resilience, subregions tend to *renew* before they *recover*. It takes longer for economic growth rates in the recovery stage of a subregion's business cycle to cumulate such that they push the subregion back above its prerecession level of output. Economic resilience is inherently a complex concept, and the heatmap in Fig. 8 illustrates the diversity of experience between geographical areas and across different recessions. Some subregions are consistently blue/red across all three measures, but many are not. Instead, on some measures, these subregions seem more/less resilient than they do on other measures.

5 Conclusions

This chapter develops a "big data" mixed-frequency Bayesian VAR model to produce historical quarterly estimates of subregional output growth in the UK. The model takes on the challenge that extant official output growth data in the UK vary in their availability both across time, regions, and subregions. This makes for a complicated and evolving data landscape. Our MF-VAR model is estimated using variational Bayesian methods with shrinkage priors. It imposes not just temporal but hierarchical aggregation constraints, such that the subregional data "add up" to the region, in which the subregions reside, and, in turn, the regional data "add up" to the UK (national) data.

We use the new quarterly subregional historical database to show that subregions of the UK differ considerably in their business cycle dynamics, in terms of when they enter and exit recessions. Such cross-subregional variation in business cycle dynamics would be obscured if we focused on extant annual regional data. Our focus in this chapter is using the new data to establish these "stylized facts" about subregional business cycle dynamics. We defer formal economic explanation of them to future research. Understanding *why* these regional and subregional business cycles differ would no doubt benefit from extending our model further to incorporate sectoral detail. One could then model how the sectoral composition of the subregional economies, and changes in this over time, explains the observed business cycle heterogeneities.

Our model could also be used to produce nowcasts, that is, more timely (out-of-sample) real-time estimates of subregional economic activity. Previous research (Koop et al., 2024) has found that conditioning quarterly nowcasts for the UK regions (ITL1) on the more timely UK GVA estimate improves their accuracy. As more real-time data accumulate and thereby enable recursive out-of-sample simulations, future research should assess whether this result continues to hold at the more disaggregated subregional (ITL2) level.

References

Chan, J. C., Poon, A., & Zhu, D. (2023). High-dimensional conditionally Gaussian state space models with missing data. *Journal of Econometrics, 236*(1), 105468.

Fingleton, B., Garretsen, H., & Martin, R. (2012). Recessionary shocks and regional employment: Evidence on the resilience of UK regions. *Journal of Regional Science, 52*(1), 109–133.

Gefang, D., Koop, G., & Poon, A. (2020). Computationally efficient inference in large Bayesian mixed frequency VARs. *Economics Letters, 191*, 109120.

Gefang, D., Koop, G., & Poon, A. (2023). Forecasting using variational Bayesian inference in large vector autoregressions with hierarchical shrinkage. *International Journal of Forecasting, 39*, 346–363.

Harding, D., & Pagan, A. (2002). Dissecting the cycle: A methodological investigation. *Journal of Monetary Economics, 49*(2), 365–381.

Koop, G., McIntyre, S., & Mitchell, J. (2020a). UK regional nowcasting using a mixed frequency vector auto-regressive model with entropic tilting. *Journal of the Royal Statistical Society: Series A (Statistics in Society), 183*(1), 91–119.

Koop, G., McIntyre, S., Mitchell, J., & Poon, A. (2020b). Regional output growth in the United Kingdom: More timely and higher frequency estimates from 1970. *Journal of Applied Econometrics, 35*(2), 176–197.

Koop, G., McIntyre, S., Mitchell, J., & Poon, A. (2024). Using stochastic hierarchical aggregation constraints to nowcast regional economic aggregates. *International Journal of Forecasting, 40*(2), 626–640.

Mariano, R. S., & Murasawa, Y. (2003). A new coincident index of business cycles based on monthly and quarterly series. *Journal of Applied Econometrics, 18*(4), 427–443.

Martin, R. (2012). Regional economic resilience, hysteresis and recessionary shocks. *Journal of Economic Geography, 12*(1), 1–32.

Martin, R., & Sunley, P. (2015). On the notion of regional economic resilience: Conceptualization and explanation. *Journal of Economic Geography, 15*(1), 1–42.

Owyang, M. T., Piger, J., & Wall, H. J. (2005). Business cycle phases in U.S. states. *The Review of Economics and Statistics, 87*(4), 604–616.

Schorfheide, F., & Song, D. (2015). Real-time forecasting with a mixed-frequency VAR. *Journal of Business & Economic Statistics, 33*(3), 366–380.

Sensier, M., Rafferty, A., & Devine, F. (2024). The economic resilience scorecard: Regional policy responses for crises recovery. *Regional Studies, 58*, 1754–1766.

Simmie, J., & Martin, R. (2010). The economic resilience of regions: Towards an evolutionary approach. *Cambridge Journal of Regions, Economy and Society, 3*(1), 27–43.

VAR Models with Fat Tails and Dynamic Asymmetry

Tamás Kiss, Stepan Mazur, Hoang Nguyen, and Pär Österholm

Abstract In this chapter, we extend the standard Gaussian stochastic volatility Bayesian VAR by employing the generalized hyperbolic skew Student's t distribution for the innovations. Allowing the skewness parameter to vary over time, our specification permits flexible modelling of innovations in terms of both fat tails and—potentially dynamic—asymmetry. In an empirical application using US data on industrial production, consumer prices and economic policy uncertainty, we find support—although to a moderate extent—for time-varying skewness. In addition, we find that shocks to economic policy uncertainty have a negative effect on both industrial production growth and CPI inflation.

1 Introduction

Many macroeconomic time series seem to be associated with properties that are inconsistent with a linear model which has error terms that are Gaussian distributed (and homoscedastic). One such property is that large shocks appear to hit the economy more frequently than what would be expected if the shocks were drawn from a Gaussian distribution. For example, it can be argued that the occurrence of the Global Financial Crisis, the crisis related to the Covid-19 pandemic and the inflation surge of 2022–2023 in a fairly short period of time warrants models with

T. Kiss · S. Mazur
School of Business, Örebro University, Örebro, Sweden
e-mail: tamas.kiss@oru.se; stepan.mazur@oru.se

H. Nguyen
Department of Management and Engineering, Linköping University, Linköping, Sweden
e-mail: hoang.nguyen@liu.se

P. Österholm (✉)
School of Business, Örebro University, Örebro, Sweden

National Institute of Economic Research, Stockholm, Sweden
e-mail: par.osterholm@oru.se

S. Mazur, P. Österholm (eds.), *Recent Developments in Bayesian Econometrics and Their Applications*, https://doi.org/10.1007/978-3-032-00110-8_5

distributions for the error terms that have fat tails. A second property is that the unconditional distribution of the variable can be characterized by skewness. This feature might be driven by disturbances whose distributions have skewness; this could potentially be time-varying, for instance, being different in expansions and contractions. Empirical work documenting and addressing fat tails and skewness include, for example, Acemoglu and Scott (1997), Ni and Sun (2005), Fagiolo et al. (2008), Ascari et al. (2015), Cross and Poon (2016), Liu (2019), Carriero et al. (2024), and Delle Monache et al. (2024). Taking these two properties into account might be beneficial from a modelling perspective—a better specified model can, for example, improve forecasting performance—and could accordingly be of interest to a fairly wide audience, including academics, forecasters, analysts and policymakers. Zooming in on policy, we note that policymakers might want to take skewness into account in policymaking (e.g. De Polis et al., 2024) and that both fat tails and skewness are relevant issues when communicating forecast uncertainty through fan charts[1] or discussing concepts such as growth-at-risk (e.g. Adrian et al., 2019).

In this chapter, we address these two issues and further the literature by specifying a Bayesian VAR model with stochastic volatility where the disturbances are drawn from a generalized hyperbolic skew Student's t distribution with time-varying skewness. This specification permits flexible modelling of innovations in terms of fat tails and—potentially dynamic—asymmetry. Our model nests specifications with Gaussian and fat-tailed innovations, as well as the specification where asymmetry is constant over time; the last specification was first defined in Karlsson et al. (2023).[2] In setting up the model, we also describe the appropriate priors and the general Markov chain Monte Carlo (MCMC) procedure used to draw inference.

We illustrate the framework with an application to US data in which we model industrial production, consumer prices and the economic policy uncertainty (EPU) index of Baker et al. (2016). The relation of the EPU index to the macroeconomy has been analysed widely over the last decade, and studies tend to show that an increasing EPU index has a dampening effect on the real economy; this includes holding back both investment (e.g. Gulen and Ion, 2016) and consumption (e.g. Nam et al., 2021), an increasing unemployment rate (e.g. Caggiano et al., 2017) and

[1] The Bank of England pioneered the concept of fan charts (Britton et al., 1998) and regularly publishes skewed fan charts for key variables in its *Monetary Policy Report*. The International Monetary Fund has provided skewed fan charts for GDP growth in many issues of its *World Economic Outlook*.

[2] Even though our framework formally nests specifications with constant volatility as well, we take stochastic volatility as a given feature of the model. There is ample evidence in the literature that many macroeconomic time series benefit from this modelling choice; see, for example, Cogley and Sargent (2005), Sims and Zha (2006), Clark (2011), Carriero et al. (2015), Chan (2017), Akram and Mumtaz (2019), Koop and Korobilis (2019), Karlsson and Österholm (2020a,b, 2023) and Kiss et al. (2023). This modelling choice is also supported in our empirical analysis. As can be seen in Fig. 2, estimated volatilities vary substantially over time.

a predictive ability for recessions (e.g. Karnizova and Li, 2014 and Balcilar et al., 2016).

The main contribution of this chapter is methodological. The most closely related research in the previous literature includes Karlsson et al. (2023) who introduced a general class of VAR model with generalized hyperbolic skew Student's t innovations and stochastic volatility. Delle Monache et al. (2024) provided evidence of time-varying conditional asymmetry of US GDP using a time-varying location, scale and asymmetry model; using the same modelling framework, De Polis et al. (2024) documented largely downside risk in inflation between the mid-1990s and the early 2020s. Iseringhausen (2024) studied the growth-at-risk of advanced economies using a univariate time-varying volatility and skewness model. Alternatively, large asymmetric VAR shocks can be modelled with nonparametric error distributions. Huber and Koop (2024) proposed an order-invariant VAR model with an additive error structure to account for the skewed and fat-tailed shocks in GDP growth, the unemployment rate and inflation. Clark et al. (2024) forecasted US inflation using a Gaussian process regression with a Dirichlet process mixture of innovations. Huber and Rossini (2022) combined VAR with Bayesian additive regression tree (BART) models for the nonlinear relations between the endogenous variables. We contribute to the literature by extending the Karlsson et al. (2023) model allowing the skewness parameters of the generalized hyperbolic skew Student's t innovations to be time-varying. Unlike the observation-driven model of Delle Monache et al. (2024) and De Polis et al. (2024), our proposed model is parameter-driven where the time-varying skewness evolves over time as a stochastic process. Furthermore, we study dynamic skewness of a multivariate distribution, where the skewness parameters follow independent random walk processes. Therefore, our framework can be seen as an extension of Tsiotas (2012) and Iseringhausen (2020). It also differs from Iseringhausen (2024) who has observable variables in the skewness equation. Our model allows us to account for the interaction between the variables included, as well as innovations that potentially feature dynamic skewness.

In addition to the methodological contribution, we also make a contribution to the empirical literature studying the relationship between the EPU index and the macroeconomy. Our results suggest that there is time variation in the asymmetry of all three variables, with the evidence being strongest for the EPU index. Furthermore, we confirm the finding of Leduc and Liu (2016) that uncertainty shocks resemble aggregate demand shocks, where an increase in uncertainty is associated with decreasing real activity and lower inflation. These effects of uncertainty shocks are also in line with the findings based on VAR models with stochastic volatility in the mean, which are frequently used to assess the macroeconomic impacts of uncertainty shocks (e.g. Mumtaz and Surico, 2018; Carriero et al., 2020; Cross et al., 2023).

The rest of this chapter is organized as follows. We present the methodological framework—including how inference is conducted—in Sect. 2. The empirical illustration of the framework is provided in Sect. 3. Finally, Sect. 4 concludes.

2 Econometric Framework

In this section, we first present a VAR model with a Gaussian distribution and stochastic volatility in the innovations. We then generalize the framework so that we allow for fat tails, dynamic asymmetry and stochastic volatility; this is done by employing a general class of generalized hyperbolic skew Student's *t* distribution for the innovations. This class of VAR model is an extension of the models developed by Karlsson et al. (2023) and allows us to capture the time-varying effects of disturbances to the conditional mean and the conditional variance of the variables.

2.1 *A VAR Model with Gaussian Innovations and Stochastic Volatility*

We assume that a VAR model with Gaussian innovations and stochastic volatility (Gaussian.SV) is given by

$$\mathbf{y}_t = \mathbf{c} + \mathbf{B}_1\mathbf{y}_{t-1} + \ldots + \mathbf{B}_p\mathbf{y}_{t-p} + \mathbf{A}^{-1}\mathbf{H}_t^{1/2}\boldsymbol{\epsilon}_t, \qquad t = 1, \ldots, T, \tag{1}$$

where $\mathbf{y}_t$ is a k-dimensional vector of time series of interest; $\mathbf{c}$ is a k-dimensional vector of constants; $\mathbf{B}_j$ is a $k \times k$ matrix of regression coefficients with $j = 1, \ldots, p$; $\mathbf{A}$ is a $k \times k$ lower triangular matrix with ones on the diagonal that describes the contemporaneous interaction of the endogenous variables; $\mathbf{H}_t = diag(h_{1t}, \ldots, h_{kt})$ is a $k \times k$ diagonal matrix that captures the heteroskedasticity and $\boldsymbol{\epsilon}_t$ is a k-dimensional vector that follows a multivariate Gaussian distribution with zero mean vector and identity covariance matrix, that is, $\boldsymbol{\epsilon}_t \sim \mathcal{N}_k(\mathbf{0}, \mathbf{I})$. We also assume that $\boldsymbol{\epsilon}_t$ is independent of $\mathbf{H}_t$. Moreover, we assume that the elements of $\mathbf{H}_t = diag(h_{1t}, \ldots, h_{kt})$ follow a random walk process such that

$$\log h_{it} = \log h_{it-1} + \sigma_{h,i}\eta_{it}, \qquad i = 1, \ldots, k, \tag{2}$$

where $\eta_{it} \sim \mathcal{N}(0, 1)$. Let us note that the VAR model with homoskedastic Gaussian innovations can be obtained by restricting the volatility to be constant, i.e. $\boldsymbol{\sigma}_h^2 = (\sigma_{h,1}^2, \ldots, \sigma_{h,k}^2)' = \mathbf{0}$ and $\log h_{it} = \log h_{i0}$ for all $i = 1, \ldots, k$ and $t = 1, \ldots, T$.

For the sake of notational simplicity, the Gaussian.SV model in (1) can be rewritten as follows:

$$\mathbf{y}_t = \mathbf{B}\mathbf{x}_t + \mathbf{u}_t, \tag{3}$$

where $\mathbf{B} = (\mathbf{c}, \mathbf{B}_1, \ldots, \mathbf{B}_p)$ is a $k \times (1 + kp)$ matrix, $\mathbf{x}_t = (1, \mathbf{y}_{t-1}', \ldots, \mathbf{y}_{t-p}')'$ is a $(1 + kp)$-dimensional vector and $\mathbf{u}_t = \mathbf{A}^{-1}\mathbf{H}_t^{1/2}\boldsymbol{\epsilon}_t$ is a k-dimensional vector of heteroskedastic innovations associated with the VAR equations. In the next subsection, we extend the innovation term $\mathbf{u}_t$ to a class of generalized hyperbolic skew Student's t distribution with stochastic volatility and dynamic asymmetry.

2.2 A VAR Model with Fat Tails, Dynamic Asymmetry and Stochastic Volatility in the Innovations

We now propose a class of generalized hyperbolic skew Student's t VAR models by assuming that the vector of innovations $\mathbf{u}_t$ in (1) follows a Gaussian variance-mean mixture and is expressed as

$$\mathbf{u}_t = (w_t - \overline{w})\boldsymbol{\gamma}_t + \sqrt{w_t}\mathbf{A}^{-1}\mathbf{H}_t^{1/2}\boldsymbol{\epsilon}_t, \tag{4}$$

where the mixing variable w_t follows an inverse gamma distribution with the same shape and rate parameters equal to $\nu/2$, that is, $w_t \sim \mathcal{IG}(\nu/2, \nu/2)$; $\overline{w} = E[w_t] = \nu/(\nu-2)$ is used for centring the innovations vector so that its mean is equal to zero and $\boldsymbol{\gamma}_t = (\gamma_{1t}, \ldots, \gamma_{kt})'$ is the k-dimensional vector of the time-varying skewness parameters that are assumed to follow independent random walk processes such that

$$\gamma_{it} = \gamma_{it-1} + \sigma_{\gamma,i}\eta_{it}, \qquad i = 1, \ldots, k, \tag{5}$$

where $\eta_{it} \sim \mathcal{N}(0, 1)$. The matrices $\mathbf{A}$ and $\mathbf{H}_t$, as well as the vector $\boldsymbol{\epsilon}_t$ in (4), are the same as in (1). Moreover, w_t, $\boldsymbol{\gamma}_t$, $\mathbf{H}_t$ and $\boldsymbol{\epsilon}_t$ are mutually independent. In what follows, the VAR model with the innovation term defined as in (4) is called *dynSkew-T.SV*.

In general, the dynSkew-T.SV VAR model imposes the fat tails and time-varying asymmetry of the innovations directly in each VAR equation. Even though there is only one parameter to control for the fat tails of all variables, it is still sufficiently flexible to capture the joint dependence of the innovations. Regarding the properties of the model, it can be noted that the following holds:

$$\mathbf{u}_t | w_t, \boldsymbol{\gamma}_t, \mathbf{H}_t \sim \mathcal{N}_k\left(\boldsymbol{\mu}_t - (w_t - \overline{w})\boldsymbol{\gamma}_t,\ \boldsymbol{\Sigma}_t = w_t\mathbf{A}^{-1}\mathbf{H}_t\mathbf{A}^{-1'}\right). \tag{6}$$

It follows that the conditional distribution of $\mathbf{u}_t$ (given $\boldsymbol{\gamma}_t$ and $\mathbf{H}_t$) follows a multivariate generalized hyperbolic skew Student's t distribution (see McNeil et al., 2015). Setting $\boldsymbol{\sigma}_\gamma^2 = (\sigma_{\gamma,1}^2, \ldots, \sigma_{\gamma,k}^2)' = \mathbf{0}$ and $\gamma_{it} = \gamma_{i0}$ for all $i = 1, \ldots, k$ and $t = 1, \ldots, T$, the dynSkew-T.SV VAR model reduces to the VAR model with generalized hyperbolic skew Student's t innovations and stochastic volatility (Skew-T.SV). Additionally, assuming $\gamma_{it} = \gamma_{i0} = 0$, we arrive to the VAR model with generalized hyperbolic Student's t innovations and stochastic volatility (T.SV). Moreover, considering $\nu \to \infty$, we obtain the VAR model with Gaussian innovations and stochastic volatility (G.SV).

From an economic point of view, the mixing variable w_t is identified from the sudden changes of the conditional volatility of variables while the stochastic volatility term $\exp(h_{it}/2)$ captures smooth changes in the volatility (Chiu et al., 2017). The data will determine whether movements that are large in magnitude come from the mixing variable or a stochastic volatility shift. It is also worth pointing out that the mixing variable itself is able to generate asymmetry in the unconditional distribution of the variables, even though innovations are drawn from a symmetric distribution. This is due to the fact that the mixing variable shifts the conditional mean of each variable. Since the uncertainty rises during recessions, these effects are different during recessions and expansions, generating asymmetry in the unconditional distribution of the variables (as described in Carriero et al., 2018). Our specification not only incorporates this feature but also directly allows for skewness in the conditional distribution of the innovations; hence, even the conditional distribution of the variables is allowed to be asymmetric.

2.3 *Bayesian Inference*

This section discusses the prior distributions for the parameters in the dynSkew-T.SV models and the general MCMC inference scheme. We also briefly comment upon some computational aspects of the model.

2.3.1 Prior Distribution

We denote $\boldsymbol{\theta} = \{vec(\mathbf{B})', \mathbf{a}', \boldsymbol{\gamma}_0', \nu, \mathbf{h}_0', \boldsymbol{\sigma}_h^{2'}, \boldsymbol{\sigma}_\gamma^{2'}, \boldsymbol{\gamma}_{1:T}', \mathbf{w}_{1:T}', \mathbf{h}_{1:T}'\}'$ as the set of the dynSkew-T.SV model parameters. We set $\mathbf{a}$ as a vector of the lower triangular elements of the matrix $\mathbf{A}$ such that $\mathbf{a} = (a_{2,1}, a_{3,1}, a_{3,2}, \ldots, a_{k,k-1})'$. Also let $\mathbf{h}_0$ and γ_0 be the vectors of initial values of the stochastic volatility and the dynamic skewness, respectively. Following the prior assumption in Koop and Korobilis (2010) and Karlsson et al. (2023), we assume that the regression coefficients $vec(\mathbf{B})$ follow a Minnesota-type prior, that is, $vec(\mathbf{B}) \sim \mathcal{N}_{k(1+kp)}(\mathbf{b}_0, \mathbf{V}_{\mathbf{b}_0})$ such that the overall shrinkage is $l_1 = 0.2$ and the cross-variable shrinkage is $l_2 = 0.5$. For other parameters, we consider vague but proper prior distributions. For example, the prior of $\mathbf{a}$ is also Gaussian, $\mathbf{a} \sim \mathcal{N}(0, 10 I_{0.5k(k-1)})$, which implies a weak assumption of no interaction among endogenous variables. The prior for the degrees of freedom parameter is $\nu \sim \mathcal{G}(2, 0.1)$ such that the prior mean of the degrees of freedom is 20. The initial skewness has zero prior mean and unit variance, while the initial log volatility is obtained from the OLS estimation of the AR(p) model for each time series. In the case of the VAR model without stochastic volatility, we consider $\log h_{i0} \sim \mathcal{N}\left(\log \hat{\Sigma}_{i,OLS}, 4\right)$, where $\hat{\Sigma}_{i,OLS}$ are variance estimators of AR(p) models using the ordinary least square method; see Clark and Ravazzolo (2015). The prior for the variance of a shock to the volatility and the time-varying skewness are $\sigma^2_{\gamma,i} \sim \mathcal{G}(\frac{1}{2}, \frac{1}{2V_\sigma})$ which is equivalent to $\pm\sqrt{\sigma^2_{\gamma,i}} \sim \mathcal{N}(0, V_\sigma)$; see Kastner and Frühwirth-Schnatter (2014). We choose $V_\sigma = 1$, noting that this prior is less influential compared to the conjugated inverse gamma prior, especially in the case of a small true value. Finally, the prior for the mixing variable w_t is based on the model assumption $w_t|\nu \sim \mathcal{IG}(\nu/2, \nu/2)$.

2.3.2 Estimation Procedure

The Bayesian inference using a Gibbs sampler is extended to make inference on model parameters. To simplify the notation, let $\boldsymbol{\Psi}$ be a set of conditioning parameters except the one that we sample from.

1. To sample $\pi(\mathbf{b}|\boldsymbol{\Psi})$ where $\mathbf{b} = vec(\mathbf{B})$, we consider Eq. (3) as

$$\begin{aligned}
\mathbf{y}_t - (w_t - \overline{w})\boldsymbol{\gamma}_t &= \mathbf{B}\mathbf{x}_t + w_t^{1/2}\mathbf{A}^{-1}\mathbf{H}_t^{1/2}\boldsymbol{\epsilon}_t, \\
w_t^{-1/2}\mathbf{A}\left(\mathbf{y}_t - (w_t - \overline{w})\boldsymbol{\gamma}_t\right) &= w_t^{-1/2}\mathbf{A}\mathbf{B}\mathbf{x}_t + \mathbf{H}_t^{1/2}\boldsymbol{\epsilon}_t \\
&= w_t^{-1/2}\mathbf{x}_t' \otimes \mathbf{A}vec(\mathbf{B}) + \mathbf{H}_t^{1/2}\boldsymbol{\epsilon}_t, \\
\widetilde{\mathbf{y}}_t &= \widetilde{\mathbf{X}}_t\mathbf{b} + \mathbf{H}_t^{1/2}\boldsymbol{\epsilon}_t,
\end{aligned}$$

where $\widetilde{\mathbf{y}}_t = w_t^{-1/2}\mathbf{A}\left(\mathbf{y}_t - (w_t - \overline{w})\boldsymbol{\gamma}_t\right)$ and $\widetilde{\mathbf{X}}_t = w_t^{-1/2}\mathbf{x}_t' \otimes \mathbf{A}$. Hence, the conditional posterior of $\mathbf{b}$ is a conjugate normal distribution

$$\pi(\mathbf{b}|\boldsymbol{\Psi}) \sim \mathcal{N}(\mathbf{b}^*, \mathbf{V}_\mathbf{b}^*),$$

where

$$\begin{aligned}
\mathbf{V}_\mathbf{b}^{*-1} &= \mathbf{V}_{\mathbf{b}_0}^{-1} + \sum_{t=1}^{T}\widetilde{\mathbf{X}}_t'\mathbf{H}_t^{-1}\widetilde{\mathbf{X}}_t, \\
\mathbf{b}^* &= \mathbf{V}_\mathbf{b}^*\left[\mathbf{V}_{\mathbf{b}_0}^{-1}\mathbf{b}_0 + \sum_{t=1}^{T}\widetilde{\mathbf{X}}_t'\mathbf{H}_t^{-1}\widetilde{\mathbf{y}}_t\right].
\end{aligned}$$

2. To sample $\pi(\boldsymbol{\gamma}_{0:T}|\boldsymbol{\Psi})$, we consider the time-varying coefficient $\boldsymbol{\gamma}_t$ on the representation of the dynSkew-T.SV model

$$\begin{aligned}
\mathbf{y}_t - \mathbf{B}\mathbf{x}_t &= (w_t - \overline{w})\boldsymbol{\gamma}_t + w_t^{1/2}\mathbf{A}^{-1}\mathbf{H}_t^{1/2}\boldsymbol{\epsilon}_t, \\
\boldsymbol{\gamma}_t &= \boldsymbol{\gamma}_{t-1} + diag(\boldsymbol{\sigma}_\gamma)\boldsymbol{\varsigma}_t.
\end{aligned}$$

We apply the forward filter backward smoothing algorithm as in Carter and Kohn (1994) to filter out the last state variable of the time-varying skewness. Then, $\boldsymbol{\gamma}_t$ is sampled from its smoothing Gaussian distribution for $t = T, \dots, 0$; see Primiceri (2005). As $\boldsymbol{\sigma}_\gamma^2$ has a non-conjugate prior distribution, we sample $\pi(\boldsymbol{\sigma}_\gamma^2|\boldsymbol{\Psi})$ using an independent Metropolis-Hastings step; see Kastner and Frühwirth-Schnatter (2014).

3. To sample $\pi(\mathbf{a}|\boldsymbol{\Psi})$, we follow Cogley and Sargent (2005) and rewrite equation (3) as a system of linear equations with equation

$$\mathbf{A}\tilde{\mathbf{u}}_t = \mathbf{H}_t^{1/2}\boldsymbol{\epsilon}_t,$$

where $\tilde{\mathbf{u}}_t = w_t^{-1/2}\left(\mathbf{y}_t - \mathbf{B}\mathbf{x}_t - (w_t - \overline{w})\boldsymbol{\gamma}_t\right)$. Then, the conditional posterior of elements in $\mathbf{a}$ can be drawn equation by equation using the conditionally Gaussian posterior distribution; see Cogley and Sargent (2005).

4. To sample $\pi(\mathbf{H}_{1:T}|\boldsymbol{\Psi})$, we follow Kim et al. (1998), Primiceri (2005) and Del Negro and Primiceri (2015). Let $\tilde{\tilde{\mathbf{u}}}_t = \mathbf{A}\tilde{\mathbf{u}}_t$, for each series $i = 1, \dots, k$, we have that $\log \tilde{\tilde{u}}_{it}^2 = \log h_{it} + \log \epsilon_{it}^2$. We follow Kim et al. (1998) who approximate the distribution of $\log(\epsilon_{it}^2)$ using a mixture of seven normal components. Then, we sample $\log \boldsymbol{h}_t$ from its smoothing Gaussian distribution using the forward filter backward smoothing algorithm in Carter and Kohn (1994). Finally, we sample $\pi(\boldsymbol{\sigma}_h^2|\boldsymbol{\Psi})$ from an independent Metropolis-Hastings; see Kastner and Frühwirth-Schnatter (2014).

5. To sample $\pi(\nu|\boldsymbol{\Psi}) \propto \mathcal{G}(\nu; 2, 0.1)\prod_{t=1}^{T} \mathcal{IG}\left(w_t; \frac{\nu}{2}, \frac{\nu}{2}\right)$ for $i = 1, \dots, k$, we use an adaptive random walk Metropolis-Hastings algorithm to accept/reject the draw $\nu^{(*)} = \nu + \eta \exp(c)$, where $\eta \sim \mathcal{N}(0, 1)$ and the adaptive variance c is adjusted automatically such that the acceptance rate is around 0.25 (Roberts and Rosenthal, 2009).

6. To sample $\pi(w_t|\boldsymbol{\Psi})$ for $t = 1, \dots, T$, we write

$$\mathbf{u}_t = \mathbf{y}_t - \mathbf{B}\mathbf{x}_t + \overline{w}\boldsymbol{\gamma}_t = w_t\boldsymbol{\gamma}_t + w_t^{1/2}\mathbf{A}^{-1}\mathbf{H}_t^{1/2}\boldsymbol{\epsilon}_t.$$

The full conditional posterior is thus generalized inverse Gaussian following Hörmann and Leydold (2014), $GIG(\lambda, \psi, \chi)$ with $\lambda = -(\nu+k)/2$, $\chi = q_t^2 + \nu$ and $\psi = p_t^2$ for $q_t^2 = \mathbf{u}_t'\mathbf{A}'\mathbf{H}_t^{-1}\mathbf{A}\mathbf{u}_t$ and $p_t^2 = \boldsymbol{\gamma}_t'\mathbf{A}'\mathbf{H}_t^{-1}\mathbf{A}\boldsymbol{\gamma}_t$ according to

$$\begin{aligned}
\pi(w_t|\boldsymbol{\Psi}) \propto w_t^{-k/2} \exp &\left(-\frac{1}{2}(\mathbf{y}_t - \mathbf{B}\mathbf{x}_t - (w_t - \overline{w})\boldsymbol{\gamma}_t)'\right. \\
&\left.\times \boldsymbol{\Sigma}_t^{-1}(\mathbf{y}_t - \mathbf{B}\mathbf{x}_t - (w_t - \overline{w})\boldsymbol{\gamma}_t)\right) \\
&\times \mathcal{IG}\left(w_t; \frac{\nu_i}{2}, \frac{\nu_i}{2}\right) \\
&\propto w_t^{-k/2} \exp\left(-\frac{1}{2}(\mathbf{u}_t - w_t\boldsymbol{\gamma}_t)'(w_t^{-1}\mathbf{A}'\mathbf{H}_t^{-1}\mathbf{A})(\mathbf{u}_t - w_t\boldsymbol{\gamma}_t)\right)
\end{aligned}$$

$$\times \mathcal{IG}\left(w_t; \frac{\nu_i}{2}, \frac{\nu_i}{2}\right)$$

$$\propto w_t^{-(k/2+\nu/2+1)} \exp\left(-\frac{1}{2}\frac{\mathbf{u}_t'\mathbf{A}'\mathbf{H}_t^{-1}\mathbf{A}\mathbf{u}_t + \nu}{w_t} - \frac{1}{2}w_t\boldsymbol{\gamma}_t'\mathbf{A}'\mathbf{H}_t^{-1}\mathbf{A}\boldsymbol{\gamma}_t\right).$$

2.3.3 Some Computational Aspects

As is well known, MCMC computation for VARs can be slow. In order to provide some information on this issue, computation times for simulated small and medium size VARs can be found in Table 3 in the Appendix. We vary the lag length (p), the number of endogenous variables (k) and sample size (T) and estimate the corresponding G.SV and dynSkew-T.SV models. Compared to the G.SV model, the dynSkew-T.SV requires about 1.5–4 times the computational cost for the estimation, where the differences are larger for the smaller number of endogenous variables.

Another computational aspect worth mentioning concerns marginal likelihood calculations. Chan and Eisenstat (2018) and Karlsson et al. (2023) computed the marginal likelihood using an importance sampling method. Since the dynSkew-T.SV VAR model contains high-dimensional latent state space parameters, integrating the latent variables in the importance sampling steps and coming up with a good proposal sampling distribution are challenging aspects. Therefore, the calculation of marginal likelihoods is beyond the scope of this chapter. It is, however, an interesting topic for future research.

3 Empirical Illustration

We next estimate models using monthly US data on industrial production, consumer prices and the economic policy uncertainty (EPU) index of Baker et al. (2016). Analysis is conducted both within- and out-of-sample.

The data—which range from 1985M1 to 2024M3—were sourced from the FRED-MD database of Federal Reserve Bank of St. Louis.[3] We calculate the year-on-year growth rate of industrial production and year-on-year inflation using the 12-month log-difference of the industrial production index and the consumer price index. Denoting industrial production at time t by IP_t, the former is defined as $i_t = 100[\log(IP_t) - \log(IP_{t-12})]$; denoting the consumer price index at time t by CPI_t, the latter is given as $p_t = 100[\log(CPI_t) - \log(CPI_{t-12})]$. For the numerical stability of the estimation procedure, we divide the EPU by 10; we accordingly define $e_t = EPU_t/10$. The data series are shown in Fig. 1.

[3] See McCracken and Ng (2016) for details.

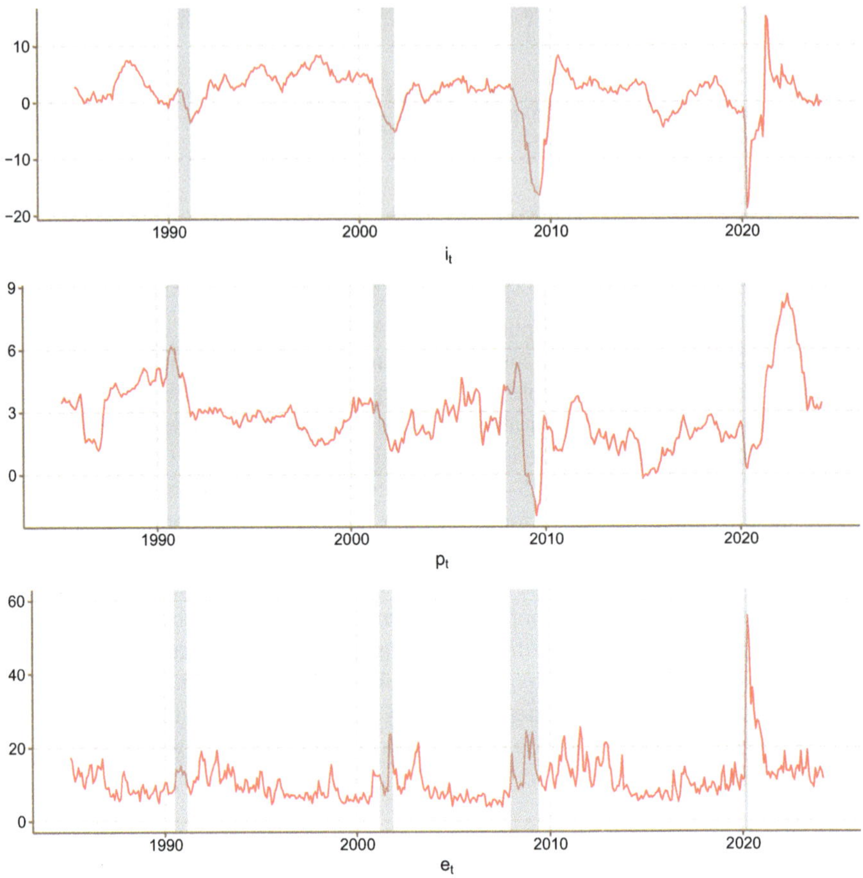

Fig. 1 Data. Industrial production growth rate (i_t) and CPI inflation (p_t) are given as 100 times the 12-month log-difference. The plotted EPU (e_t) is the original series divided by 10. The shaded areas highlight the recession periods based on the NBER indicators

From a methodological point of view, our primary interest is in the skewness estimates. But it is also relevant to see how our empirical findings relate to earlier research that has studied the relation of the EPU index to the macroeconomy. As was pointed out in the introduction, an increasing EPU index tends to be associated with a slowing real economy, where a weaker development of investment and consumption is in line with the more general notion that uncertainty should hold back the real economy; see, for example, Bernanke (1983) and Bloom (2009). Previous findings regarding the real economy are also in line with the claim of Leduc and Liu (2016) who argue that uncertainty shocks resemble aggregate demand shocks. However, Jones and Olson (2013) find that the relation between the EPU index and inflation appears ambiguous. Adding more data and new methodology to study this issue is hence relevant.

3.1 Within-Sample Analysis

Our within-sample analysis focuses on the properties of a trivariate VAR(3) model with dynamic skewness, where the vector of dependent variables is defined as $\mathbf{y}_t = (i_t, p_t, e_t)'$.[4]

Initially, we note the importance of stochastic volatility for all variables; see Fig. 2. Generally, volatility varies substantially over time, and, in particular, it rises around crisis times. This finding gives support to our choice to take stochastic volatility as a given feature. Interestingly, the estimated stochastic volatility is barely affected by allowing for non-Gaussian innovations; the estimates are slightly higher for the Gaussian model, but differences are often within the credible intervals (larger deviations are only present for some periods for the EPU index). This suggests that the dispersion of the innovation distribution is mostly captured well by stochastic volatility; potentially fat tails play a secondary role in capturing tail observations.[5]

Figure 3 shows the evolution of the skewness parameters for the trivariate VAR model with dynamic skewness. Despite the relatively wide credible intervals, we detect time variation in the asymmetry parameter for the variables. The EPU index—which typically has a positive skewness—exhibits the strongest time variation; it tends to increase in association with crises and decrease afterwards. For inflation, we observe a gradual decline in skewness, from an apparently symmetric distribution (no skewness) at the beginning of the sample to some evidence for negative skewness later on. This negative skewness is in line with the downside risk of inflation typically found between the mid-1990s and the early 2020s by De Polis et al. (2024) and for a somewhat shorter period by Le Bihan et al. (2023). The evidence for (time-varying) skewness is the weakest for industrial production, where credible intervals basically always include zero for the skewness parameter (implying a symmetric distribution), except for a short period in the mid-2010s, when skewness turns slightly positive. While this should not be over-interpreted, we note that industrial production growth can be seen as representing the evolution of the real economy, and from that perspective, the positive skewness found stands in contrast with the negative skewness that Iseringhausen (2024) tended to find for GDP growth; it is, however, in line with the finding of Karlsson et al. (2023). In addition, due to the dynamics of the model, it can still capture asymmetry in industrial production growth, such as larger negative movements during recessions (see Fig. 1).[6]

[4] The number of lags is set based on the Schwarz (1978) criterion applied to VARs assumed to have Gaussian and homoskedastic disturbances which have been estimated with maximum likelihood. As was pointed out above, no marginal likelihood calculations are carried out in this chapter and model selection based on marginal likelihoods accordingly cannot be conducted.

[5] The posterior mean of the degrees of freedom parameter is approximately 15, suggesting somewhat heavier-than-Gaussian tails.

[6] As a general caveat, it should be noted that model misspecification could affect our results. Since the model features dynamic skewness but constant parameters in the conditional mean equation, time variation in the conditional mean of one or more of the variables in the model—such as the

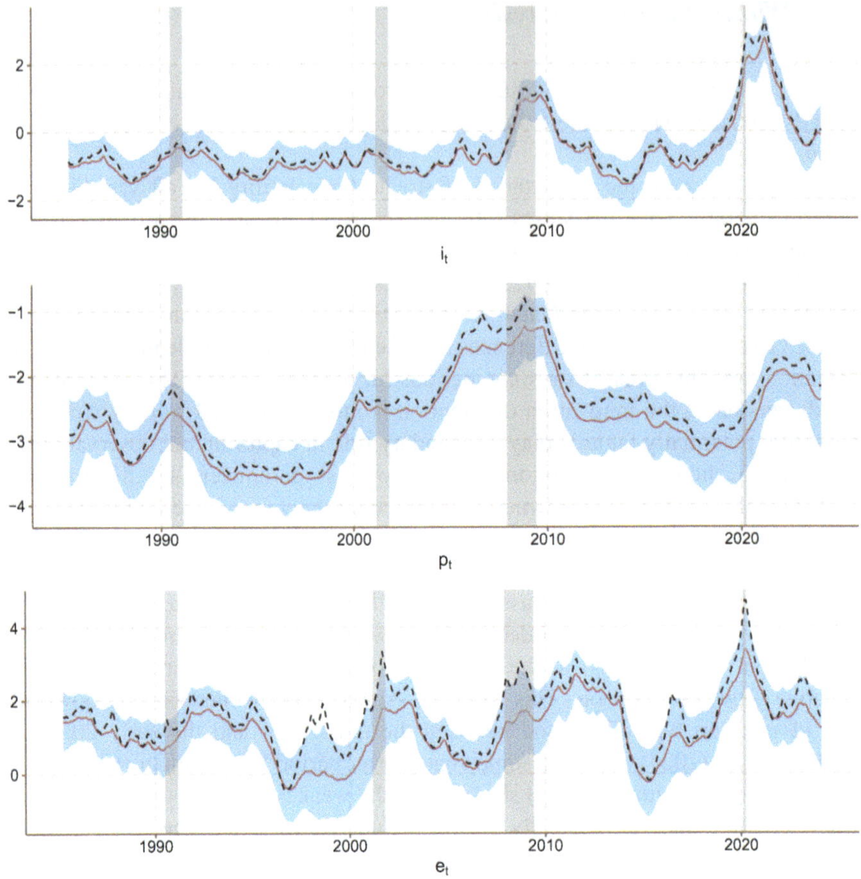

Fig. 2 Estimated log stochastic volatility. i_t is the industrial production growth rate, p_t is CPI inflation and e_t is the EPU index (divided by 10). Estimated log volatility of the VAR model with fat tails and dynamic skewness is given by the red solid line; the coloured band gives the 80% credible interval. The dashed black line shows the estimated mean log volatility from the VAR model with Gaussian error terms. The shaded areas highlight the recession periods based on the NBER indicators

Looking at the data series in Fig. 1 together with the estimates of the stochastic volatility and the conditional skewness in Figs. 2 and 3, we see that large movements in the variables around crisis times are mostly captured by an increase in the volatility and less so by an increase in the asymmetry. That is, the model attributes

time-varying "trend inflation" in, for example, Chan et al. (2018)—might partially be absorbed by the skewness parameter. To investigate this further, a model with time variation both in the skewness equation and in the conditional mean equation would have to be estimated. However, such an extension is beyond the scope of the current chapter.

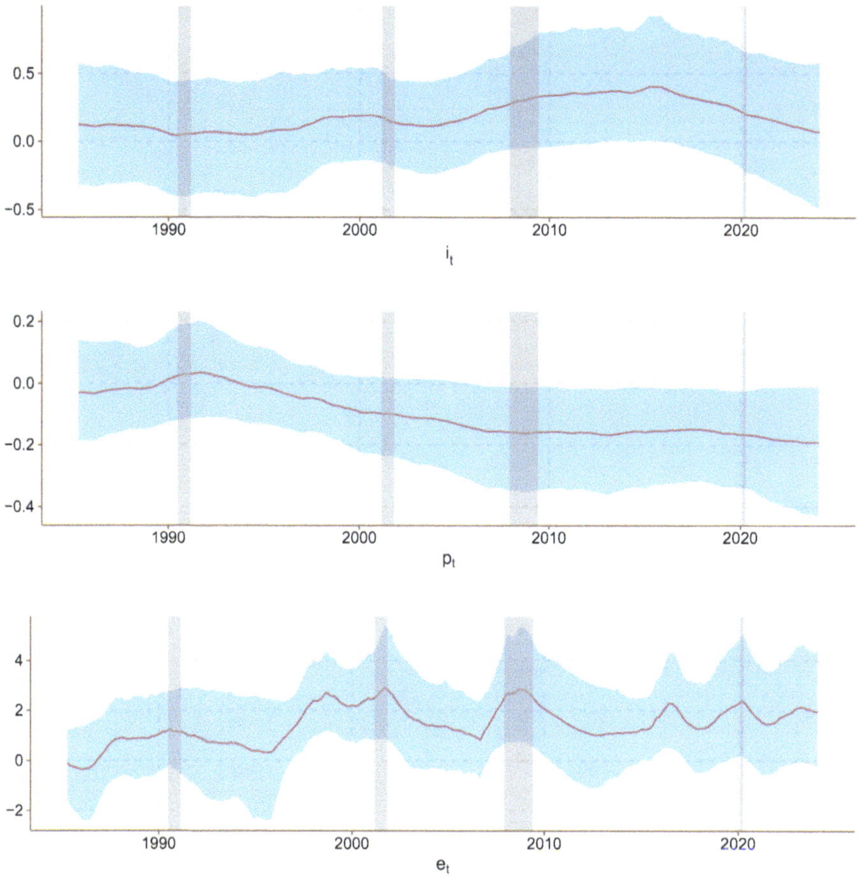

Fig. 3 Estimated time-varying skewness. i_t is the industrial production growth rate, p_t is CPI inflation and e_t is the EPU index (divided by 10). Estimated skewness (posterior mean) of the VAR model with fat tails and dynamic skewness is given by the red solid line; the coloured band gives the 80% credible interval. The shaded areas highlight the recession periods based on the NBER indicators

large movements to draws from an innovation distribution that is more spread-out, rather than draws from a distribution with increased asymmetry. The only exception is the EPU index, where not only dynamic skewness is the strongest, but also the difference in the stochastic volatility estimates for models with and without skewness is the most salient (the last panel of Fig. 2).

Next, we present the VAR dynamics using the impulse-response functions based on unit shocks. In this case, the impulse-response functions only depend on $\mathbf{A}$ and the dynamic parameters of the model ($\mathbf{B}_j$, $j = 1, \ldots, p$), which are constant over

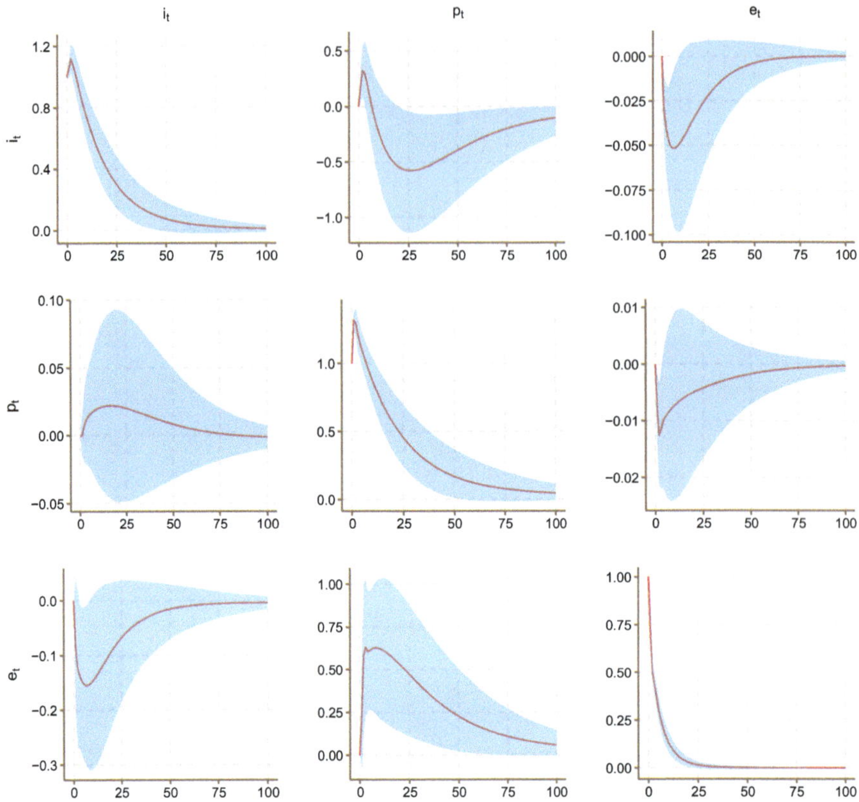

Fig. 4 The impulse-response functions of the model with time-varying skewness and fat tails. i_t is the industrial production growth rate, p_t is CPI inflation and e_t is the EPU index (divided by 10). Impulses are shown in the column headers, and response variables are given by the rows. The size of the impulse is unity. The red solid line gives the point estimate, and the coloured bands show the 80% credible interval. The size of the effect (in percentage points for i_t and p_t, and index units divided by 10 for e_t) is represented on the vertical axis. The horizontal axis shows the horizon in months

time, so responses to unit shocks are also time-invariant. These impulse-response functions, together with the 80% credible intervals, are shown in Fig. 4.[7]

The impulse-response functions are in line with what one would expect. In particular, the shock to the EPU index (the third column in Fig. 4) is in line with the findings of the previous literature in that it holds back real activity—here given

[7] Our model does feature stochastic volatility, so the impulse-response functions of a one standard deviation shock are time-varying. In this case, time variation solely comes from the time-varying standard deviation, presented in Fig. 2. By presenting responses to unit shocks, we isolate the dynamics of the conditional mean from the stochastic volatility and hence make interpretation of the impulse-response functions easier.

by industrial production growth (first row). In addition, the effect of the shock to the EPU index resembles a (negative) aggregate demand shock in that it also has a negative effect on inflation (second row). Regarding the other shocks, it can be noted that the point estimate of a shock to industrial production growth (first column) also resembles an aggregate demand shock as it is associated with an increase in inflation and a decrease in the EPU index. It should be noted though that the increase in inflation is not significant; the credible interval always includes zero. Finally, a shock to inflation (second column) looks very much like a supply-side (cost-push-type) shock; the effect on industrial production growth is negative (after a short, slightly positive effect), and the EPU increases.

Overall, the within-sample evidence shows that there is some support for time-varying skewness. And the dynamics of the model—illustrated by the impulse-response functions—are in line with mainstream findings from the previous literature on the relation between the EPU index and the macroeconomy.

3.2 Forecasting Performance

We next consider how different model features—in particular dynamic skewness—affect forecasting performance; both point and density forecasts are considered. Five different forecast horizons are assessed: one month, two months, three months, six months and one year. Our benchmark model is the univariate AR(3) model with normally distributed error terms. This is first compared to univariate specifications with error terms that allow for fat tails and asymmetry, where we consider both constant and time-varying skewness. We then also compare the benchmark model to trivariate VAR specifications, where three different trivariate specifications are used: Gaussian innovations, fat-tailed innovations with constant skewness and fat-tailed innovations with time-varying skewness. All models are estimated allowing for stochastic volatility.

The forecast evaluation is performed on an expanding sample. We use the mean squared forecast error (MSFE) for point forecast evaluation. At forecast horizon h, the MSFE is calculated as

$$\text{MSFE}_{l,h} = \frac{1}{T - T_0 - h + 1} \sum_{t=T_0}^{T-h} \left[\left(\bar{y}_{i,t+h|t} - y^o_{i,t+h} \right)^2 \right],$$

where $\bar{y}_{i,t+h|t}$ is the point forecast—the posterior mean—using all the information up to time t, and $y^o_{i,t+h}$ is the realization of the variable i, h-steps ahead. A lower MSFE indicates better model performance.

For density forecasts, we use the log predictive score (LPS). The LPS is defined such that a larger value indicates a better forecast performance of the model and is given by

$$\begin{aligned}
\mathrm{LPS}_{i,h} =& \frac{1}{T_1 - T_0 - h + 1} \sum_{t=T_0}^{T_1-h} \left[\log \int p(y^o_{i,t+h} | \boldsymbol{\theta}, \mathbf{y}_{1:t}) p(\boldsymbol{\theta} | \mathbf{y}_{1:t}) d\boldsymbol{\theta} \right] \\
=& \frac{1}{T_1 - T_0 - h + 1} \sum_{t=T_0}^{T_1-h} \left[\log \int \int p(y^o_{i,t+h} | \boldsymbol{\theta}, \mathbf{y}_{1:t}, \mathbf{y}_{t+1:t+h-1}, \boldsymbol{\theta}_{t+1:t+h}) \right. \\
& \left. p(\mathbf{y}_{t+1:t+h-1}, \boldsymbol{\theta}_{t+1:t+h} | \boldsymbol{\theta}, \mathbf{y}_{1:t}) p(\boldsymbol{\theta} | \mathbf{y}_{1:t}) d\mathbf{y}_{t+1:t+h-1} d\boldsymbol{\theta}_{t+1:t+h} d\boldsymbol{\theta} \right] \\
\approx& \frac{1}{T - T_0 - h + 1} \sum_{t=T_0}^{T-h} \left[\log \frac{1}{R} \sum_{r=1}^{R} p(y^o_{i,t+h} | \boldsymbol{\theta}^{(r)}, \mathbf{y}_{1:t}, \right. \\
& \left. \mathbf{y}^{(r)}_{t+1:t+h-1}, \boldsymbol{\theta}^{(r)}_{t+1:t+h}) \right],
\end{aligned}$$

where $\boldsymbol{\theta}^{(r)}$ is a posterior sample of the (time-invariant) parameters of the TV skewness VAR model for $r = 1, \ldots, R$; $\mathbf{y}^{(r)}_{t+1:t+h-1}$ represents a simulated path of intermediate observations between $t + 1$ and $t + h - 1$; $\boldsymbol{\theta}^{(r)}_{t+1:t+h}$ represents a simulated path of the time-varying parameters between $t + 1$ and $t + h$ and $p(y^o_{i,t+h} | \boldsymbol{\theta}^{(r)}, \mathbf{y}_{1:t}, \mathbf{y}^{(r)}_{t+1:t+h-1}, \boldsymbol{\theta}^{(r)}_{t+1:t+h})$ is the one-step ahead posterior predictive density function (the score) evaluated at the realization of the variable, conditional on the posterior sample of parameters and the simulated path.

To compare both point and density forecasting performance, we apply the one-sided test of Diebold and Mariano (1995), where the standard errors are calculated using the Newey-West estimator, in line with the previous work by Clark (2011) and Clark and Ravazzolo (2015). Forecasts are generated from January 2010 to March 2023 for all horizons, which allows for evaluating 159 forecasts uniformly over all horizons.

Tables 1 and 2 show forecast comparison results for point and density forecasts, respectively. The first row in each panel (for each variable) contains the MSFE and the LPS values for the univariate Gaussian model, and in the rest of the rows, relative values (relative MSFE and difference in LPS) are shown.

Looking at point forecasts, no quantitatively meaningful improvements can be found relative to the forecasts of the baseline univariate Gaussian model for industrial production growth or CPI inflation (even if the Diebold-Mariano test indicates that the univariate model with fat tails and dynamic skewness has a significantly lower MSFE at the 3-month horizon for industrial production growth).

Table 1 MSFEs

	1M	2M	3M	6M	12M
	(a) Industrial production growth				
Univariate AR(3)					
G.SV	4.465	11.131	14.202	16.373	23.982
Skew-T.SV	1.000	1.000	1.005	1.004	1.017
dynSkew-T.SV	0.994	0.997	0.997†	1.001	1.016
Trivariate VAR(3)					
G.SV	0.994	1.021	1.056	1.153	1.202
Skew-T.SV	0.994	1.030	1.069	1.152	1.191
dynSkew-T.SV	0.989	1.026	1.063	1.146	1.190
	(b) CPI inflation				
Univariate AR(3)					
G.SV	0.104	0.334	0.629	1.565	3.395
Skew-T.SV	1.003	1.003	1.006	1.005	1.010
dynSkew-T.SV	1.007	1.007	1.008	1.007	1.011
Trivariate VAR(3)					
G.SV	1.092	1.128	1.134	1.109	1.116
Skew-T.SV	1.094	1.123	1.127	1.099	1.101
dynSkew-T.SV	1.090	1.118	1.120	1.091	1.102
	(c) EPU/10				
Univariate AR(3)					
G.SV	16.699	33.934	43.480	59.102	68.458
Skew-T.SV	1.043	1.006	0.977	0.933*	0.883*
dynSkew-T.SV	1.046	1.010	0.981	0.939*	0.887*
Trivariate VAR(3)					
G.SV	1.004	0.999	0.989	0.986	0.986
Skew-T.SV	1.025	0.997	0.976	0.958	0.938
dynSkew-T.SV	1.024	0.992	0.970	0.949*	1.009

The first line in each panel reports the MSFE of the benchmark univariate model with Gaussian innovations based on forecasts generated from 2010M01 to 2023M03 (159 recursive estimations). The relative performance (all other rows) is computed as the ratio of the MSFE of alternative specifications over the benchmark, that is, entries less than one indicate that a given model has a lower MSFE. * denotes that the model in question significantly outperforms the benchmark model at the 5% level based on the one-sided Diebold and Mariano (1995) test where the standard errors of the test statistics are computed with the Newey–West estimator

In contrast, when forecasting the EPU index, allowing for fat tails and constant or dynamic skewness improves over Gaussian innovations in a univariate framework at the two longest horizons. Turning to the trivariate models, significant improvements over the benchmark model can only be found in one case, namely when forecasting the EPU index at the 6-month horizon using the model with fat tails and dynamic skewness; it can, however, be noted that this MSFE is higher than that of the corresponding univariate model. The overall impression is that the trivariate models

Table 2 LPSs

	1M	2M	3M	6M	12M
	(a) Industrial production growth				
Univariate AR(3)					
G.SV	−1.577	−2.149	−2.433	−2.652	−2.966
Skew-T.SV	0.009	0.109	−0.048	0.001	−0.006
dynSkew-T.SV	0.041	0.000	0.075	0.001	−0.007
Trivariate VAR(3)					
G.SV	−0.021	−0.119	−0.034	−0.209	−0.246
Skew-T.SV	−0.058	−0.138	−0.087	−0.181	−0.232
dynSkew-T.SV	−0.047	−0.116	−0.141	−0.214	−0.275
	(b) CPI inflation				
Univariate AR(3)					
G.SV	−0.294	−0.874	−1.197	−1.639	−1.999
Skew-T.SV	0.009	0.005	0.000	−0.010	−0.066
dynSkew-T.SV	0.005	−0.011	−0.030	−0.035	−0.219
Trivariate VAR(3)					
G.SV	−0.031	−0.048	−0.058	−0.037	−0.016
Skew-T.SV	−0.037	−0.058	−0.085	−0.098	−0.151
dynSkew-T.SV	−0.039	−0.076	−0.102	−0.121	−0.228
	(c) EPU/10				
Univariate AR(3)					
G.SV	−2.537	−2.816	−2.944	−3.081	−3.243
Skew-T.SV	−0.006	0.019	0.052	−0.014	0.073
dynSkew-T.SV	−0.011	0.036	0.043	−0.009	0.029
Trivariate VAR(3)					
G.SV	−0.001	0.006	0.022	0.024	0.045
Skew-T.SV	−0.002	0.018	0.042	0.028	0.078
dynSkew-T.SV	−0.002	0.010	0.004	−0.004	0.013

The first line in each panel reports the LPS of the benchmark univariate model with Gaussian innovations based on forecasts generated from 2010M01 to 2023M03 (159 recursive estimations). The relative performance is computed as the difference of the LPS of alternative specifications and the benchmark, that is, entries smaller than zero indicate that a given model has a higher LPS. * denotes that the model in question significantly outperforms the benchmark model at the 5% level based on the one-sided test Diebold and Mariano (1995) where the standard errors of the test statistics are computed with the Newey–West estimator

are not particularly successful compared to the univariate models when it comes to forecasting.

Regarding the density forecasts, we find no systematic improvements *vis-á-vis* the univariate Gaussian model. While there are some cases where the LPS is higher for the model with fat tails and dynamic skewness, these improvements are never significant. Similarly, using the trivariate VAR model never leads to significant improvements in density predictions.

4 Conclusions

In this chapter, we have extended the standard Bayesian VAR with stochastic volatility to allow for generalized hyperbolic skew Student's t distribution in the innovations, where the skewness parameter is also allowed to vary over time. This specification permits a flexible modelling of innovations in terms of fat tails and—potentially dynamic—asymmetry. These might be useful features when modelling the macroeconomy, given the recent literature documenting the non-Gaussianity of many macroeconomic variables. Our model nests specifications with Gaussian and fat-tailed innovations, as well as the specification where asymmetry is constant over time, allowing us to analyse the importance of these features in an empirical context.

In our empirical application, we revisit how aggregate uncertainty (captured by the EPU index) affects the macroeconomy. This is done by estimating a trivariate VAR model including industrial production growth, CPI inflation and the EPU index. Apart from the usual strong evidence for time-varying volatility, we also document that the estimated skewness of the variables—primarily the EPU index—changes over time. However, we also find that modelling innovations flexibly typically does not improve the point or density forecasts. Finally, we confirm previous findings in the literature that shocks to uncertainty resemble aggregate demand shocks, dampening both real activity and inflation.

Acknowledgments The authors are grateful to the reviewer for the careful reading of the manuscript and for the suggestions that have improved an earlier version of this chapter. Hoang Nguyen, Stepan Mazur and Pär Österholm acknowledge financial support from the project "Improved Economic Policy and Forecasting with High-Frequency Data" (Dnr: E47/22) funded by the Torsten Söderbergs Foundation. Stepan Mazur also acknowledges financial support from the internal research grants at Örebro University.

Appendix

Table 3 Computation time for VAR models in minutes

	$p = 1$			$p = 3$		
$T = 250$	$k = 3$	$k = 5$	$k = 10$	$k = 3$	$k = 5$	$k = 10$
G.SV	2.6	4.8	15.8	3.1	6.4	34.3
dynSkew-T.SV	10.0	14.5	29.7	10.3	16.4	48.0
$T = 500$	$k = 3$	$k = 5$	$k = 10$	$k = 3$	$k = 5$	$k = 10$
G.SV	4.4	8.3	28.5	5.1	11.2	61.2
dynSkew-T.SV	18.9	27.8	55.2	19.6	31.2	87.7

p and k are the number of lags and the number of endogenous variables in the VAR, respectively. T is the number of observations. We sample 100,000 draws with 10,000 burn-in from the posterior distributions on 4 cores of AMD Ryzen 1950X processor at 3.4 GHz

References

Acemoglu, D. & Scott, A. (1997). Asymmetric business cycles: Theory and time-series evidence. *Journal of Monetary Economics, 40*(3), 501–533.

Adrian, T., Boyarchenko, N., & Giannone, D. (2019). Vulnerable growth. *American Economic Review, 109*(4), 1263–1289.

Akram, F. Q. & Mumtaz, H. (2019). Time-varying dynamics of the Norwegian economy. *The Scandinavian Journal of Economics, 121*(1), 407–434.

Ascari, G., Fagiolo, G., & Roventini, A. (2015). Fat-tail distributions and business-cycle models. *Macroeconomic Dynamics, 19*(2), 465–476.

Baker, S. R., Bloom, N., & Davis, S. J. (2016). Measuring economic policy uncertainty. *Quarterly Journal of Economics, 131*(4), 1593–1636.

Balcilar, M., Gupta, R., & Segnon, M. (2016). The role of economic policy uncertainty in predicting us recessions: A mixed-frequency Markov-switching vector autoregressive approach. *Economics, 10*(1), 20160027.

Bernanke, B. (1983). Irreversibility, uncertainty, and cyclical investment. *Quarterly Journal of Economics, 98*(1), 85–106.

Bloom, N. (2009). The impact of uncertainty shocks. *Econometrica, 77*(3), 623–685.

Britton, E., Fisher, P., & Whitley, J. (1998). The Inflation Report projections: Understanding the fan chart. *Bank of England Quarterly Bulletin, Q1*, 30–37.

Caggiano, G., Castelnuovo, E., & Figueres, J. M. (2017). Economic policy uncertainty and unemployment in the United States: A nonlinear approach. *Economics Letters, 151*, 31–34.

Carriero, A., Clark, T. E., & Marcellino, M. (2015). Realtime nowcasting with a Bayesian mixed frequency model with stochastic volatility. *Journal of the Royal Statistical Society. Series A,(Statistics in Society), 178*(4), 837–862.

Carriero, A., Clark, T. E., & Marcellino, M. (2018). Measuring uncertainty and its impact on the economy. *Review of Economics and Statistics, 100*(5), 799–815.

Carriero, A., Clark, T. E., & Marcellino, M. (2020). Assessing international commonality in macroeconomic uncertainty and its effects. *Journal of Applied Econometrics, 35*(3), 273–293.

Carriero, A., Clark, T. E., & Marcellino, M. (2024). Capturing macro-economic tail risks with Bayesian vector autoregressions. *Journal of Money, Credit and Banking, 56*(5), 1099–1127.

Carter, C. K. & Kohn, R. (1994). On Gibbs sampling for state space models. *Biometrika, 81*(3), 541–553.

Chan, J. C. (2017). The stochastic volatility in mean model with time-varying parameters: An application to inflation modeling. *Journal of Business and Economic Statistics, 35*(1), 17–28.

Chan, J. C., Clark, T. E., & Koop, G. (2018). A new model of inflation, trend inflation, and long-run inflation expectations. *Journal of Money, Credit and Banking, 50*(1), 5–53.

Chan, J. C. & Eisenstat, E. (2018). Bayesian model comparison for time-varying parameter VARs with stochastic volatility. *Journal of Applied Econometrics, 33*(4), 509–532.

Chiu, C.-W. J., Mumtaz, H., & Pinter, G. (2017). Forecasting with VAR models: Fat tails and stochastic volatility. *International Journal of Forecasting, 33*(4), 1124–1143.

Clark, T. E. (2011). Real-time density forecasts from Bayesian vector autoregressions with stochastic volatility. *Journal of Business and Economic Statistics, 29*(3), 327–341.

Clark, T. E., Huber, F., Koop, G., & Marcellino, M. (2024). Forecasting US inflation using Bayesian nonparametric models. *Annals of Applied Statistics, 18*(2), 1421–1444.

Clark, T. E. & Ravazzolo, F. (2015). Macroeconomic forecasting performance under alternative specifications of time-varying volatility. *Journal of Applied Econometrics, 30*(4), 551–575.

Cogley, T. & Sargent, T. J. (2005). Drifts and volatilities: Monetary policies and outcomes in the post WWII US. *Review of Economic Dynamics, 8*(2), 262–302.

Cross, J. & Poon, A. (2016). Forecasting structural change and fat-tailed events in Australian macroeconomic variables. *Economic Modelling, 58*, 34–51.

Cross, J. L., Hou, C., Koop, G., & Poon, A. (2023). Large stochastic volatility in mean VARs. *Journal of Econometrics, 236*(1), 105469.

De Polis, A., Melosi, L., & Petrella, I. (2024). The taming of the skew: Asymmetric inflation risk and monetary policy. *Working Paper*. https://andreadepolis.github.io/workingpapers/inflationrisk/DPMP2024.pdf
Del Negro, M. & Primiceri, G. E. (2015). Time varying structural vector autoregressions and monetary policy: A corrigendum. *Review of Economic Studies, 82*(4), 1342–1345.
Delle Monache, D., De Polis, A., & Petrella, I. (2024). Modeling and forecasting macroeconomic downside risk. *Journal of Business and Economic Statistics, 42*(3), 1010–1025.
Diebold, F. X. & Mariano, R. S. (1995). Comparing predictive accuracy. *Journal of Business and Economic Statistics, 13*(3), 134–144.
Fagiolo, G., Napoletano, M., & Roventini, A. (2008). Are output growth-rate distributions fat-tailed? Some evidence from OECD countries. *Journal of Applied Econometrics, 23*(5), 639–669.
Gulen, H. & Ion, M. (2016). Policy uncertainty and corporate investment. *Review of Financial Studies, 29*(3), 523–564.
Hörmann, W. & Leydold, J. (2014). Generating generalized inverse Gaussian random variates. *Statistics and Computing, 24*(4), 547–557.
Huber, F. & Koop, G. (2024). Fast and order-invariant inference in Bayesian VARs with nonparametric shocks. *Journal of Applied Econometrics, 39*, 1301–1320.
Huber, F. & Rossini, L. (2022). Inference in Bayesian additive vector autoregressive tree models. *Annals of Applied Statistics, 16*(1), 104–123.
Iseringhausen, M. (2020). The time-varying asymmetry of exchange rate returns: A stochastic volatility–stochastic skewness model. *Journal of Empirical Finance, 58*, 275–292.
Iseringhausen, M. (2024). A time-varying skewness model for Growth-at-Risk. *International Journal of Forecasting, 40*(1), 229–246.
Jones, P. M. & Olson, E. (2013). The time-varying correlation between uncertainty, output, and inflation: Evidence from a DCC-GARCH model. *Economics Letters, 118*(1), 33–37.
Karlsson, S., Mazur, S., & Nguyen, H. (2023). Vector autoregression models with skewness and heavy tails. *Journal of Economic Dynamics and Control, 146*, 104580.
Karlsson, S. & Österholm, P. (2020a). A hybrid time-varying parameter Bayesian VAR analysis of Okun's law in the United States. *Economics Letters, 197*, 109622.
Karlsson, S. & Österholm, P. (2020b). A note on the stability of the Swedish Phillips curve. *Empirical Economics, 59*(6), 2573–2612.
Karlsson, S. & Österholm, P. (2023). Is the US Phillips curve stable? Evidence from Bayesian vector autoregressions. *Scandinavian Journal of Economics, 125*(1), 287–314.
Karnizova, L. & Li, J. C. (2014). Economic policy uncertainty, financial markets and probability of US recessions. *Economics Letters, 125*(2), 261–265.
Kastner, G. & Frühwirth-Schnatter, S. (2014). Ancillarity-sufficiency interweaving strategy (ASIS) for boosting MCMC estimation of stochastic volatility models. *Computational Statistics and Data Analysis, 76*, 408–423.
Kim, S., Shephard, N., & Chib, S. (1998). Stochastic volatility: Likelihood inference and comparison with ARCH models. *Review of Economic Studies, 65*(3), 361–393.
Kiss, T., Mazur, S., Nguyen, H., & Österholm, P. (2023). Modeling the relation between the US real economy and the corporate bond-yield spread in Bayesian VARs with non-Gaussian innovations. *Journal of Forecasting, 42*(2), 347–368.
Koop, G. & Korobilis, D. (2010). *Bayesian multivariate time series methods for empirical macroeconomics*. Now Publishers.
Koop, G. & Korobilis, D. (2019). Forecasting with high-dimensional panel VARs. *Oxford Bulletin of Economics and Statistics, 81*(5), 937–959.
Le Bihan, H., Leiva-Leon, D., & Pacce, M. (2023). Underlying inflation and asymmetric risks. *ECB Working Paper Series 2848*.
Leduc, S. & Liu, Z. (2016). Uncertainty shocks are aggregate demand shocks. *Journal of Monetary Economics, 82*, 20–35.
Liu, X. (2019). On tail fatness of macroeconomic dynamics. *Journal of Macroeconomics, 62*, 103154.

McCracken, M. W. & Ng, S. (2016). FRED-MD: A monthly database for macroeconomic research. *Journal of Business and Economic Statistics, 34*(4), 574–589.

McNeil, A. J., Frey, R., & Embrechts, P. (2015). *Quantitative risk management: Concepts, Techniques and Tools-revised edition.* Princeton University Press.

Mumtaz, H. & Surico, P. (2018). Policy uncertainty and aggregate fluctuations. *Journal of Applied Econometrics, 33*(3), 319–331.

Nam, E.-Y., Lee, K., & Jeon, Y. (2021). Macroeconomic uncertainty shocks and households' consumption choice. *Journal of Macroeconomics, 68*, 103306.

Ni, S. & Sun, D. (2005). Bayesian estimates for vector autoregressive models. *Journal of Business and Economic Statistics, 23*(1), 105–117.

Primiceri, G. E. (2005). Time varying structural vector autoregressions and monetary policy. *Review of Economic Studies, 72*(3), 821–852.

Roberts, G. O. & Rosenthal, J. S. (2009). Examples of adaptive MCMC. *Journal of Computational and Graphical Statistics, 18*(2), 349–367.

Schwarz, G. (1978). Estimating the dimension of a model. *Annals of Statistics, 6*, 461–464.

Sims, C. A. & Zha, T. (2006). Were there regime switches in US monetary policy? *American Economic Review, 96*(1), 54–81.

Tsiotas, G. (2012). On generalised asymmetric stochastic volatility models. *Computational Statistics and Data Analysis, 56*(1), 151–172.

International Transmission of Macroeconomic Uncertainty in Small Open Economies: An Empirical Approach

Jamie L. Cross, Chenghan Hou, and Aubrey Poon

Abstract We propose a vector autoregression with common stochastic volatility in mean (VAR-CSVM) dynamics to estimate the transmission of domestic and international sources of macroeconomic uncertainty shocks in three small open economies (SOEs): Australia, Canada, and New Zealand. We find evidence that international uncertainty spillovers shape the macroeconomic conditions in all three SOEs; that domestic uncertainty shocks have idiosyncratic transmission mechanisms in each SOE; and that accounting for uncertainty within the VAR-CSVM improves point and density forecast accuracy compared to the nested VAR-CSV and homoscedastic VAR models.

1 Introduction

Following Bloom (2009), numerous scholars have demonstrated the significance of modeling macroeconomic uncertainty in the US economy. For a recent survey, see Castelnuovo (2023). A burgeoning strand of this literature is recently focusing on the macroeconomic effects of global and international sources of uncertainty (see, e.g., Mumtaz & Theodoridis, 2015, 2017; Carriero et al., 2020; Cesa-Bianchi et al., 2020; Beckmann et al., 2023; Caggiano & Castelnuovo, 2023; Pfarrhofer, 2023, among others). However, few papers have examined the consequences of potentially different effects of domestic and international uncertainty spillovers within small

J. L. Cross
Melbourne Business School, The University of Melbourne, Parkville, VIC, USA
e-mail: j.cross@mbs.edu

C. Hou (✉)
Hunan University, Changsha, China

A. Poon
University of Kent, Canterbury, UK
e-mail: a.poon@kent.ac.uk

S. Mazur, P. Österholm (eds.), *Recent Developments in Bayesian Econometrics and Their Applications*, https://doi.org/10.1007/978-3-032-00110-8_6

open economies (SOEs).[1] This is surprising because it is well known that SOEs are highly susceptible to international macroeconomic shocks. For instance, Justiniano and Preston (2010) find that around half medium-term Canadian output growth volatility is explained by macroeconomic shocks from the US economy, and Cross and Poon (2020) find that around half of Australian business cycle fluctuations are a result of international macroeconomic spillovers. It is therefore natural to ask: What are the effects of international uncertainty spillovers on SOEs? And, does accounting for uncertainty enhance macroeconomic forecast accuracy?

In this chapter, we address these policy-relevant questions for three commonly studied inflation targeting SOEs: Australia, Canada, and New Zealand. To that end, we propose a structural VAR model for SOEs that allows for both domestic and international sources of uncertainty. As in studies on the US economy (e.g., Jurado et al., 2015; Carriero et al., 2018, 2020), our model-based measures of domestic uncertainty are defined as the common component in the second moment of a particular country's macroeconomic variables. Unanticipated changes in the common second moment of domestic and international variables, i.e., domestic/international uncertainty shocks, are modeled with common stochastic volatility terms that directly impact the conditional mean (CSVM). The resulting VAR-CSVM model is estimated for each of the three SOEs individually, using the USA as a representative large economy.

The results suggest that domestic and international sources of macroeconomic uncertainty shape the economic climate in each SOE. In all three cases, international uncertainty shocks have a persistent negative impact on real GDP (RGDP) growth, while increasing both inflation and interest rates. Domestic uncertainty shocks also have a similar effect on inflation and interest rates; however, the RGDP growth responses are idiosyncratic. The transmission of domestic uncertainty shocks is negative in Canada and positive in New Zealand, while the Australian response is initially negative and becomes positive over time. These results are somewhat at odds with some finding for the US economy where uncertainty can be viewed as an aggregate demand shock. For instance, using a VAR model (Leduc & Liu, 2016) finds that uncertainty shocks elicit demand shock type of effects—a finding which (Haque & Magnusson, 2021) show is robust to potential parameter instability as modeled by time-varying parameters and stochastic volatility. However, they are in line with (Fasani & Rossi, 2018) who show that the negative response of inflation in their model can become positive when modeling interest rate inertia—as a result uncertainty shocks look more like cost-push shocks. A more detailed discussion of the conflict in the literature is provided in the recent survey by Castelnuovo (2023), who notes that while uncertainty is generally counter cyclical in nature, the impact of uncertainty on inflation is uncertain.

[1] Two exceptions are Cross et al. (2017) and Caggiano et al. (2020). The former use a structural macromodel to investigate the effects of various uncertainties between the USA and Canada. The latter use an empirical model to investigate the effects of international uncertainty spillovers from the USA to Canada in booms and busts.

After establishing these results, we then ask whether accounting for uncertainty can improve upon the forecast accuracy of a homeoscedastic VAR model, and a VAR model with common stochastic volatility that does not impact the conditional mean (VAR-CSV). Since each of the SOEs has an inflation targeting mandate, any improvements in forecast performance would allow for more informed policy decisions. The major difficulty in this task is that commonly used estimation procedures are computationally intensive, making forecasting exercises infeasible. To overcome this hurdle, we adapt a recently proposed efficient Markov chain Monte Carlo (MCMC) algorithm for large VAR-SVM models proposed in Cross et al. (2023), to our VAR-CSVM model. The main idea in this algorithm is to sample each of the common log-volatility terms across all dates in a single step, without the need for iterative procedures, like the particle filter used elsewhere in the literature. The key feature of the algorithm that makes it fast is that the log-conditional densities of the log volatilities have a block-banded Hessian matrix, thereby facilitating the use of computationally efficient band and sparse matrix algorithms. The main result is that accounting for uncertainty improves upon the accuracy of a conventional VAR with and without stochastic volatility. Taken together with results from the in-sample analysis, this suggest that uncertainty plays a key role in shaping the macroeconomies of each SOE, and accounting for this uncertainty can aid in making better informed policy decisions.

This chapter is structured as follows. Section 2 details the methodology. Results from the in-sample and out-of-sample analyses are presented in Sects. 3 and 4, respectively. Section 5 concludes.

2 Common Stochastic Volatility in Mean VAR

This section contains three subsections with distinct methodological details. In Sect. 2.1, we introduce the VAR with common stochastic volatility in mean (VAR-CSVM) model used in our empirical analysis. It is essentially an extended version of the common stochastic volatility VAR (VAR-CSV) model developed by Carriero et al. (2016), in which we allow for two additional features. First, as in conventional VAR-SVM models, the second moment shocks directly enter the observation equation allowing them to impact the levels of the variables (see, e.g., Mumtaz & Zanetti, 2013; Mumtaz & Theodoridis, 2015, 2017; Carriero et al., 2018, 2020). Second, for reasons explained later, we deviate from the state equation specification of an independent AR(1) process in Carriero et al. (2016) and instead specify a general VAR(1) process with full-covariance matrix.

To determine whether including the CSVM terms is supported by the data, we use the Bayes factor to conduct an in-sample model comparison exercise. Since the marginal likelihood for the VAR-CSVM model is not available analytically, we use the fact that it can be decomposed into the product of one-step-ahead predictive likelihoods evaluated at the observed data (Geweke & Amisano, 2011). The same procedure is also used to test various hypotheses about the flow of domestic and international uncertainties between the large and small economies in the model. These details are in Sect. 2.2.

Finally, in relation to the aforementioned VAR-SVM models, our VAR-CSVM model is most similar to the two-factor SVM model in Carriero et al. (2018). The main difference is that we focus solely on common variation in the macroeconomic variables, as opposed to idiosyncratic volatility. Since we are interested in the effects of country-specific macroeconomic uncertainty, we believe that nothing is lost in the context to our research question. Instead, a key advantage of our proposed specification is that it allows for the development of an efficient MCMC algorithm which facilitates our forecasting exercise in Sect. 4. The essential details of this algorithm are given in Sect. 2.3.

2.1 The Model

To set the stage, let $\mathbf{y}_t = (y_{1,t}, \ldots, y_{2n,t})' = (\mathbf{y}_t^{L'}, \mathbf{y}_t^{S'})'$ denote a vector of variables of interest, where the superscripts respectively denote the set of variables in the large and SOE. In our study, both $\mathbf{y}_t^L$ and $\mathbf{y}_t^S$ are $n \times 1$ vectors; however, the model can also accommodate vectors of distinct size. We consider the following structural VAR-SVM model[2]

$$\mathbf{B}_0\mathbf{y}_t = \mathbf{c} + \sum_{i=1}^{p} \mathbf{B}_i\mathbf{y}_{t-i} + \mathbf{A}\begin{pmatrix} e^{h_t^L} \\ e^{h_t^S} \end{pmatrix} + \boldsymbol{\epsilon}_t^y, \quad \boldsymbol{\epsilon}_t^y \sim \mathcal{N}(\mathbf{0}, \boldsymbol{\Sigma}_t), \tag{1}$$

where $\mathcal{N}(\cdot, \cdot)$ denotes the Gaussian distribution, $\mathbf{c}$ is a $2n \times 1$ vector, $\mathbf{B}_i, i = 1, \ldots, p$ are conditional mean coefficients of size $2n \times 2n$, and $\mathbf{B}_0$ is a lower triangular "structural impact matrix" with ones on the main diagonal.[3] The $2n \times 2$ "uncertainty

[2] In this general setup, we allow the small open economy to affect the large economy. This is motivated by the small open economy having some predictive power for the large economy (evidence of this is provided in Sect. 3.1). In future research, it would be interesting to consider placing priors on the dynamics that make the large economy block exogenous as in Villani and Warne (2003) and Österholm and Zettelmeyer (2008).

[3] The lower triangular structure imposed on $\mathbf{B}_0$ is to identify the VAR structural shocks $\boldsymbol{\epsilon}_t^y$, which are required for the estimation of the uncertainty measures of interest. This identifying scheme assumes that the VAR structural shocks of the large economy have contemporaneous effects on the small economy, while the small economy reacts to the large economy with a lag. A drawback of this setting is that the estimation results might be sensitive to the ordering of the VAR variables (see Davidson et al., 2025 for a recent discussion of this issue in the context of VAR-SVM models more generally). However, a key advantage of this modeling specification is that it allows us to simulate posterior draws of $(\mathbf{B}_0, \mathbf{B}_1, \ldots, \mathbf{A})$ in one step (see Appendix A.2), which can reduce the correlation among posterior draws.

impact matrix" $\mathbf{A}$ and the $2n \times 2n$ covariance matrix $\boldsymbol{\Sigma}_t$ are given by

$$\mathbf{A} = \begin{pmatrix} \mathbf{a}_{11} & \mathbf{a}_{12} \\ \mathbf{a}_{21} & \mathbf{a}_{22} \end{pmatrix} \quad \text{and} \quad \boldsymbol{\Sigma}_t = \begin{pmatrix} e^{h_t^L}\boldsymbol{\Sigma}_L & \mathbf{0} \\ \mathbf{0} & e^{h_t^S}\boldsymbol{\Sigma}_S \end{pmatrix},$$

where each $\mathbf{a}_{ij}$ with $i, j = 1, 2$ is a $n \times 1$ vector, and both $\boldsymbol{\Sigma}_L$ and $\boldsymbol{\Sigma}_S$ are diagonal matrices of size $n \times n$, i.e., $\boldsymbol{\Sigma}_L = \text{diag}(\sigma^2_{1,L}, \ldots, \sigma^2_{n,L})$ and $\boldsymbol{\Sigma}_S = \text{diag}(\sigma^2_{n+1,S}, \ldots, \sigma^2_{2n,S})$.[4] We note that the covariance matrix $\boldsymbol{\Sigma}_t$ is changing over time and that this time variation is driven by the common stochastic volatilities from both the large economy $e^{h_t^L}$ and the small economy $e^{h_t^S}$. Since they capture the common component in the volatility of the country-specific macroeconomic variables, these terms provide measures of uncertainty, and we refer to unanticipated changes in these volatilities as *uncertainty shocks*.

The state equations for the log volatilities $\mathbf{h}_t = (h_t^L, h_t^S)'$ are assumed to follow a stationary VAR(1) process

$$\mathbf{h}_t = \boldsymbol{\Phi}\mathbf{h}_{t-1} + \boldsymbol{\epsilon}_t^h, \quad \boldsymbol{\epsilon}_t^h \sim \mathcal{N}(\mathbf{0}, \boldsymbol{\Sigma}_h), \tag{2}$$

where $\boldsymbol{\Phi}$ is a full 2×2 coefficient matrix and the initial condition is set to be $\mathbf{h}_t \sim \mathcal{N}(\mathbf{0}, \mathbf{V}_h)$. In order to allow the log volatilities of the small and large open economies to be correlated with each other, $\boldsymbol{\Sigma}_h$ is assumed to be a full positive-definite matrix which is free to be estimated.

We use a stationary vector autoregressive law of motion for the log volatilities for both economic and econometric reasons. Economically, it allows for spillovers between the uncertainty measures of the large and small economies. In contrast, a simpler autoregressive specification would rule this out a priori. For instance, it is well documented that US uncertainty ($e_t^{h^L}$) increased during the 2008 Global Financial Crisis (Jurado et al., 2015; Carriero et al., 2018, 2020). If we were to instead adopt an independent AR specification, then there would be no channel through which this US originated uncertainty shock can affect the level of domestic uncertainty in Canada, for example. Econometrically, stationarity is important as it ensures the existence of closed-form solutions for the first and second moments (see Andreasen, 2010 for a discussion of this point in the univariate context).

---

[4] Note that the coefficient matrix $\mathbf{A}$ should not be interpreted as the contemporaneous impulse responses of uncertainty shocks; it only specifies the relationship between the uncertainty measures in the structural form of our model. By premultiplying both sides of equation (1) by $\mathbf{B}_0^{-1}$, we obtain the reduced form model:

$$\mathbf{y}_t = \bar{\mathbf{c}} + \sum_{i=1}^{p} \bar{\mathbf{B}}_i \mathbf{y}_{t-i} + \bar{\mathbf{A}} \begin{pmatrix} e^{h_t^L} \\ e^{h_t^S} \end{pmatrix} + \bar{\boldsymbol{\epsilon}}_t^y, \quad \bar{\boldsymbol{\epsilon}}_t^y \sim \mathcal{N}(\mathbf{0}, \bar{\boldsymbol{\Sigma}}_t),$$

where $\bar{\mathbf{c}} = \mathbf{B}_0^{-1}\mathbf{c}$, $\bar{\mathbf{B}}_i = \mathbf{B}_0^{-1}\mathbf{B}_i$, $\bar{\mathbf{A}} = \mathbf{B}_0^{-1}\mathbf{A}$, and $\bar{\boldsymbol{\Sigma}}_t = \mathbf{B}_0^{-1}\boldsymbol{\Sigma}_t\mathbf{B}_0^{-1'}$. It can be seen that the contemporaneous impulse responses of uncertainty shocks are characterized by $\bar{\mathbf{A}}$.

To identify the structural uncertainty shocks, we adopt a simple, yet economically plausible, identifying assumption. That is, uncertainty in the large economy does not respond contemporaneously to uncertainty shocks in the small economy. To be specific, we assume $\boldsymbol{\epsilon}_t^h = \mathbf{C}\mathbf{e}_t$, where $\mathbf{C}$ is a lower triangular matrix and $\mathbf{e}_t$ collects the structural shocks that follow $\mathcal{N}(\mathbf{0}, \mathbf{I}_2)$. We also note that since the uncertainty measures are estimated within the model, uncertainty shocks are, by construction, orthogonal to the VAR shocks. In this sense, our identification strategy is very similar to that in Carriero et al. (2018, 2020).

To complete the model specification, we assume independent prior distributions for each of the model parameters. Let $\boldsymbol{\beta} = \text{vec}([\mathbf{c}, \mathbf{B}_1, , \ldots, \mathbf{B}_p]')$, $\boldsymbol{\gamma}$ be a vector stacking the nonzero elements in each row $\mathbf{B}_0$ and $\mathbf{a} = \text{vec}(\mathbf{A})'$. For the measurement equation, we specify Gaussian priors for the VAR coefficients and the uncertainty impact matrix $\boldsymbol{\beta} \sim \mathcal{N}(\boldsymbol{\beta}_0, \mathbf{V}_\beta)$, $\boldsymbol{\gamma} \sim \mathcal{N}(\boldsymbol{\gamma}_0, \mathbf{V}_\gamma)$, $\mathbf{a} \sim \mathcal{N}(\mathbf{a}_0, \mathbf{V}_a)$, and inverse-gamma priors for the variable-specific variances:

$$\sigma_{i,L}^2 \sim \mathcal{IG}(\eta_{i,L}, \omega_{i,L}), \quad i = 1, \ldots, n,$$

$$\sigma_{j,S}^2 \sim \mathcal{IG}(\eta_{j,S}, \omega_{j,S}), \quad j = n+1, \ldots, 2n.$$

For the state equation, the coefficient and covariance matrices respectively follow truncated normal and inverse-Wishart distributions:

$$\text{vec}(\boldsymbol{\Phi}') = \boldsymbol{\phi} \sim \mathcal{N}(\boldsymbol{\phi}_0, \mathbf{V}_\phi) 1(\boldsymbol{\Phi} \in A), \quad \boldsymbol{\Sigma}_h \sim \mathcal{IW}(\boldsymbol{\Xi}_h, \nu_h),$$

in which $1(Q)$ is the indicator function which equals one if statement Q is true and zero otherwise, and the set A is the region in which the VAR(1) process is stationary.

2.2 Model Comparison

It is easy to see that our proposed VAR-CSVM model nests both the traditional SVAR in Sims (1980) and the common stochastic volatility VAR (VAR-CSV) model in Carriero et al. (2016). In particular, the VAR-CSV model can be viewed as a restricted version of our VAR-CSVM with $\mathbf{A} = \mathbf{0}$ and only one common stochastic volatility component, i.e., $h_t^L = h_t^S$ for all dates $t = 1, \ldots, T$. Next, we get the traditional SVAR by making the additional restriction that $h_t^L = h_t^S = 0$ for all dates. By testing whether these restrictions are supported by the data, we can readily investigate the importance of uncertainty shocks in the proposed economies. Given that the uncertainty measures are important, we can further address our research question by imposing additional economically motivated restrictions on the uncertainty impact matrix $\mathbf{A}$. To facilitate our discussion of the results, we defer exact details for these restrictions to the empirical section.

To test these hypotheses, we utilize a formal Bayesian model comparison method, via the *Bayes factor*—a special case of the *posterior odds ratio*. The posterior odds ratio can be written as

$$\mathrm{PO}_{1,2} = \frac{p\left(\mathbf{y^o}|M_1\right)}{p\left(\mathbf{y^o}|M_2\right)} \times \frac{\mathbb{P}\left(M_1\right)}{\mathbb{P}\left(M_2\right)},$$

where $p\left(\mathbf{y^o}|M_i\right)$ and $\mathbb{P}\left(M_i\right)$ respectively denote the marginal likelihood and prior model probability for M_i, $i = 1, 2$. The marginal likelihood is defined as

$$p(\mathbf{y^o}|M_i) = \int_{\boldsymbol{\Theta}_i} p(\mathbf{y^o}|\boldsymbol{\theta}_i, M_i)p(\boldsymbol{\theta}_i|M_i)d\boldsymbol{\theta}_i,$$

where $\boldsymbol{\theta}_i$ is a vector of the parameters in model M_i, and $\boldsymbol{\Theta}_i$ is the associated parameter space. The ratio of two such likelihoods is known as the *Bayes factor*. It can easily be seen that the posterior odds ratio reduces to the Bayes factor of M_1 against M_2, denoted $\mathrm{BF}_{1,2}$, under the assumption of equal prior model probabilities (i.e., $\mathbb{P}\left(M_1\right) = \mathbb{P}\left(M_2\right)$). Moreover, since it is a special case of the posterior odds ratio, the Bayes factor takes on a probabilistic interpretation. For instance, if $\mathrm{BF}_{1,2} = 2$, then M_1 is twice as likely as M_2.

One difficulty in using Bayes factor to compare models is that the analytical solution of the marginal likelihood for many nonlinear models does not exist in general. For this reason, it is common to approximate it through a product of one-step-ahead predictive likelihoods evaluated at the observed data (Geweke & Amisano, 2011). That is

$$p(\mathbf{y^o}|M_i) = p(\mathbf{y}_1^o|M_i)\prod_{t=2}^{T} p(\mathbf{y}_t^o|\mathbf{y}_1^o, \ldots, \mathbf{y}_{t-1}^o, M_i).$$

Since it requires estimating the model at each date, the computation of this one-step-ahead predictive likelihood is computationally intensive for nonlinear models such as stochastic volatility in mean VARs. Fortunately, we can overcome this hurdle through the use of our efficient posterior sampler which we detail in the next subsection.

2.3 Bayesian Estimation

This section discusses the MCMC algorithm we use for estimating the VAR-CSVM model defined in Eqs. (1) and (2). For notational convenience, we define $\mathbf{y} = (\mathbf{y}_1, \ldots, \mathbf{y}_T)'$, $\mathbf{h}^L = (h_1^L, \ldots, h_T^L)'$, and $\mathbf{h}^S = (h_1^S, \ldots, h_T^S)'$.

Posterior draws of the model parameters can be obtained by sequentially sampling from:

1. $p(\mathbf{h}^L|\mathbf{h}^S, \mathbf{a}, \boldsymbol{\beta}, \boldsymbol{\gamma}, \boldsymbol{\Sigma}_L, \boldsymbol{\Sigma}_S, \boldsymbol{\Phi}, \boldsymbol{\Sigma}_h, \mathbf{y})$.
2. $p(\mathbf{h}^S|\mathbf{h}^L, \mathbf{a}, \boldsymbol{\beta}, \boldsymbol{\gamma}, \boldsymbol{\Sigma}_L, \boldsymbol{\Sigma}_S, \boldsymbol{\Phi}, \boldsymbol{\Sigma}_h, \mathbf{y})$.

3. $p(\mathbf{a}, \boldsymbol{\beta}, \boldsymbol{\gamma}|\mathbf{h}^L, \mathbf{h}^S, \boldsymbol{\Sigma}_L, \boldsymbol{\Sigma}_S, \boldsymbol{\Phi}, \boldsymbol{\Sigma}_h, \mathbf{y})$.
4. $p(\boldsymbol{\Sigma}_L|\mathbf{h}^L, \mathbf{h}^S, \mathbf{a}, \boldsymbol{\beta}, \boldsymbol{\gamma}, \boldsymbol{\Sigma}_S, \boldsymbol{\Phi}, \boldsymbol{\Sigma}_h, \mathbf{y})$.
5. $p(\boldsymbol{\Sigma}_S|\mathbf{h}^L, \mathbf{h}^S, \mathbf{a}, \boldsymbol{\beta}, \boldsymbol{\gamma}, \boldsymbol{\Sigma}_L, \boldsymbol{\Phi}, \boldsymbol{\Sigma}_h, \mathbf{y})$.
6. $p(\boldsymbol{\Sigma}_h|\mathbf{h}^L, \mathbf{h}^S, \boldsymbol{\Sigma}_L, \boldsymbol{\Sigma}_S, \mathbf{a}, \boldsymbol{\beta}, \boldsymbol{\gamma}, \boldsymbol{\Phi}, \mathbf{y})$.
7. $p(\boldsymbol{\Phi}|\mathbf{h}^L, \mathbf{h}^S, \boldsymbol{\Sigma}_L, \boldsymbol{\Sigma}_S, \mathbf{a}, \boldsymbol{\beta}, \boldsymbol{\gamma}, \boldsymbol{\Sigma}_h, \mathbf{y})$.

The main difficulty in estimating VAR-SVM models arises in sampling from the nonstandard conditional distributions of the log volatilities in Steps 1 and 2. Since the common stochastic volatilities, $e^{h_t^L}$ and $e^{h_t^S}$, appear in both the conditional mean and the conditional variance of the model, this nonlinear feature makes the auxiliary mixture sampler (Kim et al., 1998), often used for estimating SV-VAR models, not applicable.

In recent studies, Mumtaz and Theodoridis (2017) show that it is possible to sample such states using the single-move Metropolis-Hasting algorithm proposed in Jacquier et al. (2002), while (Carriero et al., 2018, 2020) employ a particle Gibbs sampler developed in Andrieu et al. (2010). However, Cross et al. (2023) has indicated that these existing methods might produce highly correlated posterior draws and are computationally intensive when estimating high-dimensional models, making posterior inference for the VAR-SVM model difficult.

To overcome these issues, Cross et al. (2023) develop an efficient single-block sampler for drawing the log volatilities in Steps 1 and 2, which has been shown to substantially improve MCMC mixing. This is done by first constructing a Gaussian proposal distribution with its mean and covariance matrix set respectively at the mode and the negative inverse Hessian matrix evaluated at the mode of the log-conditional posterior of the log volatilities. One key feature of the algorithm proposed by Cross et al. (2023) is based on the fact that the Hessian matrix of the log-conditional densities for the log volatilities is block-banded. Hence, the recent advances in precision sampling algorithm (Rue et al., 2009; Chan & Jeliazkov, 2009; McCausland et al., 2011) can be used to efficiently draw from the Gaussian proposal distribution, and posterior draws of the log volatilities can then be obtained using an independence-chain ARMH algorithm (Chib & Greenberg, 1995).

In what follows, we discuss how the algorithm of Cross et al. (2023) can be used to implement Steps 1–2. Since the other steps utilize standard techniques, we defer remaining details to the Appendix. Readers who are only interested in the empirical application may skip this section and go straight to the results.

To sample from the conditional distribution in Step 1, first note that Eq. (1) can be expressed as

$$\mathbf{y}_t = \widetilde{\mathbf{X}}_t\boldsymbol{\beta} + \mathbf{W}_t\boldsymbol{\gamma} + \mathbf{A}\begin{pmatrix} e^{h_t^L} \\ e^{h_t^S} \end{pmatrix} + \boldsymbol{\epsilon}_t^y, \quad \boldsymbol{\epsilon}_t^y \sim \mathcal{N}(\mathbf{0}, \boldsymbol{\Sigma}_t), \tag{3}$$

where $\widetilde{\mathbf{X}}_t = \mathbf{I}_{2n} \otimes (1, \mathbf{y}'_{t-1}, \ldots, \mathbf{y}'_{t-p})$ and

$$\mathbf{W}_t = \begin{pmatrix} 0 & 0 & 0 & \cdots & \cdots & \cdots & 0 \\ -y_{1,t} & 0 & 0 & \cdots & \cdots & \cdots & 0 \\ 0 & -y_{1,t} & -y_{2,t} & \cdots & \cdots & \cdots & 0 \\ \vdots & \vdots & \ddots & \vdots & \cdots & \cdots & 0 \\ 0 & \cdots & \cdots & -y_{1,t} & -y_{2,t} & \cdots & -y_{2n-1,t} \end{pmatrix}.$$

It can be seen that the Eq. (3) above can be written as

$$\mathbf{y}_t = \widetilde{\mathbf{X}}_t\boldsymbol{\beta} + \mathbf{W}_t\boldsymbol{\gamma} + e^{h_t^L}\begin{pmatrix}\mathbf{a}_{11}\\ \mathbf{a}_{21}\end{pmatrix} + e^{h_t^S}\begin{pmatrix}\mathbf{a}_{12}\\ \mathbf{a}_{22}\end{pmatrix} + \boldsymbol{\epsilon}_t^y, \quad \boldsymbol{\epsilon}_t^y \sim \mathcal{N}(\mathbf{0}, \boldsymbol{\Sigma}_t).$$

After rearranging the equation above, we obtain

$$\widetilde{\mathbf{y}}_t = e^{h_t^L}\begin{pmatrix}\mathbf{a}_{11}\\ \mathbf{a}_{21}\end{pmatrix} + \boldsymbol{\epsilon}_t^y, \quad \boldsymbol{\epsilon}_t^y \sim \mathcal{N}(\mathbf{0}, \boldsymbol{\Sigma}_t), \tag{4}$$

where

$$\widetilde{\mathbf{y}}_t = \mathbf{y}_t - \widetilde{\mathbf{X}}_t\boldsymbol{\beta} - \mathbf{W}_t\boldsymbol{\gamma} - e^{h_t^S}\begin{pmatrix}\mathbf{a}_{12}\\ \mathbf{a}_{22}\end{pmatrix}. \tag{5}$$

Thus, by a change of variable, it follows that

$$p(\mathbf{h}^L|\mathbf{h}^S, \mathbf{a}, \boldsymbol{\beta}, \boldsymbol{\Sigma}_L, \boldsymbol{\Sigma}_S, \rho_L, \sigma_L^2, \mathbf{y}) \propto p(\mathbf{y}|\mathbf{h}^L)p(\mathbf{h}^L) \propto p(\widetilde{\mathbf{y}}|\mathbf{h}^L)p(\mathbf{h}^L).$$

Here we have suppressed the conditional parameters except $\mathbf{h}^L$ for notational convenience. Note that the log likelihood with respect to $\widetilde{\mathbf{y}}$ can be written as $\log p(\widetilde{\mathbf{y}}|\mathbf{h}^L) = \sum_{t=1}^T \log p(\widetilde{\mathbf{y}}_t|h_t^L)$. Taking the second-order Taylor expansion of the log likelihood around a given point $\widetilde{\mathbf{h}}^L = (\widetilde{h}_1^L, \ldots, \widetilde{h}_T^L)'$, which yields the approximation

$$\begin{aligned} \log p(\widetilde{\mathbf{y}}|\mathbf{h}^L) &\approx \log p(\widetilde{\mathbf{y}}|\widetilde{\mathbf{h}}^L) + (\mathbf{h}^L - \widetilde{\mathbf{h}}^L)'\mathbf{f} - \frac{1}{2}(\mathbf{h}^L - \widetilde{\mathbf{h}}^L)'\mathbf{G}(\mathbf{h}^L - \widetilde{\mathbf{h}}^L), \\ &= -\frac{1}{2}\left(\mathbf{h}^{L\prime}\mathbf{G}\mathbf{h}^L - 2\mathbf{h}^{L\prime}(\mathbf{f} + \mathbf{G}\widetilde{\mathbf{h}}^L)\right) + c_1, \end{aligned} \tag{6}$$

where c_1 is a constant independent of $\mathbf{h}^L$, $\mathbf{f}$ is the gradient vector and $\mathbf{G}$ is the negative Hessian matrix of the log-likelihood function evaluated at $\widetilde{\mathbf{h}}^L$. Specifically, if we write $\mathbf{f} = (f_1, \ldots, f_T)'$ and $\mathbf{G} = \mathrm{diag}(G_1, \ldots, G_T)$, then

$$f_t = \frac{\partial}{\partial h_t^L}\log p(\widetilde{\mathbf{y}}_t|h_t^L)\Big|_{\mathbf{h}^L=\widetilde{\mathbf{h}}^L}, \quad G_t = -\frac{\partial^2}{\partial {h_t^L}^2}\log p(\widetilde{\mathbf{y}}_t|h_t^L)\Big|_{\mathbf{h}^L=\widetilde{\mathbf{h}}^L}.$$

Given the expression in (4), the log-conditional density of $\widetilde{\mathbf{y}}_t$ is

$$\log p(\widetilde{\mathbf{y}}_t|\mathbf{h}_t^L) = -\frac{nh_t^L}{2} - \frac{1}{2}e^{-h_t^L}\left(\widetilde{\mathbf{y}}_{1,t} - e^{h_t^L}\mathbf{a}_{11}\right)' \boldsymbol{\Sigma}_L^{-1}\left(\widetilde{\mathbf{y}}_{1,t} - e^{h_t^L}\mathbf{a}_{11}\right)$$
$$- \frac{1}{2}e^{-h_t^S}\left(\widetilde{\mathbf{y}}_{2,t} - e^{h_t^L}\mathbf{a}_{21}\right)' \boldsymbol{\Sigma}_S^{-1}\left(\widetilde{\mathbf{y}}_{2,t} - e^{h_t^L}\mathbf{a}_{21}\right). \quad (7)$$

Therefore, it is straightforward to check that

$$\frac{\partial}{\partial h_t^L}\log p(\widetilde{\mathbf{y}}_t|h_t^L) = -\frac{1}{2}\left(n - e^{-h_t^L}\widetilde{\mathbf{y}}_{1,t}'\boldsymbol{\Sigma}_L^{-1}\widetilde{\mathbf{y}}_{1,t} + e^{h_t^L}\mathbf{a}_{11}'\boldsymbol{\Sigma}_L^{-1}\mathbf{a}_{11}\right)$$
$$+ e^{h_t^L - h_t^S}\widetilde{\mathbf{y}}_{2,t}'\boldsymbol{\Sigma}_S^{-1}\mathbf{a}_{21} - e^{2h_t^L - h_t^S}\mathbf{a}_{21}'\boldsymbol{\Sigma}_S^{-1}\mathbf{a}_{21}, \quad (8)$$

$$\frac{\partial^2}{\partial {h_t^L}^2}\log p(\widetilde{\mathbf{y}}_t|h_t^L) = -\frac{1}{2}\left(e^{-h_t^L}\widetilde{\mathbf{y}}_{1,t}'\boldsymbol{\Sigma}_L^{-1}\widetilde{\mathbf{y}}_{1,t} + e^{h_t^L}\mathbf{a}_{11}'\boldsymbol{\Sigma}_L^{-1}\mathbf{a}_{11}\right)$$
$$+ e^{h_t^L - h_t^S}\widetilde{\mathbf{y}}_{2,t}'\boldsymbol{\Sigma}_S^{-1}\mathbf{a}_{21} - 2e^{2h_t^L - h_t^S}\mathbf{a}_{21}'\boldsymbol{\Sigma}_S^{-1}\mathbf{a}_{21}. \quad (9)$$

Next, we will find the log-conditional prior density for $\mathbf{h}^L$ and also its gradient vector and Hessian matrix. To this end, we first stack the state equation (2) over $t = 1, \dots, T$ and obtain

$$\mathbf{H}\mathbf{h} = \boldsymbol{\epsilon}^h, \quad \boldsymbol{\epsilon}^h \sim \mathcal{N}(\mathbf{0}, \mathbf{S}_h),$$

where $\mathbf{h} = (\mathbf{h}_1', \dots, \mathbf{h}_T')'$, $\mathbf{S}_h = \text{diag}(\mathbf{V}_h, \boldsymbol{\Sigma}_h, \dots, \boldsymbol{\Sigma}_h)$, and

$$\mathbf{H} = \begin{pmatrix} \mathbf{I} & \mathbf{0} & \cdots & \cdots & \mathbf{0} \\ -\boldsymbol{\Phi} & \mathbf{I} & \ddots & \ddots & \vdots \\ \mathbf{0} & -\boldsymbol{\Phi} & \mathbf{I} & \ddots & \vdots \\ \vdots & \ddots & \ddots & \ddots & \mathbf{0} \\ \mathbf{0} & \cdots & \cdots & -\boldsymbol{\Phi} & \mathbf{I} \end{pmatrix}.$$

This implies that $\mathbf{h} \sim \mathcal{N}(\mathbf{0}, \widetilde{\mathbf{K}}_h^{-1})$ with $\widetilde{\mathbf{K}}_h = \mathbf{H}'\mathbf{S}_h^{-1}\mathbf{H}$. Let $\mathbf{P}$ be an $2n \times 2n$ selection matrix such that $\mathbf{P}\mathbf{h} = (\mathbf{h}^{L'}, \mathbf{h}^{S'})'$, and then we have $(\mathbf{h}^{L'}, \mathbf{h}^{S'})' \sim \mathcal{N}(\mathbf{0}, \mathbf{K}_h^{-1})$ with $\mathbf{K}_h = \mathbf{P}\widetilde{\mathbf{K}}_h\mathbf{P}'$. Partition $\mathbf{K}_h$ as

$$\mathbf{K}_h = \begin{pmatrix} \mathbf{K}_L & \mathbf{K}_{L,S} \\ \mathbf{K}_{L,S}' & \mathbf{K}_S \end{pmatrix},$$

where $\mathbf{K}_L$, $\mathbf{K}_S$, and $\mathbf{K}_{L,S}$ are $T \times T$ matrices. Using the standard property of Gaussian distribution, we have that $\mathbf{h}^L \sim \mathcal{N}(\boldsymbol{\mu}_L, \mathbf{K}_L^{-1})$ conditional on $\mathbf{h}^S$ with

$\boldsymbol{\mu}_L = -\mathbf{K}_L^{-1}\mathbf{K}_{L,S}\mathbf{h}^S$, which gives the log-conditional prior density

$$\log p(\mathbf{h}^L) = c_2 - \frac{1}{2}(\mathbf{h}^L - \boldsymbol{\mu}_L)'\mathbf{K}_L(\mathbf{h}^L - \boldsymbol{\mu}_L), \tag{10}$$

where c_2 is the normalization constant that is independent of $\mathbf{h}^L$. Combining the results obtained in Eqs. (6) and (10), we obtain the following approximation for the conditional posterior of $\mathbf{h}^L$:

$$\begin{aligned}\log p(\mathbf{h}^L|\mathbf{h}^S, \mathbf{A}_0, \boldsymbol{\beta}, \boldsymbol{\Sigma}_L, \boldsymbol{\Sigma}_S, \rho_L, \sigma_L^2, \mathbf{y}) &= \log p(\widetilde{\mathbf{y}}|\mathbf{h}^L) + \log p(\mathbf{h}^L),\\ &\approx -\frac{1}{2}\left(\mathbf{h}^{L'}\mathbf{G}\mathbf{h}^L - 2\mathbf{h}^{L'}(\mathbf{f} + \mathbf{G}\widetilde{\mathbf{h}}^L)\right) - \frac{1}{2}(\mathbf{h}^L - \boldsymbol{\mu}_L)'\mathbf{K}_L(\mathbf{h}^L - \boldsymbol{\mu}_L) + c_3,\\ &= -\frac{1}{2}\left(\mathbf{h}^{L'}(\mathbf{G} + \mathbf{K}_L)\mathbf{h}^L - 2\mathbf{h}^{L'}(\mathbf{f} + \mathbf{G}\widetilde{\mathbf{h}}^L + \mathbf{K}_L\boldsymbol{\mu}_L)\right).\end{aligned}$$

It can be seen that the above equation is the log kernel of the Gaussian distribution, which shows that the conditional posterior $p(\mathbf{h}^L|\mathbf{h}^S, \mathbf{A}_0, \boldsymbol{\beta}, \boldsymbol{\Sigma}_L, \boldsymbol{\Sigma}_S, \rho_L, \sigma_L^2, \mathbf{y})$ can be approximated using a Gaussian distribution $\mathcal{N}(\widehat{\mathbf{h}}^L, \widehat{\mathbf{K}}_L^{-1})$, where $\widehat{\mathbf{K}}_L = \mathbf{G} + \mathbf{K}_L$ and $\widehat{\mathbf{h}}^L = \widehat{\mathbf{K}}_L^{-1}(\mathbf{f} + \mathbf{G}\widetilde{\mathbf{h}}^L + \mathbf{K}_L\boldsymbol{\mu}_L)$. Note also that both $\widehat{\mathbf{h}}^L$ and $\widehat{\mathbf{K}}_L^{-1}$ are dependent on $\widetilde{\mathbf{h}}^L$. We set $\widetilde{\mathbf{h}}^L$ to be the mode of the conditional posterior of $\mathbf{h}^L$, which can be obtained numerically using the Newton-Raphson algorithm.[5] Thus, we use the resulting $\mathcal{N}(\widehat{\mathbf{h}}^L, \widehat{\mathbf{K}}_L^{-1})$ as our proposal distribution and implement the ARMH algorithm of Chib and Greenberg (1995) to draw from the conditional posterior of $\mathbf{h}^L$. We refer readers to Cross et al. (2023) for more details about the ARMH algorithm and the implementation of the Newton-Raphson algorithm in the SVM-SVAR framework.

Since Step 2 is symmetric to Step 1, sampling can be accomplished through straightforward relabeling of terms in the above equations. We close by making a few remarks on the computation. First, in contrast to the multi-move samplers used in existing studies, our common stochastic volatilities are drawn in a single-block sampler. In general, single-block samplers have been proven to be more numerically efficient than multi-move samplers when the posterior samples are highly correlated. Second, due to the availability of the first- and second-order derivatives of the log-conditional density, the Newton-Raphson method can be used to obtain the mode of the log density efficiently. This idea is similar to the single-block version of Shephard and Pitt (1997) and Chan (2017) in that the proposal distribution is based on the second-order Taylor expansion of the likelihood function in terms of

[5] Note that, in general, the Hessian of the conditional posterior of $\mathbf{h}_L$, i.e., $-\mathbf{G} - \mathbf{K}_L$, is not guaranteed to be negative definite. If $-\mathbf{G} - \mathbf{K}_L$ is not negative definite, the Newton-Raphson method might not work well. One way to avoid this issue is to impose a tight prior on $\boldsymbol{\Sigma}_h$ that restricts the diagonal elements of $\boldsymbol{\Sigma}_h$ to be small values. In our application, we assume $\boldsymbol{\Sigma}_h \sim \mathcal{IW}(\boldsymbol{\Xi}_h, \nu_h)$ with $\nu_h = 40$ and $\boldsymbol{\Xi}_h = 0.1^2(\nu_h - n - 1)\mathbf{I}_n$. With this setting, we find that the Newton-Raphson method can effectively locate the mode of the conditional posterior.

stochastic volatilities. Of course, they are different in that we consider a multivariate model and exploit techniques from band and sparse matrix algorithms. Specifically, since the precision matrix $\widehat{\mathbf{K}}_L$ is a band matrix, we can efficiently draw from the proposal distribution by applying the precision sampler in Chan and Jeliazkov (2009).[6]

3 Empirical Results

In this section, we present our main empirical results on the effects of macroeconomic uncertainty in three commonly studied SOEs: Australia, Canada, and New Zealand. We use a lag length of $p = 2$ to maintain consistency with the broader uncertainty literature (e.g., Caggiano et al., 2017; Carriero et al., 2018; Mumtaz & Theodoridis, 2017; Carriero et al., 2020). All estimates are based on 50,000 posterior draws after a burn-in period of 5000. The priors are in the Appendix.

We use quarterly data on real GDP (RGDP) growth, CPI inflation, and a short-term interest rate—taken to be the respective country's 3-month (90-day) rate on domestic bank bills—from 1978Q3-2016Q4. All series were sourced from the International Monetary Fund's (IMF's) International Financial Statistics (IFS) database. We then converted the RGDP and CPI indexes to annualized growth measures, i.e., $400([X_t/X_{t-1}] - 1)$, where X_t is the variable of interest at date t. Since RGDP data in New Zealand starts in 1987:M5, from 1978Q1 to 1987Q1, we use industrial production growth rate for this period. The data is plotted in Fig. 1.

3.1 Hypothesis Testing via Model Selection

In this section, we conduct a model comparison exercise using the marginal likelihood as a selection criterion. The main objective is to provide some statistical evidence that helps determine whether uncertainty plays a role in shaping the macroeconomic environments of the countries in question. To this end, we focus on the two major issues concerned in this chapter. First, we ask whether allowing the macroeconomic variables in a specific country to share a common volatility is consistent with the data. Second, we investigate the relevance of domestic and international sources of uncertainty within each of the SOEs. To that end, we impose

[6] The attentive reader will note that we have referred to two distinct notions of efficiency here. The first refers to reduced autocorrelation in the posterior draws. The second refers to sampling speed. For this particular model, our proposed algorithm is superior to existing samplers in that it is efficient in both respects.

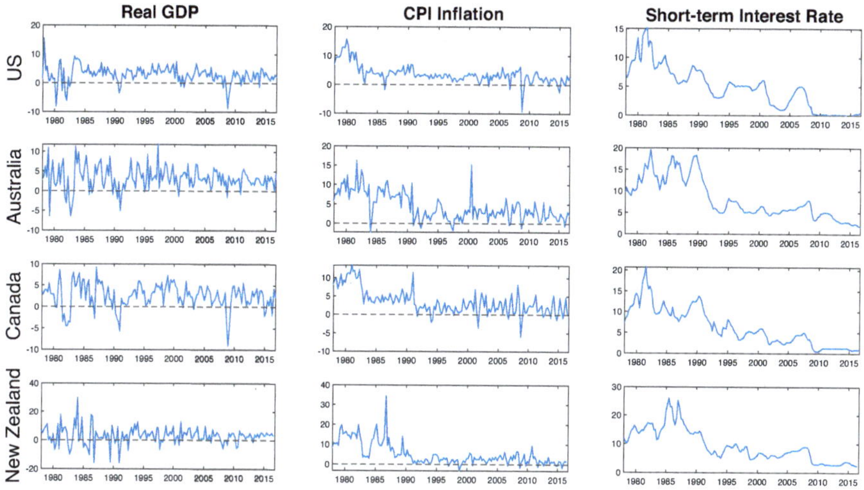

Fig. 1 Real GDP growth, CPI inflation, and short-term interest rates from 1978Q1–2016Q4. Note: The RGDP growth rate and CPI inflation rate are expressed as annualized growth rates, i.e., $400([X_t/X_{t-1}] - 1)$, where X_t is the variable of interest at date t. The short-term interest rate refers to each country's 3-month (90-day) rates on domestic bank bills. Since RGDP data in New Zealand starts in 1987:M5, from 1978Q1 to 1987Q1, we use industrial production growth rate for this period

Table 1 A list of models

VAR	Constant VAR
VAR-CSV	With common SV
VAR-CSVM	With common SVM
VAR-CSVM-R1	With common SVM with $\mathbf{a}_{12} = \mathbf{0}$
VAR-CSVM-R2	With common SVM with $\mathbf{a}_{21} = \mathbf{0}$
VAR-CSVM-R3	With common SVM with $\mathbf{a}_{12} = \mathbf{a}_{21} = \mathbf{0}$
VAR-CSVM-R4	With common SVM with $\mathbf{a}_{12} = \mathbf{a}_{22} = \mathbf{0}$

four economically motivated restrictions on the impact matrix $\mathbf{A}$. To ease exposition, recall that

$$\mathbf{A} = \begin{pmatrix} \mathbf{a}_{11} & \mathbf{a}_{12} \\ \mathbf{a}_{21} & \mathbf{a}_{22} \end{pmatrix}.$$

The restrictions are: (1) $\mathbf{a}_{12} = \mathbf{0}$; (2) $\mathbf{a}_{21} = \mathbf{0}$; (3) $\mathbf{a}_{12} = \mathbf{a}_{21} = \mathbf{0}$; (4) $\mathbf{a}_{12} = \mathbf{a}_{22} = \mathbf{0}$. Restriction (1) hypothesizes that uncertainty spillovers in the SOE do not transmit to the large economy; restriction (2) that uncertainty spillovers in the large economy do not transmit to the small economy; restriction (3) that uncertainty is purely idiosyncratic (i.e., no spillovers); restriction (4) that the only source of global uncertainty is from the large economy. We summarize each of the model specifications in Table 1, and the estimated marginal likelihood of each model is reported in Table 2.

Table 2 Estimated log marginal likelihoods for various models in Table 1

	Australia	Canada	New Zealand
VAR	−990.02	−923.65	−1187.20
VAR-CSV	−932.83	−889.78	−1077.74
VAR-CSVM	**−920.24**	**−874.55**	**−1045.58**
VAR-CSVM-R1	−931.26	−878.28	−1052.43
VAR-CSVM-R2	−922.12	−874.66	−1063.20
VAR-CSVM-R3	−934.46	−875.83	−1059.59
VAR-CSVM-R4	−925.28	−879.11	−1055.71

Note: The best model for each country is in bold

The general conclusion is that the log marginal likelihoods provide overwhelming support in favor of the unrestricted VAR-CSVM model against the VAR, VAR-CSV, and restricted VAR-CSVM models across all countries. In the first instance, we thus conclude that there is a statistically relevant country-specific common component in the macroeconomic variables of interest. For instance, in the case of Australia, our results indicate that the difference of the log marginal likelihood of the VAR-CSV is about 57 larger than the standard VAR, highlighting strong evidence in favor of the VAR-CSV.

Finally, the fact that the unrestricted VAR-CSVM model is preferred to the VAR-CSV and each of the restricted VAR-CSVMs suggests that both domestic and international sources of macroeconomic uncertainty are a key feature of the economic environment in each of the SOEs.

3.2 *Uncertainty Measures*

Having identified the significance of both domestic and international sources of uncertainty across each of the SOEs, we now discuss the qualitative behavior of the uncertainty measures. Figures 2 and 3 respectively present the macroeconomic uncertainty estimates for the USA and the SOEs, i.e., $e^{h_t^L}$ and $e^{h_t^S}$, respectively. In each plot, the blue line represents the posterior means, and the red lines represent the associated 90% credible intervals.

If the model is well-specified, then all three versions of estimated US uncertainty measures should be similar to each other. The plots in Fig. 2 show that the qualitative behavior of the resulting measures is almost identical across the three panels. We are therefore confident that the model is well-specified.[7] Moreover, the general pattern of the US uncertainty measures is consistent with those in Carriero et al. (2018). In particular, they show significant increases around some of the political and economic events that were first highlighted in Bloom (2009). For instance, uncertainty is high in the 1980s and then declines, with some spikes around the

[7] Since the estimated value of the uncertainty measure has no direct economic interpretation, we are not concerned about the quantitative differences in these figures across estimates.

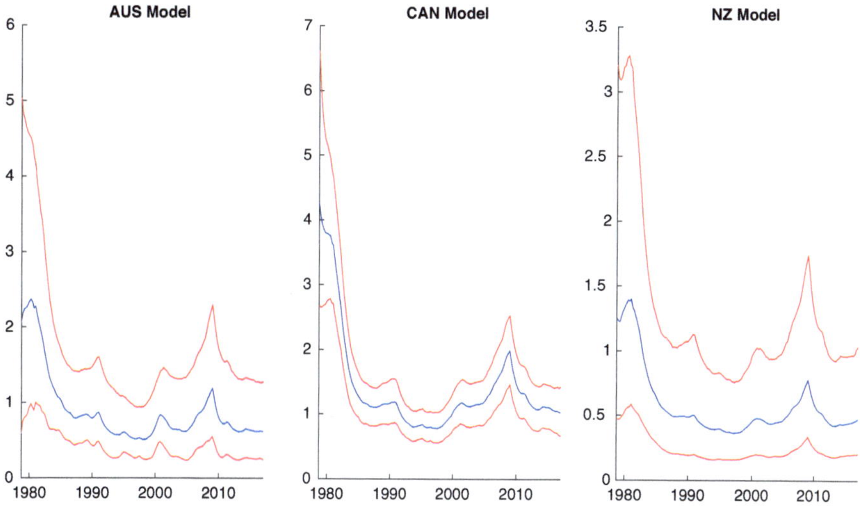

Fig. 2 Uncertainty estimates for the USA in each model: Australia (left), Canada (center), and New Zealand (right). Note: The macroeconomic uncertainty estimates for the USA are $e^{h_t^L}$ in each model. In each plot, the blue line represents the posterior means, and the red lines represent the associated 90% credible intervals

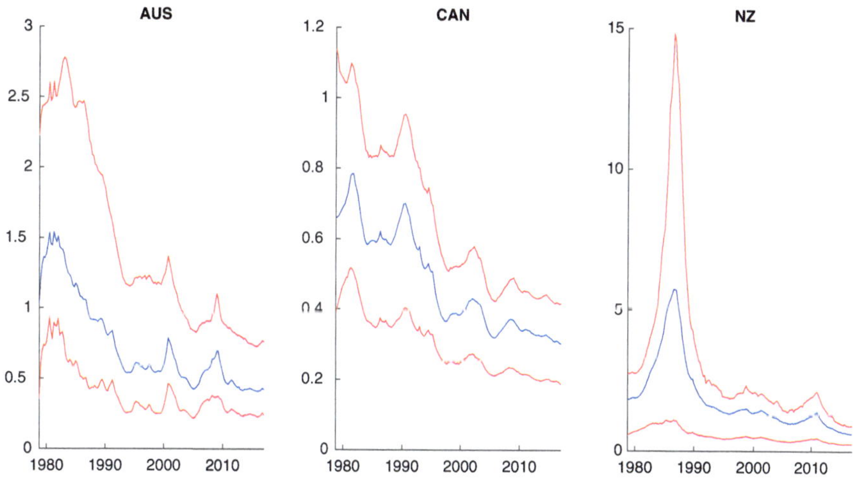

Fig. 3 Uncertainty estimates for the small open economies in each model: Australia (left), Canada (center), and New Zealand (right). Note: The macroeconomic uncertainty estimates for the small open economies are $e^{h_t^S}$ in each model. In each plot, the blue line represents the posterior means, and the red lines represent the associated 90% credible intervals

time of the Gulf War around 1990s and the "Dot-Com bubble" in the early 2000s. The estimates also capture the heightened economic uncertainty around the 2008 Global Financial Crisis.

As for the SOEs, the first plot in Fig. 3 shows that uncertainty in Australia (left panel) initially increases following the 1979 energy crisis, before declining after the movement to a flexible exchange rate regime in 1983. Uncertainty was then relatively low and stable over the next two decades. The stabilization of uncertainty commencing in the early 1990s coincides with the adoption of inflation targeting around 1992/93. The notable spikes in the early 2000s and 2008 relate to the "Dot-Com bubble" and the Global Financial Crisis, respectively.

Similar to Australia, uncertainty in Canada increases following the 1979 energy crisis before returning to baseline by the mid-1980s. Uncertainty then peaks again around the 1990 recession before steadily declining after the adoption of inflation targeting in 1991. As in the Australian case, the notable spike in the early 2000s likely relates to the "Dot-Com bubble." In contrast to the Australian case, there is a spike in uncertainty during the GFC; however, it is clear that most of the uncertainty during that period stemmed from the US economy.

Finally, the New Zealand experience is distinct from the similar fluctuations in both Australia and Canada. In particular, aside from the peak in uncertainty around 1985, uncertainty exhibits a constant mean, with relatively modest increases around the crisis periods of 2000 and 2008. The large spike in the mid-1980s is associated with the nine-month-long recession in 1982–83 and the year-long recession of 1987–88 recessions. Finally, as in the case of Australia and Canada, a possible explanation of the post-1980s reduction in uncertainty is the adoption of inflation targeting in the early 1990s.

3.3 *Transmission of Uncertainty Shocks*

How do macroeconomic uncertainty shocks affect the SOEs? We address this question by computing impulse response functions (IRFs) of each variable following a one standard deviation uncertainty shock. Since the model is nonlinear in the uncertainty terms, we address this question using generalized impulse response functions (GIRFs) as in Koop et al. (1996). The difference between GIRFs and traditional IRFs is that future shocks are not "zeroed-out" by assumption, but instead "integrated-out" through a Monte Carlo integration procedure, details which are provided in Koop et al. (1996). In the context of our model, the GIRF can be expressed as

$$\text{GIRF}(y_t, h, \epsilon_t) = \mathbb{E}[y_{t+h} \mid \epsilon_t = \delta] - \mathbb{E}[y_{t+h}],$$

where $\text{GIRF}(y_t, h, \epsilon_t)$ represents the response of variable y at horizon h to the shock ϵ_t, $\mathbb{E}[y_{t+h} \mid \epsilon_t = \delta]$ is the expected value of y at time $t + h$ given a one standard deviation uncertainty shock $\epsilon_t = \delta$, and $\mathbb{E}[y_{t+h}]$ is the baseline expected value of y at time $t + h$ without the shock.

We show how the SOEs respond to one standard deviation uncertainty shocks stemming from the US economy in Fig. 4, and the respective SOEs in Fig. 5. In both figures, the solid line represents the average mean point estimates to a one standard

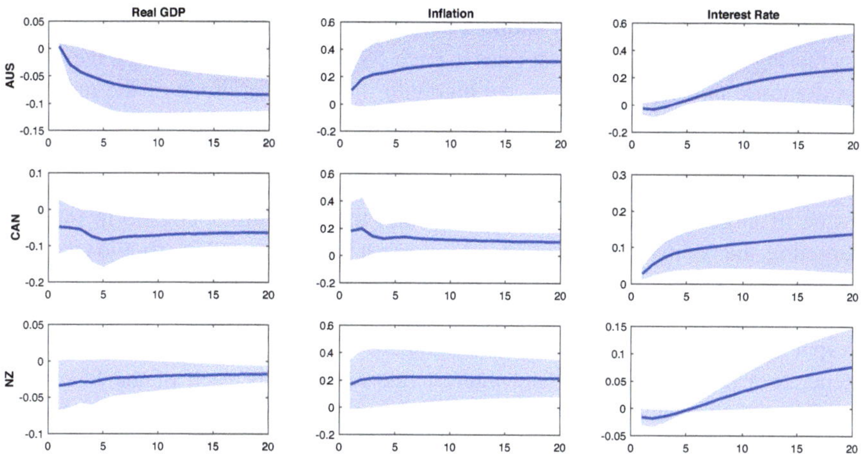

Fig. 4 GIRF: 1 standard deviation uncertainty shock from the USA: Australia (top row), Canada (center row), and New Zealand (bottom row). Note: The uncertainty shocks are 1 standard deviation shocks to the log volatility, h_t^L, in (2). In each plot, the thick blue line represents the posterior mean GIRF, and the shaded area represents the associated GIRFs between the 5th and 95th percentiles of the responses

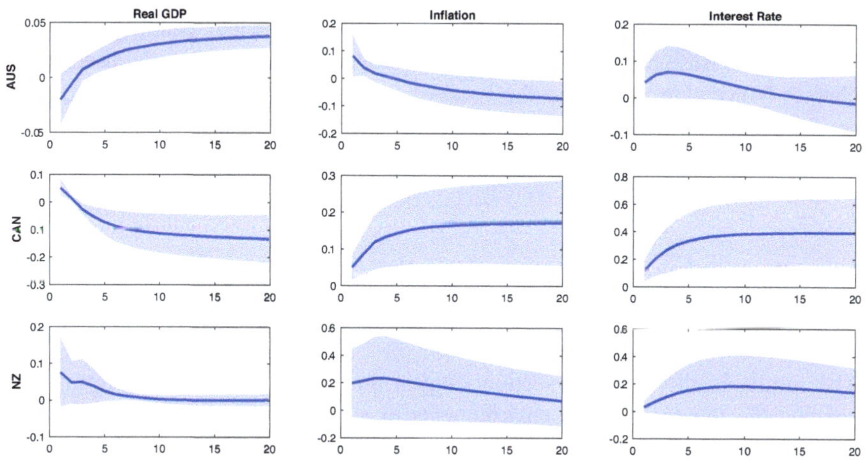

Fig. 5 GIRF: 1 standard deviation domestic uncertainty shock from the SOE: Australia (top row), Canada (center row), and New Zealand (bottom row). Note: The uncertainty shocks are 1 standard deviation shocks to the log volatility h_t^S in (2). In each plot, the thick blue line represents the posterior mean GIRF, and the shaded area represents the associated GIRFs between the 5th and 95th percentiles of the responses

deviation shock to the log volatilities in (2). The shaded regions are the 5th and 95th percentiles of the responses. The horizontal axis represents an impulse horizon of 20 quarters (5 years), while the vertical axis displays percentage point changes.

The responses in Fig. 4 show that the transmission of international uncertainty spillovers from the USA is qualitatively similar across each of the SOEs. That is, the rise in US uncertainty induces persistent declines in RGDP growth, while increasing inflation and interest rates in each country. Before discussing the effects of domestic uncertainty shocks, we highlight that the magnitude of the responses to international uncertainty shocks differs across the economies. In particular, the output responses in Australia and New Zealand are about half the size of that in Canada. Conversely, the medium run inflation response in these countries is about twice as large as that in Canada. Given that these countries engage in inflation targeting, the positive inflation responses suggest that policymakers should take such shocks into consideration when making interest rate decisions.

Interestingly, Fig. 5 shows that domestic uncertainty shocks elicit different responses relative to their international counterparts. In particular, while such shocks generate a positive inflation and interest rate response in each of the SOEs, the RGDP growth responses are idiosyncratic. For instance, the response in Canada is initially positive but then becomes persistently negative over time. In contrast, the New Zealand response is initially positive and gradually reverts to zero, and the Australian response is initially negative but then becomes positive over time. This suggests that each country has a different transmission mechanism for domestic uncertainty shocks, at least with respect to output. In particular, the responses in New Zealand are potentially consistent with growth options (Paddock et al., 1988), and Oi-Hartman-Abel effects (Abel, 1983; Hartman, 1972; Oi, 1961), while the Australian response is consistent with a precautionary savings channel (Bansal & Yaron, 2004; Basu & Bundick, 2017).[8] Of course, the empirical model used in this chapter is not able to address the issue of the underlying behavioral foundations of these mechanisms since a Lucas, 1976 critique issue arises in this type of counterfactual analysis. It is therefore not pursued in this chapter, but highlights an important area of future research.

4 Forecasting

The results in the previous section suggest that uncertainty plays a key role in shaping the macroeconomic environments in all SOEs. Given the importance of these shocks, along with the fact that each of the SOEs has an inflation targeting mandate, an important question is: *Can accounting for uncertainty improve the forecast accuracy of traditional econometric models*? To address this question, we compare the point and density forecast accuracy of the VAR-CSVM compared to

[8] Bloom (2014) provides excellent general overview of these mechanisms.

the VAR and VAR-CSV specifications. In this exercise, we evaluate the iterated h-step-ahead forecast of each model with $h = 1, 2, 4, 8$, and the forecast evaluation period is from 1990Q1–2016Q4. The starting date is roughly equal to the period when each country adopted inflation targeting. Thus, our exercise resembles that of the forecasting departments in the respective countries' central banks.

To assess the point forecast accuracy, we report both the root mean-squared forecast error (RMSFE) and the mean absolute forecast error (MAFE):

$$\text{RMSFE} = \sqrt{\frac{\sum_{t=t_0}^{T-h}\left(\mathbf{y}_{t+h}^{o} - \mathbb{E}(\mathbf{y}_{t+h}|\mathbf{y}_{1:t})\right)^2}{T-h-t_0+1}},$$

$$\text{MAFE} = \frac{\sum_{t=t_0}^{T-h}|\mathbf{y}_{t+h}^{o} - \hat{\mathbf{y}}_{t+h}^{M}|}{T-h-t_0+1},$$

where $\mathbb{E}(\mathbf{y}_{t+h}|\mathbf{y}_{1:t})$ is the posterior mean of the predictive density and $\hat{\mathbf{y}}_{t+h}^{M}$ is the posterior median of the predictive density.

To assess the density forecast accuracy, we report the average log-predictive likelihoods (ALPL) and the average continuous rank probability score (ACRPS):

$$\text{ALPL} = \frac{\sum_{t=t_0}^{T-k}\log p_{t+h}(\mathbf{y}_{t+k} = \mathbf{y}_{t+h}^{o}|\mathbf{y}_1^{o}, \ldots, \mathbf{y}_{t-1}^{o})}{T-h-t_0+1},$$

$$\text{ACRPS} = \frac{1}{T-h-t_0+1}\sum_{t=t_0}^{T-h}\text{CRPS}_t,$$

where $\text{CRPS}_t = \int_{-\infty}^{\infty}\left(F_{t+h}(z) - 1(\mathbf{y}_{t+h}^{o} < z)\right)^2 dz = \mathbb{E}_{p_{t+h}}|\mathbf{y}_{t+h} - \mathbf{y}_{t+h}^{o}| - 0.5\mathbb{E}_{p_{t+h}}|\mathbf{y}_{t+h} - \mathbf{y}_{t+h}^{\prime}|$ and F_{t+h} is the cumulative distribution of the predictive density at time $t+h$ given all information up to time t. A small value of the ACRPS indicates a better forecasting performance.

The point and density forecast results for each country are reported across Table 3. To facilitate comparison, we report relative scores to a VAR benchmark. The relative RMSFEs, MAFEs, and ACRPSs are ratios, whereas the relative ALPL is a difference as this is measured on a logarithmic scale. Set in this manner, a relative RMSFE, MAFE, and ACRPS of less than one indicates that the given model provides a better forecast than the VAR benchmark. Conversely, a positive value for the relative ALPL indicates a better forecasting performance than the benchmark.

The general trend is that the VAR-CSVM forecasts better than the alternatives. Two notable exceptions are the point forecasts of Canadian interest rates and New Zealand RGDP growth. In the former case, the VAR-CSV gives the best point forecasts at all forecast horizons; however, the VAR-CSVM provides better density forecasts at short-term horizons under the ALPS metric. In the latter case, the VAR-CSVM gives the best one-step-ahead point forecasts, but then the conventional

Table 3 Forecasting results for Australia, Canada, and New Zealand

			RMSFE				MAE				ALPS				ACRPS			
Country	Variable	Model	h = 1	h = 2	h = 4	h = 8	h = 1	h = 2	h = 4	h = 8	h = 1	h = 2	h = 4	h = 8	h = 1	h = 2	h = 4	h = 8
Australia	RGDP growth	VAR	1.00	1.00	1.00	1.00	1.00	1.00	1.00	1.00	0.00	0.00	0.00	0.00	1.00	1.00	1.00	1.00
		VAR-CSV	0.97	0.95	0.96	0.99	0.95	0.92**	0.94**	0.98	0.06***	0.10***	0.13***	0.11***	0.96	0.93**	0.93**	0.95
		VAR-CSVM	0.97	0.93	0.95	0.99	0.96	0.90**	0.93	0.98	0.11***	0.15***	0.16***	0.16***	0.97	0.93**	0.94**	0.95
	CPI inflation	VAR	1.00	1.00	1.00	1.00	1.00	1.00	1.00	1.00	0.00	0.00	0.00	0.00	1.00	1.00	1.00	1.00
		VAR-CSV	0.99	0.97	0.97	0.97	0.96	0.96	0.96	0.97	0.01***	0.01***	0.00***	0.00***	0.97	0.96	0.97	0.98
		VAR-CSVM	0.97	0.95	0.95	0.98	0.94	0.93*	0.90*	0.95	0.12***	0.14***	0.13***	0.16***	0.97	0.96	0.96	0.96
	Interest rate	VAR	1.00	1.00	1.00	1.00	1.00	1.00	1.00	1.00	0.00	0.00	0.00	0.00	1.00	1.00	1.00	1.00
		VAR-CSV	0.90	0.95	0.97	0.97	0.86***	0.90	0.98	1.00	0.28***	0.20***	0.14***	0.14***	0.84***	0.90**	0.96	0.99
		VAR-CSVM	0.89**	0.93**	0.94	1.02	0.85	0.88**	0.92**	0.93**	0.34***	0.24***	0.19***	0.24***	0.88**	0.95	1.00	0.97
Canada	RGDP growth	VAR	1.00	1.00	1.00	1.00	1.00	1.00	1.00	1.00	0.00	0.00	0.00	0.00	1.00	1.00	1.00	1.00
		VAR-CSV	0.94	0.92	0.96	1.01	0.94	0.91	0.95	1.01	0.09	0.13 ***	0.10***	0.06***	0.94	0.92	0.96	1.01
		VAR-CSVM	0.92	0.90	0.94	1.03	0.94	0.91	0.90	1.03	0.09***	0.11***	0.05***	− 0.01***	0.95	0.93	0.99	1.09

	CPI Inflation	VAR	1.00	1.00	1.00	1.00	1.00	1.00	1.00	1.00	0.00	0.00	0.00	0.00	1.00	1.00	1.00	1.00
		VAR-CSV	0.96	0.92*	0.96	0.91	0.95	0.91*	0.95	0.91	0.07***	0.11***	0.05***	0.10***	0.96	0.92*	0.96	0.89 **
		VAR-CSVM	0.90	0.87***	0.94*	0.95	0.88***	0.86***	0.89**	0.91*	0.12***	0.14***	0.00***	0.03***	0.92**	0.89**	0.96	0.96
	Interest rate	VAR	1.00	1.00	1.00	1.00	1.00	1.00	1.00	1.00	0.00	0.00	0.00	0.00	1.00	1.00	1.00	1.00
		VAR-CSV	0.97	0.87**	0.79**	0.76**	0.95	0.89**	0.76**	0.72**	0.10***	0.14***	0.19***	0.27***	0.96	0.89	0.82**	0.77
		VAR-CSVM	1.00	0.97	0.96	0.96	0.97	0.96	0.90	0.86***	0.19 ***	0.19***	0.19***	0.22****	1.00	0.99	0.97	0.95
New Zealand	RGDP growth	VAR	1.00	1.00	1.00	1.00	1.00	1.00	1.00	1.00	0.00	0.00	0.00	0.00	1.00	1.00	1.00	1.00
		VAR-CSV	0.98	1.02	1 02	1.02	0.98	1.02	1.02	1.03	0.22***	0.22	0.26***	0.22***	0.90*	0.91*	0.89 **	0.90
		VAR-CSVM	0.97	1.01	1.01	1.02	0.97	1.02	1.02	1.03	0.27***	0.27 ***	0.32***	0.33***	0.90*	0.91*	0.88**	0.89**
	CPI inflation	VAR	1.00	1.00	1.00	1.00	1.00	1.00	1.00	1.00	0.00	0.00	0.00	0.00	1.00	1.00	1.00	1.00
		VAR-CSV	0.96	0.97	0.90*	0.84**	0.96	0.93	0.86**	0.81**	0.24***	0.23***	0.30***	0.33***	0.88**	0.89**	0.85**	0.81**
		VAR-CSVM	0.83***	0.79***	0.72***	0.64***	0.84**	0.76***	0.68***	0.65***	0.33***	0.37***	0.46***	0.52***	0.83***	0.80***	0.75***	0.70***
	Interest rate	VAR	1.00	1.00	1.00	1.00	1.00	1.00	1.00	1.00	0.00	0.00	0.00	0.00	1.00	1.00	1.00	1.00
		VAR-CSV	0.91*	0.92*	0.90*	0.87**	0.92	0.94	0.94	0.92	0.25***	0.20***	0.18***	0.19***	0.87**	0.89*	0.89*	0.88*
		VAR-CSVM	0.89**	0.89*	0.82***	0.76***	0.96	0.96	0.91*	0.82**	0.27***	0.23***	0.25***	0.30***	0.90**	0.92 *	0.87**	0.80***

Note: *RMSFE* Root Mean-Squared Forecast Error, *MAE* Mean Absolute Error, *ALPS* Average Log Predictive Score, *ACRPS* Average Continuous Ranked Probability Score. All metrics are reported relative to the VAR baseline. The relative RMSFEs, MAFEs, and ACRPSs are ratios, whereas the relative ALPL is a difference as this is measured on a logarithmic scale. A relative RMSFE, MAFE, and ACRPS of less than one indicates that the given model provides a better forecast than the VAR benchmark. A positive value for the relative ALPL indicates a better forecasting performance than the benchmark. Finally, *,**,*** denotes the 10, 5, and 1 percent significant level of the Diebold-Mariano predictability test

VAR forecasts quite well at longer horizons. This is unsurprising given that the impulse response functions in Figs. 4 and 5 suggested that the New Zealand RGDP growth response to uncertainty shocks is short-lived. The finding that the VAR-CSVM is superior only at a one-step-ahead horizon is therefore consistent with this result. Since each of these countries has an explicit inflation targeting mandate, the results are of practical significance to central bankers. By providing more accurate forecasts, accounting for uncertainty can aid policymakers in making better informed decisions.

5 Conclusion

Our objective in the chapter was to estimate the effects of domestic and international sources of macroeconomic uncertainty in three commonly studied small open economies (SOEs): Australia, Canada, and New Zealand. To this end, we proposed a common stochastic volatility in mean VAR (VAR-CSVM) model. To estimate the model, we developed an efficient algorithm that built upon recent advances in band and sparse matrix algorithms. Our primary result was that both domestic and foreign sources of macroeconomic uncertainty shape the economic climate in each of the SOEs. In particular, international macroeconomic uncertainty spillovers were found to elicit a persistent negative effect on RGDP growth, while increasing both inflation and interest rates. Domestic uncertainty shocks were found to have a similar effect on inflation and interest rates; however, the RGDP growth responses were idiosyncratic. In particular, the transmission of domestic uncertainty shocks in New Zealand was persistently positive, while the Australian response is initially negative and becomes positive over time. This suggests that these economies exhibit different transmission mechanisms compared to the US economy. More precisely, the New Zealand experience is suggestive of *growth options* or *Oi-Hartman-Abel* effects, while the Australian response is suggestive of a *precautionary savings* channel. This highlights a clear direction of future research: the investigation of such a channel within an estimated and small open economy DSGE model. Finally, in a forecasting exercise, the proposed model generally provided more accurate point and density forecasts compared to nested VAR specifications. Taken together, our results suggested that international uncertainty spillovers play a key role in shaping the macroeconomic environments in all SOEs, and accounting for this uncertainty can aid policymakers in making better informed decisions.

In future work, it would be interesting to extend the model specification to account for idiosyncratic volatilities in addition to those common volatilities estimated here. An example of such a specification is provided in two-factor SVM model in Carriero et al. (2018). While such an extension is mathematically trivial, it induces a nontrivial computation problem regarding how to efficiently sample these volatilities in a way that facilitates timely forecasting exercises. Another interesting direction for future research is to consider the transmission of uncertainty to possibly asymmetric tail risks. This could potentially be done by adapting the model specification to include a non-normal error distribution in the measurement equation, such as the skewed-t distribution in Karlsson et al. (2023). Since our efficient sampling approach is based on a conditionally Gaussian posterior distribution for the log volatilities, this extension would require nontrivial changes to the algorithm. We invite other scholars to undertake these ideas as future research possibilities.

Appendix

Priors

In this Appendix, we present the priors used in the empirical analysis in Sect. 3. To conduct the analysis, we set a Minnesota type prior for the VAR coefficients. In particular, the prior mean is set to a zero vector, $\boldsymbol{\beta}_0 = \mathbf{0}$, and the prior covariance matrix is diagonal with its corresponding elements set as follows:

$$\text{Var}(\mathbf{c}) = 10 \times \mathbf{I}_{2n},$$

$$\text{Var}(\mathbf{B}_l^{ij}) = \begin{cases} \frac{\lambda_1^2 \lambda_2}{l^{\lambda_3}} \frac{\sigma_i^2}{\sigma_j^2} & \text{for } l = 1, \ldots, p \text{ and } i \neq j, \\ \frac{\lambda_1^2}{l^{\lambda_3}} & \text{for } l = 1, \ldots, p \text{ and } i = j, \end{cases}$$

where $\mathbf{B}_l^{i,j}$ denotes the $(i, j)th$ element of the matrix $\mathbf{B}_l$ and σ_r^2 is set equal to the variance of the residual from AR(p) model for the variable r. The hyperparameters are set to be $\lambda_1 = 0.2$, $\lambda_2 = 0.5$, $\lambda_3 = 2$. We assume a relatively non-informative prior on the uncertainty impact matrix. To be specific, we set $\mathbf{a}_0 = \mathbf{0}$ and $\mathbf{V}_a = 5 \times \mathbf{I}_{2n}$. For the covariance matrix of the log volatilities, we set $\nu_h = 40$ and $\boldsymbol{\Xi}_h = 0.1^2(\nu_h - n - 1) \times \mathbf{I}_n$. These values imply that the expected value of $\boldsymbol{\Sigma}_h$ is equal to $0.1^2 \times \mathbf{I}_n$. For the VAR coefficient matrix of the log volatilities, we set $\boldsymbol{\phi}_0 = (0.9, 0, 0.9, 0)'$ and $\mathbf{V}_\phi = \mathbf{I}_2$. Lastly, for the variable-specific variance, we set $\eta_{i,L} = 10$, $\omega_{i,S} = 9$ for $i = 1, \ldots, n$, and $= \eta_{j,L} = 10$, $\omega_{i,L} = 9$ for $j = n+1, \ldots, 2n$. This implies that $\mathbb{E}(\sigma_{i,L}^2) = \mathbb{E}(\sigma_{j,S}^2) = 1$.

Gibbs Sampler

In this appendix, we explain how to obtain draws from the conditional distributions in Steps 3–7 of the posterior simulator in Sect. 2.3. To this end, first note that Eq. (2) can be written as

$$\mathbf{y}_t = \mathbf{X}_t \boldsymbol{\theta} + \boldsymbol{\epsilon}_t^y, \quad \boldsymbol{\epsilon}_t^y \sim \mathcal{N}(\mathbf{0}, \boldsymbol{\Sigma}_t), \tag{11}$$

where $\mathbf{X}_t = \left(\widetilde{\mathbf{X}}_t, \mathbf{W}_t, \mathbf{I}_{2n} \otimes (e^{h_t^L}, e^{h_t^S})\right)$ and $\boldsymbol{\theta} = \left(\boldsymbol{\beta}', \boldsymbol{\gamma}', \text{vec}(\mathbf{A}')'\right)'$. Stacking the above equation over $t = 1, \dots, T$, we get

$$\mathbf{y} = \mathbf{X}\boldsymbol{\theta} + \boldsymbol{\epsilon}^y, \quad \boldsymbol{\epsilon}^y \sim \mathcal{N}(\mathbf{0}, \boldsymbol{\Sigma}),$$

where $\mathbf{y} = (\mathbf{y}_1, \dots, \mathbf{y}_T)'$, $\boldsymbol{\Sigma} = \mathbf{I}_{2n} \otimes \boldsymbol{\Sigma}_t$, and

$$\mathbf{X} = \begin{pmatrix} \mathbf{X}_1 \\ \vdots \\ \mathbf{X}_T \end{pmatrix}.$$

For Step 3 of the sampler, let $\mathbf{V}_{\boldsymbol{\theta}} = \text{diag}(\mathbf{V}_\beta, \mathbf{V}_\gamma, \mathbf{V}_a)$ and $\boldsymbol{\theta}_0 = (\boldsymbol{\beta}_0', \boldsymbol{\gamma}_0, \mathbf{a}_0')'$. Then, by standard results from linear regression, we have

$$\left(\boldsymbol{\beta}, \boldsymbol{\gamma}, \mathbf{A} | \mathbf{h}^L, \mathbf{h}^S, \boldsymbol{\Sigma}_L, \boldsymbol{\Sigma}_S, \mathbf{y}\right) \equiv \left(\boldsymbol{\theta} | \mathbf{h}^L, \mathbf{h}^S, \boldsymbol{\Sigma}_L, \boldsymbol{\Sigma}_S, \mathbf{y}\right) \sim \mathcal{N}\left(\widehat{\boldsymbol{\theta}}, \mathbf{D}_{\boldsymbol{\theta}}\right), \tag{12}$$

where $\mathbf{D}_{\boldsymbol{\theta}}^{-1} = \mathbf{X}'\boldsymbol{\Sigma}^{-1}\mathbf{X} + \mathbf{V}_{\boldsymbol{\theta}}^{-1}$ and $\widehat{\boldsymbol{\theta}} = \mathbf{D}_{\boldsymbol{\theta}}\left(\mathbf{X}'\boldsymbol{\Sigma}^{-1}\mathbf{y} + \mathbf{V}_{\boldsymbol{\theta}}^{-1}\boldsymbol{\theta}_0\right)$.

In Step 4, the full conditional distributions of the variance terms are standard. In particular, it follows that

$$\sigma_{i,L}^2 \sim \mathcal{IG}\left(\frac{T}{2} + \eta_{i,L}, \omega_{i,L} + \sum_{t=1}^{T} dy_{i,t}^2 e^{-h_t^L}\right), \quad i = 1, \dots, n,$$

$$\sigma_{j,S}^2 \sim \mathcal{IG}\left(\frac{T}{2} + \eta_{j,S}, \omega_{j,S} + \sum_{t=1}^{T} dy_{j,t}^2 e^{-h_t^S}\right), \quad j = n+1, \dots, 2n,$$

where $dy_{i,t}$ is the jth element of the vector $\mathbf{y}_t - \mathbf{X}_t\boldsymbol{\theta}$. Step 5 is essentially the same as Step 4. Step 6 is also standard. In particular, given the log volatilities, it follows that

$$(\boldsymbol{\Sigma}_h | \mathbf{h}^L, \mathbf{h}^S, \boldsymbol{\Phi}) \sim \mathcal{IW}(\widehat{\mathbf{S}}_h, \widehat{\nu}_h),$$

with $\widehat{\mathbf{S}}_h = \sum_{t=2}^{T} \mathbf{e}_t \mathbf{e}_t' + \boldsymbol{\Xi}_h$ and $\widehat{\nu}_h = \nu_h + T - 1$, where

$$\mathbf{e}_t = \begin{pmatrix} h_t^L \\ h_t^S \end{pmatrix} - \boldsymbol{\Phi} \begin{pmatrix} h_{t-1}^L \\ h_{t-1}^S \end{pmatrix}.$$

For Step 7, we again apply standard linear regression results. In particular, we first note that the state equation can be expressed as

$$\mathbf{h}_t = \mathbf{X}_t^h \boldsymbol{\phi} + \boldsymbol{\epsilon}_t^h, \quad \boldsymbol{\epsilon}_t^h \sim \mathcal{N}(\mathbf{0}, \boldsymbol{\Sigma}_h),$$

where $\mathbf{h}_t = (h_t^L, h_t^S)'$. Stacking all the state equations from $t = 2, \ldots, T$ gives

$$\mathbf{h} = \mathbf{X}^h \boldsymbol{\phi} + \boldsymbol{\epsilon}^h, \quad \boldsymbol{\epsilon}^h \sim \mathcal{N}(\mathbf{0}, \mathbf{I}_2 \otimes \boldsymbol{\Sigma}_h).$$

Thus, the Step 7 can be implemented as follows:

$$(\boldsymbol{\Phi} | \mathbf{h}^L, \mathbf{h}^S, \boldsymbol{\Sigma}_h) \equiv (\boldsymbol{\phi} | \mathbf{h}^L, \mathbf{h}^S, \boldsymbol{\Sigma}_h) \sim \mathcal{N}(\widehat{\boldsymbol{\phi}}, \mathbf{D}_\phi) 1(\boldsymbol{\Phi} \in A),$$

where $\mathbf{D}_\phi^{-1} = \mathbf{X}^{h'} (\mathbf{I}_2 \otimes \boldsymbol{\Sigma}_h)^{-1} \mathbf{X}^h + \mathbf{V}_\phi^{-1}$ and $\widehat{\boldsymbol{\phi}} = \mathbf{D}_\phi \left(\mathbf{X}^{h'} (\mathbf{I}_2 \otimes \boldsymbol{\Sigma}_h)^{-1} \mathbf{h} + \mathbf{V}_\phi^{-1} \boldsymbol{\phi}_0 \right)$. Since this full conditional distribution is not standard, we implement a Metropolis-Hasting step using $\mathcal{N}(\widehat{\boldsymbol{\phi}}, \mathbf{D}_\phi)$ as a proposal distribution for drawing from the conditional posterior of $\boldsymbol{\phi}$.

Acknowledgments We thank Gary Koop, Dimitris Korobilis, Joshua Chan, Knut Are Aastveit, Todd Clark, Benjamin Wong, Joscha Beckmann, Yiqiao Sun, and members of the 12th RCEA Bayesian Workshop and The 27th Annual SNDE Symposium for their comments in the development of this research.

References

Abel, A. B. (1983). Optimal investment under uncertainty. *American Economic Review, 73*(1), 228–233.

Andreasen, M. M. (2010). Stochastic volatility and DSGE models. *Economics Letters, 108*(1), 7–9.

Andrieu, C., Doucet, A., & Holenstein, R. (2010). Particle Markov chain Monte Carlo methods. *Journal of the Royal Statistical Society: Series B, 72*(3), 269–342.

Bansal, R., & Yaron, A. (2004). Risks for the long run: A potential resolution of asset pricing puzzles. *The Journal of Finance, 59*(4), 1481–1509.

Basu, S., & Bundick, B. (2017). Uncertainty shocks in a model of effective demand. *Econometrica, 85*(3), 937–958.

Beckmann, J., Davidson, S. N., Koop, G., & Schüssler, R. (2023). Cross-country uncertainty spillovers: Evidence from international survey data. *Journal of International Money and Finance, 130*, 102760.

Bloom, N. (2009). The impact of uncertainty shocks. *Econometrica, 77*(3), 623–685.

Bloom, N. (2014). Fluctuations in uncertainty. *Journal of Economic Perspectives, 28*(2), 153–76.

Caggiano, G., & Castelnuovo, E. (2023). Global financial uncertainty. *Journal of Applied Econometrics, 38*(3), 432–449.

Caggiano, G., Castelnuovo, E., & Figueres, J. M. (2017). Economic policy uncertainty spillovers in booms and busts. *Working Paper* (2017)

Caggiano, G., Castelnuovo, E., & Figueres, J. M. (2020). Economic policy uncertainty spillovers in booms and busts. *Oxford Bulletin of Economics and Statistics, 82*(1), 125–155.

Carriero, A., Clark, T. E., & Marcellino, M. (2016). Common drifting volatility in large Bayesian VARs. *Journal of Business and Economic Statistics, 34*(3), 375–390.

Carriero, A., Clark, T. E., & Marcellino, M. (2018). Measuring uncertainty and its impact on the economy. *Review of Economics and Statistics, 100*(5), 799–815.

Carriero, A., Clark, T. E., & Marcellino, M. (2020). Assessing international commonality in macroeconomic uncertainty and its effects. *Journal of Applied Econometrics, 35*(3), 273–293.

Castelnuovo, E. (2023). Uncertainty before and during covid-19: A survey. *Journal of Economic Surveys, 37*(3), 821–864.
Cesa-Bianchi, A., Pesaran, M. H., & Rebucci, A. (2020). Uncertainty and economic activity: A multicountry perspective. *The Review of Financial Studies, 33*(8), 3393–3445.
Chan, J. C. (2017). The stochastic volatility in mean model with time-varying parameters: An application to inflation modeling. *Journal of Business and Economic Statistics, 35*(1), 17–28.
Chan, J. C., & Jeliazkov, I. (2009). Efficient simulation and integrated likelihood estimation in state space models. *International Journal of Mathematical Modelling and Numerical Optimisation, 1*(1–2), 101–120.
Chib, S., & Greenberg, E. (1995). Understanding the Metropolis-Hastings algorithm. *The American Statistician, 49*(4), 327–335.
Cross, J., Chan, J., Kam, T., & Poon, A. (2017). Economic, fiscal- or monetary-policy uncertainty shocks: What matters for a small open economy? *Working Paper*, 2017.
Cross, J. L., Hou, C., Koop, G., & Poon, A. (2023). Large stochastic volatility in mean VARs. *Journal of Econometrics, 236*(1), 105469.
Cross, J. L., & Poon, A. (2020). On the contribution of international shocks in Australian business cycle fluctuations. *Empirical Economics, 59*(6), 2613–2637.
Davidson, S. N., Hou, C., & Koop, G. (2025). Investigating economic uncertainty using stochastic volatility in Mean VARs: The importance of model size, order-invariance and classification. *Journal of Business and Economic Statistics*, forthcoming.
Fasani, S., & Rossi, L. (2018). Are uncertainty shocks aggregate demand shocks? *Economics Letters, 167*, 142–146.
Geweke, J., & Amisano, G. (2011). Hierarchical Markov normal mixture models with applications to financial asset returns. *Journal of Applied Econometrics, 26*(1), 1–29.
Haque, Q., & Magnusson, L. M. (2021). Uncertainty shocks and inflation dynamics in the US. *Economics Letters, 202*, 109825.
Hartman, R. (1972). The effects of price and cost uncertainty on investment. *Journal of Economic Theory, 5*(2), 258–266.
Jacquier, E., Polson, N. G., & Rossi, P. E. (2002). Bayesian analysis of stochastic volatility models. *Journal of Business and Economic Statistics, 20*(1), 69–87.
Jurado, K., Ludvigson, S. C., & Ng, S. (2015). Measuring uncertainty. *American Economic Review, 105*(3), 1177–1216.
Justiniano, A., & Preston, B. (2010). Can structural small open-economy models account for the influence of foreign disturbances? *Journal of International Economics, 81*(1), 61–74.
Karlsson, S., Mazur, S., & Nguyen, H. (2023). Vector autoregression models with skewness and heavy tails. *Journal of Economic Dynamics and Control, 146*, 104580.
Kim, S., Shephard, N., & Chib, S. (1998). Stochastic volatility: Likelihood inference and comparison with ARCH models. *The Review of Economic Studies, 65*(3), 361–393.
Koop, G., Pesaran, M. H., & Potter, S. M. (1996). Impulse response analysis in nonlinear multivariate models. *Journal of Econometrics, 74*(1), 119–147.
Leduc, S., & Liu, Z. (2016). Uncertainty shocks are aggregate demand shocks. *Journal of Monetary Economics, 82*, 20–35.
Lucas, R. E. (1976). Econometric policy evaluation: A critique. In *Carnegie-Rochester Conference Series on Public Policy* (Vol. 1, pp. 19–46). Elsevier.
McCausland, W. J., Miller, S., & Pelletier, D. (2011). Simulation smoothing for state–space models: A computational efficiency analysis. *Computational Statistics and Data Analysis, 55*(1), 199–212.
Mumtaz, H., & Theodoridis, K. (2015). The international transmission of volatility shocks: An empirical analysis. *Journal of the European Economic Association, 13*(3), 512–533.
Mumtaz, H., & Theodoridis, K. (2017). Common and country specific economic uncertainty. *Journal of International Economics, 105*, 205–216.
Mumtaz, H., & Zanetti, F. (2013). The impact of the volatility of monetary policy shocks. *Journal of Money, Credit and Banking, 45*(4), 535–558.

Oi, W. Y. (1961). The desirability of price instability under perfect competition. *Econometrica, 29*, 58–64.

Österholm, P., & Zettelmeyer, J. (2008). The effect of external conditions on growth in Latin America. *IMF Staff Papers, 55*(4), 595–623.

Paddock, J. L., Siegel, D. R., & Smith, J. L. (1988). Option valuation of claims on real assets: The case of offshore petroleum leases. *The Quarterly Journal of Economics, 103*(3), 479–508.

Pfarrhofer, M. (2023). Measuring international uncertainty using global vector autoregressions with drifting parameters. *Macroeconomic Dynamics, 27*(3), 770–793.

Rue, H., Martino, S., & Chopin, N. (2009). Approximate Bayesian inference for latent Gaussian models by using integrated nested Laplace approximations. *Journal of the Royal Statistical Society: Series B, 71*(2), 319–392.

Shephard, N., & Pitt, M. K. (1997). Likelihood analysis of non-Gaussian measurement time series. *Biometrika, 84*(3), 653–667.

Sims, C. A. (1980). Macroeconomics and reality. *Econometrica, 48*, 1–48.

Villani, M., & Warne, A. (2003). Monetary policy analysis in a small open economy using Bayesian cointegrated structural VARs. *Available at SSRN 487477.*

Modeling Local Predictive Ability Using Power-Transformed Gaussian Processes

Oscar Oelrich and Mattias Villani

Abstract A Gaussian process (GP) is proposed as a model for the posterior distribution of the local predictive ability of a model or expert, conditional on a vector of covariates, from historical predictions in the form of log predictive scores. Assuming Gaussian expert predictions and a Gaussian data-generating process, a linear transformation of the predictive score follows a noncentral chi-squared distribution with one degree of freedom. Motivated by this, we develop a noncentral chi-squared Gaussian process regression to flexibly model local predictive ability, with the posterior distribution of the latent GP function and kernel hyperparameters sampled by Hamiltonian Monte Carlo (HMC). We show that a cube root transformation of the log scores is approximately Gaussian with homoscedastic variance, making it possible to estimate the model much faster by marginalizing the latent GP function analytically. A multi-output Gaussian process regression is also introduced to model the dependence in predictive ability between experts, both for inference and for prediction purposes. Linear pools based on learned local predictive ability are applied to predict daily bike usage in Washington DC.

1 Introduction

Combining predictions from several models or experts, be they point forecasts (Bates & Granger, 1969) or forecast distributions (Yao et al., 2018), has a long history in statistics and econometrics; see Wang et al. (2023) for a recent review. A forecast combination can better quantify total uncertainty (Draper, 1995) and improve accuracy (Wang et al., 2023). Some prominent examples of aggregation methods include weighted averages with weights based on past forecast errors (Bates & Granger, 1969), posterior model probabilities (Hoeting et al., 1999), and historical performance of the experts' density forecasts (Hall & Mitchell, 2007; Geweke & Amisano, 2011).

O. Oelrich (✉) · M. Villani
Department of Statistics, Stockholm University, Stockholm, Sweden
e-mail: oscar.oelrich@stat.su.se; mattias.villani@stat.su.se

S. Mazur, P. Österholm (eds.), *Recent Developments in Bayesian Econometrics and Their Applications*, https://doi.org/10.1007/978-3-032-00110-8_7

The out-of-sample predictive accuracy of an expert, which could be a formal model or a human, is typically treated as a fixed quantity to be estimated (Gelman et al., 2014). Our aim here is to model *local predictive ability*, i.e., predictive ability as a function of a vector of covariates $\mathbf{z}$, to capture how the predictive abilities of a set of experts vary over the space of $\mathbf{z}$ (Yao et al., 2021; Li et al., 2023; Oelrich et al., 2024). We aim to obtain the posterior distribution of the local predictive ability to properly quantify the uncertainty of the expert's predictive ability.

Since one use of measures of predictive ability is to combine, or pool, expert predictions, we call the variables in $\mathbf{z}$ *pooling variables* and the space of $\mathbf{z}$ the *pooling space* (Oelrich et al., 2024). The choice of pooling variables is determined by the *decision-maker* (Lindley et al., 1979) making the aggregated prediction, based on what she believes are important determinants of the experts' local predictive abilities. In an economic application, a measure of the business cycle or a survey of expectations could, for example, be a pooling variable, if the forecasting experts are expected to have different forecasting accuracies depending on where the economy is in the business cycle.

Our focus is on the accuracy of an expert's forecast *density* (Gelman et al., 2014). Density forecasts are typically part of the output from probabilistic models, but are also increasingly reported from human experts. Examples include the histogram densities for key macroeconomic variables from the Survey of Professional Forecasters conducted by the Federal Reserve Bank of Philadelphia for the USA (Croushore & Stark, 2019) and the European Central Bank for the Euro Area (de Vincent-Humphreys et al., 2019), and probabilistic forecasts of temperature and precipitation in meteorology from a combination of subjective experts, historical data, and complex deterministic scientific models (AMS Council, 2008).

A theoretically motivated measure of predictive ability is the expected log predictive density (ELPD), which can be unbiasedly estimated by average log predictive density evaluations (log scores) on a test dataset. The distribution of the log scores is typically complex however (Sivula et al., 2022b), making inferences for ELPD challenging. The likelihood of the observed log scores depends not only on the model/expert but also on the data-generating process (DGP). For example, using a simple model with additive noise will typically work poorly, as when the predictive distribution and the DGP are normal, the log scores will follow a scaled and translated noncentral χ_1^2 distribution (Sivula et al., 2022a).

Further, specifying the relationship between the set of covariates and predictive ability is a challenging problem in itself. Certain variables may be expected to be important for predictive ability, but the exact functional form may be harder to pinpoint. In a situation of little prior information, it makes sense for the decision-maker to use a nonparametric model that can capture variations in predictive ability over an arbitrary pooling space in a flexible manner. Oelrich et al. (2024) propose the nonparametric caliper estimator of local predictive ability, which averages log scores locally in pooling space. The caliper estimator is flexible, but has the potential drawback that the inferred local ability surface is discontinuous, and the lack of smoothness may result in large variability in the local estimates.

We here propose a flexible noncentral χ^2 Gaussian process regression, where the latent GP surface models the conditional ELPD over the space of pooling variables in $\mathbf{z}$. A Gaussian process explicitly models the correlation properties, but makes otherwise minimal assumptions about the specific relationship between the covariates and the ELPD. The smoothness and other hyperparameters in the covariance kernel are assumed unknown and are inferred jointly with the latent GP surface. A Gaussian process model for local ELPD is attractive since it gives a direct quantification of the uncertainty about the underlying ELPD surface for each expert.

Separate GPs can be estimated for each expert, implicitly assuming independence when forecasts are pooled based on local predictive ability. However, correctly capturing any dependence structure between experts is useful both from a pure inference standpoint, as it allows us to better understand the relationship between the predictions of the experts, and for creating better aggregate predictions (McAlinn et al., 2020). The latter point is clearly illustrated by Winkler (1981), who shows that an expert with relatively low predictive ability can nevertheless get a substantial weight in a combined forecast if the expert's forecast error is negatively correlated with the forecast errors of the other experts. We therefore propose a multi-output GP to *jointly* model the local predictive ability of a set of K experts over the pooling space. The inferred dependence between the experts is exploited to improve the forecast combination in a linear pool.

The proposed multi-output GP uses a kernel function that is constructed by correlating a set of underlying independent GPs, one for each expert. This construction makes it possible to use separate smoothness kernel parameters, or even completely different classes of kernels, for each expert. It also gives us access to interpretable parameters that describe the correlation between experts. The multi-output model allows us to jointly infer the latent GP surfaces of all experts, which gives a more full quantification of the uncertainty of the ELPD surface of prediction pools based on the set of experts.

The joint posterior distribution of the conditional ELPD and kernel hyperparameters are sampled by HMC using Stan (Stan Development Team, 2024). The dimension of the sampled conditional ELPD grows with the sample size however, and inferences are time-consuming even for moderately large datasets. We show empirically that a cube root transformation of the log scores is approximately Gaussian with homoscedastic variance. This makes it possible to integrate out the latent GP surface analytically and sample the low-dimensional marginal posterior kernel hyperparameters by HMC, which is much faster and scalable for larger datasets. We show that the ELPD surface can then be easily sampled by standard GP methods (Rasmussen & Williams, 2006, Chapter 2) combined with analytically available third moments of Gaussian processes. In the more general case when either the expert or the data-generating process is non-Gaussian, we propose to use a general power transformation of the ELPD and to estimate the power transformation using HMC jointly with the kernel hyperparameters.

The chapter proceeds as follows. Section 2 defines local predictive ability and describes the noncentral χ^2 GP regression for estimating it, along with a computationally more efficient single- and multi-output GP regression on power-transformed log scores. Section 3 consists of an empirical example analyzing local predictive ability for a set of models in a bike-sharing dataset, along with several methods to pool predictive distributions based on estimated predictive ability. Section 4 concludes.

2 A Framework for Estimating Local Predictive Ability

This section defines local predictive ability within a decision-maker framework (Oelrich et al., 2024), discusses the implications of using log scores as data to estimate local predictive ability, and proposes Gaussian process regression as a model for estimating both joint and marginal local predictive ability.

2.1 *Local Predictive Ability*

We consider the problem where a decision-maker is presented with a set of K experts, either in the form of models or opinionated humans, and wants to estimate how well the experts make predictions, based on their historical performance. We assume that the decision-maker has access to historic predictions made by the experts in the form of log predictive density scores $\ell_{ik} = \log p_k(y_i \mid \mathbf{y}_k)$, assessing the prediction of observation y_i in the kth expert's predictive density given a set of training data $\mathbf{y}_k$. While nothing prevents us from considering other measures, such as mean-squared error, the derived model for predictive ability depends on the chosen measure. The log predictive density score is considered the gold standard for measuring predictive performance since it has the unique advantage of being both local and proper (Gneiting & Raftery, 2007), and is therefore commonly used in model selection. Further, the ELPD of a predictive distribution is proportional to the Kullback-Leibler divergence with regard to the DGP (Hall & Mitchell, 2007).

The decision-maker uses the ELPD for a new data point $\tilde{y}$ from the DGP $p_*(\cdot)$ to define the predictive ability of expert k

$$\eta_k \equiv \mathrm{E}_{p_*(\tilde{y})}\left[\log p_k(\tilde{y} \mid \mathbf{y}_k)\right] = \int_{-\infty}^{\infty} \left[\log p_k(\tilde{y} \mid \mathbf{y}_k)\right] p_*(\tilde{y})\, d\tilde{y}, \tag{1}$$

where $\tilde{y}$ is a new (out-of-sample) observation, $p_*(\cdot)$ is the density of the data-generating process, and $\mathbf{y}_k$ is the training data the expert has access to. The variable of interest can be univariate or multivariate, but we will assume for simplicity that it is univariate here.

The decision-maker further believes that the predictive ability of the experts varies over a set of pooling variables $\mathbf{z}$ (Oelrich et al., 2024). She is therefore not only interested in determining how good the experts are at making predictions in general, but how good they are at making predictions conditional on the pooling variables, $\mathbf{z}$. We call this conditional predictive ability the *local predictive ability* of the expert at $\mathbf{z}$

$$\eta_k(\mathbf{z}) \equiv \mathrm{E}_{p_*(\tilde{y}|\mathbf{z})}\left[\log p(\tilde{y} \mid \mathbf{y}_k, \mathbf{z})\right] = \int_{-\infty}^{\infty} \left[\log p(\tilde{y} \mid \mathbf{y}_k, \mathbf{z})\right] p_*(\tilde{y}|\mathbf{z})\, d\tilde{y}, \tag{2}$$

where the data-generating process $p_*(\tilde{y}|\mathbf{z})$ may or may not depend on the pooling variables $\mathbf{z}$.

2.2 *Distribution of the Log Scores*

The log scores of the experts, ℓ_{ik} for $k = 1, \ldots, K$ and $i = 1, \ldots, n$, are treated as data by the decision-maker (Lindley et al., 1979) from which she infers how the ELPD of each expert varies over the pooling space. The choice of pooling variables is therefore a pure modeling problem from the viewpoint of the decision-maker. The decision-maker does not need to know if the experts have access to the pooling variables when making their predictions, but this information can of course affect the modeling choices of the decision-maker.

The distribution of the local predictive ability is determined by the predictive distribution of the expert, which is typically known, and the data-generating process, which is typically unknown. In order to specify a parametric model for the log scores, we therefore need access to the predictive distribution of the expert. Different predictive distributions will lead to different models of local predictive ability.

Consider the practically important scenario where the expert predictive distributions, as well as the data-generating process, are all Gaussian. Specifically, let the predictive distribution of expert k for the data point y_i be $y_i \mid \mathbf{y}_k \sim \mathcal{N}(\mu_{ik}, \sigma_{ik}^2)$, where $\mathbf{y}_k$ is the training data used by the kth expert. The log score is then

$$\ell_{ik} \equiv \log p_k(y_i \mid \mathbf{y}_k) = -\frac{1}{2}\log(2\pi\sigma_{ik}^2) - \frac{1}{2\sigma_{ik}^2}(y_i - \mu_{ik})^2 \,. \tag{3}$$

If we let the data-generating process be given by $y_i \sim \mathcal{N}(\mu_{i*}, \sigma_{i*}^2)$, we can trivially rewrite Eq. (3) as

$$\ell_{ik} \equiv \log p_k(y_i \mid \mathbf{y}_k) = -\frac{1}{2}\log(2\pi\sigma_{ik}^2) - \frac{\sigma_{i*}^2}{2\sigma_{ik}^2}\left(\frac{y_i - \mu_{ik}}{\sigma_{i*}}\right)^2 . \tag{4}$$

Since $(y_i - \mu_{ik})/\sigma_{i*} \sim \mathcal{N}(\mu_{i*} - \mu_{ik}, 1)$, it follows that (Sivula et al., 2022a)

$$\ell_{ik} \stackrel{d}{=} a_{ik} - b_{ik} \cdot X_{ik}, \tag{5}$$

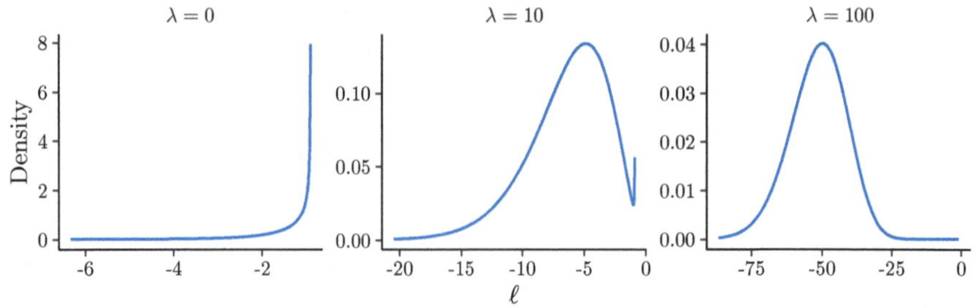

Fig. 1 Distribution of log scores for a Gaussian expert and DGP with equal variance for three different $\lambda_k = (\mu_* - \mu_k)^2$

where $X_{ik} \sim \chi_1^2(\lambda_{ik})$ with noncentrality parameter $\lambda_{ik} = (\mu_{i*} - \mu_{ik})^2/\sigma_{i*}^2$, $a_{ik} = -\frac{1}{2}\log(2\pi\sigma_{ik}^2)$, and $b_{ik} = \sigma_{i*}^2/(2\sigma_{ik}^2)$. Hence, the noncentrality parameter λ_{ik} depends on the difference between the predictive mean of the expert and the mean of the data-generating process; the shape of the distribution of the log scores therefore varies dramatically with $\mu_{i*} - \mu_{ik}$ and is typically highly non-Gaussian for all b, see Fig. 1 for an illustration.

Since $a_{ik} = -\frac{1}{2}\log(2\pi\sigma_{ik}^2)$ only depends on the expert's predictive distribution and is therefore assumed known, we can define $\ell'_{ik} \equiv -(\ell_{ik} - a_{ik})$ and write the model in (5) as

$$\ell'_{ik} \mid \lambda_{ik}, b_{ik} \sim \chi_1^2(\lambda_{ik}, b_{ik}), \tag{6}$$

where $\chi_1^2(\lambda, b)$ denotes a *scaled* noncentral χ^2 with one degree of freedom, noncentrality parameter λ, and scale parameter b. A useful simplification of the model in (6) is to assume that $b_{ik} = b_k$ for all $i = 1, \ldots, n$, i.e., that the variance ratio $\sigma_{i*}^2/\sigma_{ik}^2$ is constant over observations for a given expert. Note that this assumption does not mean that the DGP needs to be homoscedastic, only that the variance in the expert's predictive distribution is proportional to the variance in the DGP. Put differently, an expert with a homoscedastic variance cannot fulfill this restriction if the DGP is truly heteroscedastic; for such an expert, we would need to model also b_{ik} as a function of the pooling variables (see next subsection), which is more complicated, but certainly possible. Alternatively, such a poorly calibrated expert may be excluded in the forecast combination altogether.

2.3 Scaled Noncentral χ^2 Gaussian Process Regression

The setup with Gaussian data-generating process and experts in the previous subsection suggests a scaled noncentral χ^2 regression for modeling the local predictive ability of experts. We will here propose a nonparametric Bayesian model where the log of the noncentrality parameter is modeled by a Gaussian process over the pooling space.

A Gaussian process generalizes the Gaussian distribution from a random variable to a stochastic process (Rasmussen & Williams, 2006). In simple terms, a Gaussian process is an infinite sequence of function values with different inputs $\{f(\mathbf{z}_1), f(\mathbf{z}_2), f(\mathbf{z}_3), \dots\}$, any subset of which follows a multivariate Gaussian distribution. Each GP has a covariance function, or kernel, which depends on the input values and specifies how the covariance matrix for each subset of the sequence should be constructed. The specific covariance function we select determines our prior regarding how smoothly the outputs vary over the input space. The covariance function we use in the examples and applications in this chapter is the squared exponential with automatic relevance determination (ARD) (Neal, 1996)

$$k_{\mathrm{SE}}(\mathbf{z}_i, \mathbf{z}_j) = \alpha^2 \exp\left(-\frac{1}{2}(\mathbf{z}_i - \mathbf{z}_j)(\mathrm{diag}(\boldsymbol{l})^{-2})(\mathbf{z}_i - \mathbf{z}_j)\right) + \delta_{ij}\sigma_n^2, \tag{7}$$

where $\boldsymbol{l}$ is a vector of *length scales*, α denotes the signal standard deviation, σ_n^2 denotes the noise variance, and δ_{ij} is the Kronecker function, which takes the value 1 when $i = j$ and 0 otherwise. Kernels where the length scales of different inputs are allowed to differ are referred to as automatic relevance determination kernels, as the length scales are inversely related to relevance (Rasmussen & Williams, 2006). Any prior can be used for the prior hyperparameters since our sampling scheme does not rely on conjugacy; in the application in Sect. 3, we use inverse gamma for the length scales and zero-truncated normal distributions for the signal and noise variances.

Our noncentral χ^2 Gaussian process model for the log scores is

$$\begin{aligned}
\ell'_{ik} \mid \lambda_k(\mathbf{z}_i), b_k &\overset{\text{indep.}}{\sim} \chi_1^2(\lambda_k(\mathbf{z}_i), b_k) \\
\log \lambda_k(\mathbf{z}) &\sim \mathrm{GP}\left(\mu_k(\mathbf{z}), g_k(\mathbf{z}, \mathbf{z}')\right) \\
b_k &\sim \mathcal{N}^+(1/2, \psi_b^2),
\end{aligned} \tag{8}$$

where $\mathcal{N}^+(\cdot, \cdot)$ denotes a normal distribution truncated at 0. Note that $b_k > 0$, and $b_k = 1/2$ when the variance of the DGP and the expert match. Further, note that the GP is on the log of the noncentrality parameter, ensuring that $\lambda_k(\mathbf{z})$ stays nonnegative. Since the mean of a $\chi_1^2(\lambda, b)$ variable is $b(1+\lambda)$, the Gaussian process in (8) directly implies a flexible prior for the mean function of the log scores, i.e., the local ELPD. We will refer to the model in (8) as $\mathrm{GP}(\chi_1^2)$. The mean function $\mu_k(\cdot)$ and the covariance kernel $g_k(\cdot, \cdot)$ are subscripted by k since their hyperparameters are allowed to be different across the K experts. Any valid kernel function can be used. For notational simplicity, we let $\boldsymbol{\theta}_k$ be the vector with all the hyperparameters in the mean and kernel function of the GP prior.

The ultimate goal of the decision-maker is to compute the posterior distribution of the local ELPD $\eta_k(\tilde{\mathbf{z}}) = \mathrm{E}_{p_*}(\tilde{\ell}_k) = \tilde{a}_k - b_k(1+\lambda_k(\tilde{\mathbf{z}}))$ at a new point $\tilde{\mathbf{z}}$ in pooling space for each expert. This is achieved by sampling from the joint posterior

$$p\big(\lambda_k(\tilde{\mathbf{z}}), \boldsymbol{\lambda}_k, b_k, \boldsymbol{\theta}_k \mid \boldsymbol{\ell}'_k\big), \tag{9}$$

where $\boldsymbol{\ell}'_k = (\ell'_{1k}, \ldots, \ell'_{nk})^\top$ and $\boldsymbol{\lambda}_k = (\lambda_k(\mathbf{z}_1), \ldots, \lambda_k(\mathbf{z}_n))^\top$ is the vector of noncentrality parameters in the training data. We sample the potentially high-dimensional posterior in (9) using HMC in Stan.

2.4 Cube-Root-Transformed Gaussian Process Regression

HMC sampling for the scaled noncentral χ^2 Gaussian process regression is efficient, but necessarily slow in even moderately large datasets since $\boldsymbol{\lambda}_k$ is high dimensional. This subsection presents a cube root transformation of the log scores, which makes the log scores approximately normal with constant variance.[1] This allows us to analytically integrate out the high-dimensional latent ELPD surfaces from the posterior, thereby substantially speeding up the computation of the ELPD posterior at a new point in the pooling space. We will refer to this model in tables and figures as GP (1/3).

Starting from the linear transformation $\ell'_{ik} = -(\ell_{ik} - a_{ik})$, we further transform the log scores ℓ'_{ik} by taking the cube root, which is motivated by Abdel-Aty (1954) who shows that a version of the Wilson-Hilferty transformation of the χ^2_v distribution approximately holds for $X \sim \chi^2_v(\lambda)$ as well. That is,

$$\left(\frac{X}{v+\lambda}\right)^{1/3} \mathrel{\dot\sim} \mathcal{N}, \tag{10}$$

where the mean and variance of the normal distribution depend on the noncentrality parameter λ as well as the degrees of freedom. Since the noncentrality parameter λ is

[1] The cube root transformation is suitable when the log scores follow a scaled noncentral χ^2_1 distribution, which is the correct model in the important case when both the DGP and the expert are Gaussian. Deviations from Gaussianity lead to new models that need to be derived on a case-by-case basis. The posterior distribution of the ELPD can be sampled with HMC as in the noncentral χ^2_1 model described above. There may be new transformations that convert the log scores into something that can be modeled by a Gaussian process regression with Gaussian errors, but also this needs to be determined for each particular model.

Moderate departures from Gaussianity can however be handled with a more general power transformation than the cube root transformation. Let $\ell_{ik}^{(\alpha)} = (\ell'_{ik})^\alpha$ for $\alpha > 0$ be a general power transformation. The ELPD will then be $\eta_k(\tilde{\mathbf{z}}) = \tilde{a}_k - \mathrm{E}_{p_*}\left[(\tilde{\ell}_k^{(\alpha)})^{1/\alpha}\right]$. The fractional moments $\mathrm{E}_{p_*}\left[(\tilde{\ell}_k^{(\alpha)})^{1/\alpha}\right]$ can again be expressed in terms of $f_k(\tilde{\mathbf{z}})$ and σ_k^2 as in (14) using the formulas in Haldane (1942). The power parameter, α, can be sampled in the HMC step along with σ_k and the kernel hyperparameters in $\boldsymbol{\theta}_k$. The prior for α can be centered on $\alpha = 1/3$, i.e., the cube root transformation.

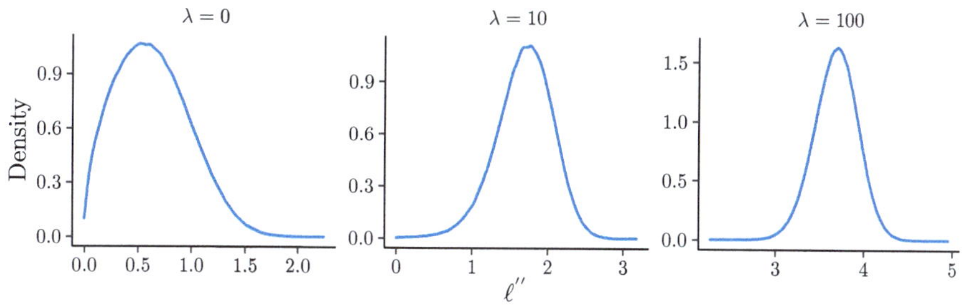

Fig. 2 Approximate distribution of transformed log scores for a Gaussian expert and DGP with equal variance for three different $\lambda_k = (\mu_* - \mu_k)^2$

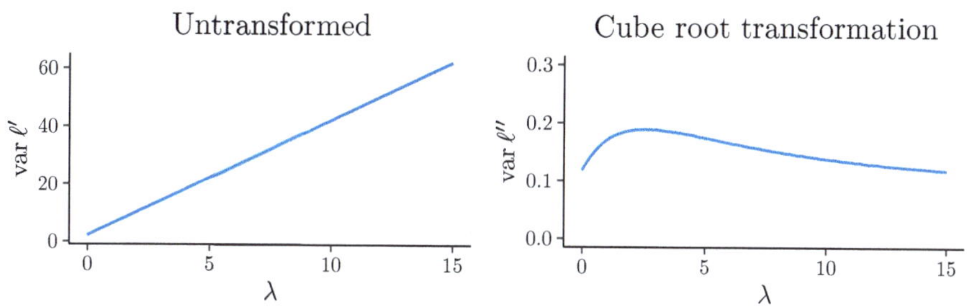

Fig. 3 Variance of log scores as function of $\lambda = (\mu_* - \mu_k)^2$ for a Gaussian expert and DGP with equal variance

not known, we use a plain cube root $X^{1/3}$ transformation when transforming the log scores. Figure 2 shows that this transformation makes the log scores approximately Gaussian; compare with Fig. 1. As λ goes toward infinity, the variance goes to zero, but Fig. 3 shows that the variance is approximately constant in a range of reasonable λ values. We use the notation $\ell'' = (\ell')^{1/3}$ for these twice transformed log scores.

Once our log scores have been transformed to approximate normality with constant variance, we can use a Gaussian process with homoscedastic Gaussian noise to model the predictive ability of each expert:

$$\begin{aligned} \ell''_{ik} \mid f_k(\mathbf{z}_i), \sigma_k &\overset{\text{indep.}}{\sim} \mathcal{N}(f_k(\mathbf{z}_i), \sigma_k^2) \\ f_k(\mathbf{z}) &\sim \mathrm{GP}\left(\mu_k(\mathbf{z}), g_k(\mathbf{z}, \mathbf{z}')\right) \\ \sigma_k^2 &\sim \mathcal{N}^+(0, \psi_\sigma^2) \, . \end{aligned} \tag{11}$$

Again, let $\boldsymbol{\theta}_k$ denote the vector of hyperparameters in the mean and kernel functions of the GP for the kth expert.

Since $\ell_{ik} = a_{ik} - \left(\ell''_{ik}\right)^3$, the ELPD at the test point $\tilde{\mathbf{z}}$ is $\eta_k(\tilde{\mathbf{z}}) = \tilde{a}_k - \mathrm{E}_{p*}\left[\left(\tilde{\ell}''_k\right)^3\right]$. For a general $X \sim \mathcal{N}(\mu, \sigma^2)$, we have

$$\mathrm{E}(X^3) = \mu^3 + 3\mu\sigma^2, \tag{12}$$

and hence

$$\begin{aligned}\mathrm{E}_{p*}\left[(\tilde{\ell}''_k)^3\right] &= [\mathrm{E}_{p*}(\tilde{\ell}''_k)]^3 + 3\,\mathrm{E}_{p*}(\tilde{\ell}''_{*k}) \cdot \mathrm{var}_{p*}(\tilde{\ell}''_k) \\ &= f_k^3(\tilde{\mathbf{z}}) + 3 f_k(\tilde{\mathbf{z}})\sigma_k^2\,. \end{aligned}\tag{13}$$

We therefore have the following expression for the local ELPD at $\tilde{\mathbf{z}}$

$$\eta_k(\tilde{\mathbf{z}}) = \tilde{a}_k - f_k^3(\tilde{\mathbf{z}}) - 3 f_k(\tilde{\mathbf{z}})\sigma_k^2\,. \tag{14}$$

To obtain the posterior for $\eta_k(\tilde{\mathbf{z}})$, we need the joint posterior $p\big(f_k(\tilde{\mathbf{z}}), \sigma_k, \boldsymbol{\theta}_k \mid \boldsymbol{\ell}''_k\big)$. Due to the approximate normality induced by the cube root transformation, we can analytically integrate out $\boldsymbol{f}_k = (f_k(\mathbf{z}_1), \ldots, f_k(\mathbf{z}_n))^\top$ in the training data and obtain the marginal posterior of σ_k and the kernel hyperparameters $\boldsymbol{\theta}_k$ in closed form (Rasmussen & Williams, 2006, Eq. 2.28)

$$p\big(\sigma_k, \boldsymbol{\theta}_k \mid \boldsymbol{\ell}''_k\big) = \int p\big(\boldsymbol{f}_k, \sigma_k, \boldsymbol{\theta}_k \mid \boldsymbol{\ell}''_k\big) d\boldsymbol{f}_k, \tag{15}$$

where $\boldsymbol{\ell}''_k$ is a vector with transformed log scores in the training data. This low-dimensional marginal posterior can be sampled with HMC using Stan in a fraction of the time that it takes to sample with full joint posterior for the GP(χ_1^2) model in (9).

For each HMC sampled $(\sigma_k, \boldsymbol{\theta}_k)$, we can now easily sample $f_k(\tilde{\mathbf{z}})$ from (Rasmussen & Williams, 2006)

$$f_k(\tilde{\mathbf{z}}) \mid \boldsymbol{\ell}''_k, \mathbf{Z}, \tilde{\mathbf{z}}, \sigma_k, \boldsymbol{\theta}_k \sim \mathcal{N}\big[\bar{f}(\tilde{\mathbf{z}}), \mathrm{var}(f(\tilde{\mathbf{z}}))\big], \tag{16}$$

where $\mathbf{Z}$ is the matrix with pooling variables in the training data, and

$$\bar{f}(\tilde{\mathbf{z}}) = \mu_k(\tilde{\mathbf{z}}) + \mathrm{G}(\tilde{\mathbf{z}}, \mathbf{Z})\left[\mathrm{G}(\mathbf{Z}, \mathbf{Z}) + \sigma_k^2 I\right]^{-1}(\boldsymbol{\ell}''_k - \boldsymbol{\mu}_k),$$

$$\mathrm{var}(f(\tilde{\mathbf{z}})) = \mathrm{G}(\tilde{\mathbf{z}}, \tilde{\mathbf{z}}) - \mathrm{G}(\tilde{\mathbf{z}}, \mathbf{Z})\left[\mathrm{G}(\mathbf{Z}, \mathbf{Z}) + \sigma_k^2 I\right]^{-1}\mathrm{G}(\mathbf{Z}, \tilde{\mathbf{z}}),$$

where, for example, $G(\mathbf{Z}, \mathbf{Z})$ is the $n \times n$ matrix with covariances among the $f(\mathbf{z})$ in the training data given by the kernel function $g_k(\cdot, \cdot)$. The posterior draws of σ_k from (15) and $f_k(\tilde{\mathbf{z}})$ from (16) can finally be inserted into (14) to obtain draws from the posterior of the local ELPD $\eta_k(\tilde{\mathbf{z}})$. The procedure is summarized in Algorithm 1.

Algorithm 1: Posterior sampling for the ELPD at $\tilde{\mathbf{z}}$ in the GP (1/3) model

Input: Number of posterior draws M, vector of pooling variables $\tilde{\mathbf{z}}$ and $\tilde{a}_k = -(1/2)\log(2\pi\tilde{\sigma}_k^2)$ for expert k.

Generate M draws $\{\sigma_k^{(j)}, \boldsymbol{\theta}_k^{(j)}\}_{j=1}^M$ from $p(\sigma_k, \boldsymbol{\theta}_k \mid \boldsymbol{\ell}_k'')$ in (15) using HMC.

for $j = 1, \ldots, M$ **do**

  Sample from the posterior predictive $f_k^{(j)}(\tilde{\mathbf{z}}) \mid \boldsymbol{\ell}_k'', \mathbf{Z}, \tilde{\mathbf{z}}, \sigma_k^{(j)}, \boldsymbol{\theta}_k^{(j)}$ in (16)

  Compute the ELPD draw: $\eta_k^{(j)}(\tilde{\mathbf{z}}) = \tilde{a}_k - (f_k^{(j)}(\tilde{\mathbf{z}}))^3 - 3f_k^{(j)}(\tilde{\mathbf{z}})(\sigma_k^{(j)})^2$

end

Output: Sample from ELPD posterior: $\eta_k^{(1)}(\tilde{\mathbf{z}}), \eta_k^{(2)}(\tilde{\mathbf{z}}), \ldots, \eta_k^{(M)}(\tilde{\mathbf{z}})$.

As the GP(1/3) adds an approximation layer on top of the original model, we want to ensure that the gains in computational speed do not come at the cost of a drop in performance. To this end, we conduct a simulation study that shows good performance of the GP(1/3) model compared to the GP(χ_1^2) model. For details, see the Appendix.

2.5 *The Multi-output* GP(1/3) *Model*

The GP(1/3) model proposed in Sect. 2.4 can be used to model the predictive ability of each of a set of experts. If these estimates are then used to create a pooled predictive distribution, two independence assumptions are implied: independence in the predictive abilities $\eta_1(\mathbf{z}_i), \ldots, \eta_K(\mathbf{z}_i)$, and independence in the errors, $\ell_{ik}'' - f_k(\mathbf{z}_i)$ for $i = 1, \ldots, n$, conditional on the predictive ability. In this section, we propose a generalization of the GP(1/3) model to a multi-output GP that relaxes both of these assumptions.

Let $\ell_{ik}'' = (a_{ik} - \ell_{ik})^{1/3}$ denote the cube-root-transformed log scores. The distribution of ℓ_{ik}'' for K experts jointly, $\boldsymbol{\ell}_i'' = (\ell_{i1}'', \ldots, \ell_{iK}'')^\top$, can then be modeled using a multi-output GP with Gaussian noise

$$\begin{aligned} \boldsymbol{\ell}_i'' \mid \mathbf{f}(\mathbf{z}_i), \boldsymbol{\Sigma} &\sim \mathcal{N}(\mathbf{f}(\mathbf{z}_i), \boldsymbol{\Sigma}) \\ \mathbf{f}(\mathbf{z}) \mid \boldsymbol{\Theta}, \mathbf{C} &\sim \mathrm{GP}\left(\boldsymbol{\mu}(\mathbf{z}), g(\mathbf{z}, \mathbf{z}')\right), \end{aligned} \tag{17}$$

where $\mathbf{f}(\mathbf{z}) = (f_1(\mathbf{z}), \ldots, f_K(\mathbf{z}))^\top$ is a multi-output GP with hyperparameters $\boldsymbol{\Theta} = \{\boldsymbol{\theta}_1,\ldots,\boldsymbol{\theta}_K\}$ including a vector of parameters $\boldsymbol{\theta}_k$ for the mean and covariance function of each expert as well, and $\mathbf{C}$ is a $K \times K$ matrix that will be used below to correlate the experts' abilities $\mathbf{f}$. The noise covariance matrix $\boldsymbol{\Sigma}$ is here taken to be a general positive-definite matrix, but may also be restricted to a diagonal matrix. The goal of the model in (17) is to capture correlation of the log scores between experts. This correlation is coming from two sources: in the noise distribution via $\boldsymbol{\Sigma}$, and in the underlying GP prior for $\mathbf{f}(\mathbf{z})$ with correlation generated by $\mathbf{C}$, which we now describe.

The model in (17) for all log scores in the training sample can be written

$$\underset{n\times K}{\boldsymbol{\ell}''} = \underset{n\times K}{\mathbf{F}} + \underset{n\times K}{\mathbf{E}}, \tag{18}$$

where $\boldsymbol{\ell}''$ is the matrix consisting of the transformed log scores of K experts at n time points, $\mathbf{F}$ is determined by the multi-output GP, and $\mathbf{E}$ is a matrix of Gaussian noise terms.

Reformulating the model in vectorized form, we can write the vector of errors as $\operatorname{vec}\mathbf{E} = \boldsymbol{\Sigma} \otimes \mathbf{I}_n$. By letting $\boldsymbol{\Sigma}$ be an arbitrary covariance matrix, we allow correlation in the error terms across experts, but assume independence between observations.

Modeling the correlation in the underlying predictive ability is done via the GP prior. A first attempt for a model for $\mathbf{F}$ is a matrix normal

$$\mathbf{F} \sim \mathcal{MN}(\mathbf{M}, \mathbf{G}, \boldsymbol{\Omega}), \tag{19}$$

i.e., that $\operatorname{vec}\mathbf{F}$ follows a multivariate normal with Kronecker-structured covariance matrix $\boldsymbol{\Omega} \otimes \mathbf{G}$. The matrix $\mathbf{G}$ models the correlation between rows of $\mathbf{F}$, which correspond to observations, and is therefore typically a covariance matrix generated from a kernel function, for example the squared exponential in (7). The matrix $\boldsymbol{\Omega}$ models the correlation between columns, in our case experts, and can most naturally be taken as a general positive-definite covariance matrix. The signal variances in the squared exponential kernel, α, would then be set to 1 for each expert.

While the matrix normal model in (19) has some intuitive appeal and is easy to work with from a computational perspective, it assumes that the columns of $\mathbf{F}$ all follow the same multivariate normal distribution, which puts the restriction on our model that all experts have identical covariance function, i.e., the same covariance matrix $\mathbf{G}$. To be able to specify separate covariance functions for each expert, we instead induce a prior for $\mathbf{F}$ by correlating a set of K underlying independent GPs, $\mathbf{h}_1, \ldots, \mathbf{h}_K$, each with its own kernel function (Teh et al., 2005)

$$\mathbf{F} = \mathbf{HC}, \qquad \mathbf{H} = [\mathbf{h}_1 \ldots \mathbf{h}_K], \qquad \mathbf{h}_k \overset{\text{indep}}{\sim} \mathcal{N}[\mathbf{0}, \mathrm{G}_k(\mathbf{Z}, \mathbf{Z})], \tag{20}$$

where $\mathrm{G}_k(\mathbf{Z}, \mathbf{Z}')$ is the covariance matrix for expert k generated from kernel function $g_k(\mathbf{z}, \mathbf{z}')$ and $\mathbf{C}$ is a $K \times K$ matrix that generates the dependence between the $\mathbf{f}_k$ for the experts. We set the signal standard deviation to unity in all kernels, i.e., $\alpha_1 = \cdots = \alpha_K = 1$ in the squared exponential kernel in (7) so that the scales of $\mathbf{F}$ are generated by the $\mathbf{C}$ matrix. While the decision-maker does not need to use the same pooling variables for each expert, we let the matrix of pooling variables $\mathbf{Z}$ contain the pooling variables for all experts to simplify notation.

The model in (20) allows the decision-maker to specify completely different covariance functions for the experts. Importantly, this allows the length scales to differ between the experts in the case where an ARD kernel is used, but it is also possible to use, for example, a Matérn kernel (Rasmussen & Williams, 2006) for some of the experts and squared exponential for others. It is also straightforward to force a subset of experts to share the exact same covariance function.

The covariance function for $\mathbf{F}$ in (20) is a weighted sum of the covariance functions of the K experts

$$\text{cov}(f_k(\mathbf{z}_i), f_l(\mathbf{z}_j)) = \sum_{s=1}^{K} c_{sk} c_{sl} g_s(\mathbf{z}_i, \mathbf{z}_j), \tag{21}$$

where c_{kl} is the element (k, l) of the matrix $\mathbf{C}$. To fully specify the multi-output GP(1/3) model—which we will denote by multi-GP(1/3)—we need to specify the covariance function of each expert, including priors for any unknown hyperparameters, as well as a prior for the matrix $\mathbf{C}$ that describes the between-expert covariances.

Note that while we choose to select a number of underlying processes equal to the number of experts, there is nothing that prevents us from creating our multi-output process by correlating an arbitrary number of latent GPs. A common justification for using several models is that it gives a more holistic quantification of uncertainty by capturing the structural uncertainty of the parametric form of the models, rather than just the parametric uncertainty within models (Draper, 1995). From this perspective, including several models that share the majority of structural DNA can be problematic, for example, in Bayesian model averaging (Hoeting et al., 1999), where adding an effective "copy" of an already existing model while using a uniform prior will essentially double the posterior model weight of that model. When modeling the joint predictive ability of a large pool of highly correlated experts, it is possible that many of the elements of $\mathbf{C}$ will be close to zero, in which case restricting the number of underlying GPs could simplify the model and speed up calculations.

2.5.1 Inference for the Multi-output GP(1/3) Model

When modeling local predictive ability jointly, the end goal is the joint posterior of the ELPD for all experts at a new point $\tilde{\mathbf{z}}$ in the pooling space, $\boldsymbol{\eta}(\mathbf{z}) = (\eta_1(\tilde{\mathbf{z}}), \ldots, \eta_K(\tilde{\mathbf{z}}))^\top$, i.e., the mean of $\boldsymbol{\ell}$ at $\tilde{\mathbf{z}}$. Since $\text{E}(\boldsymbol{\ell}_i) = (\text{E}(\ell_{i1}), \ldots, \text{E}(\ell_{iK}))^\top$ and $\ell_{ik} = a_{ik} - \left(\ell''_{ik}\right)^3$, we can use the same formula for the third moments of Gaussians as in the univariate case and obtain the local predictive ability from model (17) pointwise as

$$\eta_k(\tilde{\mathbf{z}}) = \tilde{a}_k - f_k^3(\tilde{\mathbf{z}}) - 3 f_k(\tilde{\mathbf{z}}) \boldsymbol{\Sigma}_{k,k}, \tag{22}$$

where $\boldsymbol{\Sigma}_{k,k}$ denotes the kth diagonal element of the noise covariance matrix, and $f_k(\tilde{\mathbf{z}})$ the kth output of the GP at $\tilde{\mathbf{z}}$.

According to Eq. (22), we obtain the posterior for $\eta(\tilde{\mathbf{z}})$ through the joint posterior $p\left(\mathbf{f}(\tilde{\mathbf{z}}), \boldsymbol{\Sigma}, \boldsymbol{\Theta}, \mathbf{C} \mid \boldsymbol{\ell}''\right)$, where $\boldsymbol{\ell}''$ is the $n \times K$ matrix of observed (transformed) log scores for all experts. This posterior is expensive to sample from. However, the cube root transformation makes it possible to extend the results from the univariate case to analytically integrate out $\mathbf{F}$ in the training data, and obtain the marginal posterior of $\boldsymbol{\Sigma}$ and the GP hyperparameters $\boldsymbol{\Theta}$ and $\mathbf{C}$ in closed form using a multivariate version of Eq. 2.8 in Rasmussen and Williams (2006)

$$p\left(\boldsymbol{\Sigma}, \boldsymbol{\Theta}, \mathbf{C} \mid \boldsymbol{\ell}''\right) = \int p\left(\mathbf{F}, \boldsymbol{\Sigma}, \boldsymbol{\Theta}, \mathbf{C} \mid \boldsymbol{\ell}''\right) d\mathbf{F} . \tag{23}$$

This lower-dimensional marginal posterior can be sampled with HMC using Stan in a fraction of the time that it takes to sample the full joint posterior.

For each HMC sampled $(\boldsymbol{\Theta}, \mathbf{C}, \boldsymbol{\Sigma})$ from (23), we can now easily sample from the distribution of $\mathbf{f}(\tilde{\mathbf{z}})$ using (Rasmussen & Williams, 2006)

$$\mathbf{f}(\tilde{\mathbf{z}}) \mid \boldsymbol{\ell}'', \mathbf{Z}, \tilde{\mathbf{z}}, \boldsymbol{\Sigma}, \boldsymbol{\Theta}, \mathbf{C} \sim \mathcal{N}\left[\bar{\tilde{\mathbf{f}}}, \operatorname{var}(\tilde{\mathbf{f}})\right], \tag{24}$$

where $\mathbf{Z}$ is the matrix of pooling variables in the training data, and

$$\begin{aligned} \bar{\tilde{\mathbf{f}}} &= \boldsymbol{\mu}(\tilde{\mathbf{z}}) + \mathrm{G}(\tilde{\mathbf{z}}, \mathbf{Z})\,[\mathrm{G}(\mathbf{Z}, \mathbf{Z}) + \boldsymbol{\Sigma} \otimes \mathbf{I}_n]^{-1}\,(\operatorname{vec}\boldsymbol{\ell}'' - \boldsymbol{\mu}), \\ \operatorname{var}(\tilde{\mathbf{f}}) &= \mathrm{G}(\mathbf{z}, \tilde{\mathbf{z}}) - \mathrm{G}(\tilde{\mathbf{z}}, \mathbf{Z})\,[\mathrm{G}(\mathbf{Z}, \mathbf{Z}) + \boldsymbol{\Sigma} \otimes \mathbf{I}_n]^{-1}\,\mathrm{G}(\mathbf{Z}, \tilde{\mathbf{z}}), \end{aligned}$$

where, for example, $\mathrm{G}(\mathbf{Z}, \mathbf{Z})$ is the $Kn \times Kn$ matrix with covariances of the form $\operatorname{cov}(f_k(\mathbf{z}_i), f_l(\mathbf{z}_j))$ for experts k and l and observations i and j in the training data. The posterior draws of $\boldsymbol{\Sigma}$ from (23) and $\mathbf{f}(\tilde{\mathbf{z}})$ from (24) can finally be inserted into (22) to obtain draws from the posterior of the local ELPD $\eta(\tilde{\mathbf{z}})$. The method is summarized in Algorithm 2.

Algorithm 2: Posterior sampling for ELPD at $\tilde{\mathbf{z}}$ in the multi- GP (1/3) model

---

Input: Number of posterior draws M, a vector of pooling variables $\tilde{\mathbf{z}}$ and $\tilde{a}_k = -(1/2)\log(2\pi\tilde{\sigma}_k^2)$ for expert k.

Generate M draws $\{\boldsymbol{\Sigma}^{(j)}, \boldsymbol{\Theta}^{(j)}, \mathbf{C}^{(j)}\}_{j=1}^{M}$ from $p\left(\boldsymbol{\Sigma}, \boldsymbol{\Theta}, \mathbf{C} \mid \boldsymbol{\ell}''\right)$ in (23) using HMC.

for $j = 1, \ldots, M$ **do**

  Sample the posterior $\mathbf{f}^{(j)}(\tilde{\mathbf{z}}) \mid \boldsymbol{\ell}'', \mathbf{Z}, \tilde{\mathbf{z}}, \boldsymbol{\Sigma}^{(j)}, \boldsymbol{\Theta}^{(j)}, \mathbf{C}^{(j)}$ in (24)

  Compute joint local predictive ability $\eta^{(j)}(\tilde{\mathbf{z}})$ from (22)

end

Output: Sample from the joint local predictive ability posterior:

$\eta^{(1)}(\tilde{\mathbf{z}}), \eta^{(2)}(\tilde{\mathbf{z}}), \ldots, \eta^{(M)}(\tilde{\mathbf{z}})$.

---

3 Predicting Bike-Sharing Utilization Rates in Washington D.C.

The last decade has seen an explosion in companies offering short-term rentals of final mile transportation, such as bikes or electric scooters. Key for the success of such a business is to be able to correctly predict demand. In this section, we use single- and multi-output GP models to estimate the joint one-step-ahead local predictive ability of three experts predicting bike rentals based on the data in Fanaee-T and Gama (2014). We then propose novel methods for how to use the estimates of local predictive ability to form aggregate predictions and evaluate the single- and multi-output versions with regard to out-of-sample performance.

3.1 Data and Models

The dataset in Fanaee-T and Gama (2014) includes the daily number of bike rentals over the two-year period from January 1, 2011, to December 31, 2012. The dataset further contains several covariates relating to the weather, as well as indicators for official US holidays and an indicator for if the day in question is a workday. Figure 4 illustrates the full dataset together with the last two months. Throughout this application, the last 330 observations (February 6th to December 31st) are used for evaluation, leaving us a training set of 400 observations after losing one observation due to the inclusion of lagged covariates in some of the models.

We use the same three experts as in Oelrich et al. (2024) to generate predictions: a Bayesian regression model (BREG), a Bayesian additive regression tree model (BART), and a Bayesian linear regression model with stochastic volatility (SVBREG). These experts use several weather-related variables, an indicator for season, the number of rentals the previous day, and indicators for workday and holiday as covariates. For further discussion of the data, see Oelrich et al. (2024).

As pooling variables, the decision-maker uses humidity, wind speed, and temperature from the dataset in Fanaee-T and Gama (2014). The decision-maker also decides to use a variable she calls *family holiday* which is not included in the dataset, or in the estimation of any of the experts, which takes the value 1 at Thanksgiving and Christmas (Eve and Day) (Oelrich et al., 2024). This variable is included to represent the decision-makers belief that there are certain stay-at-home holidays where the demand for rental bikes is almost nonexistent. As neither of the experts have access to this variable, she expects their predictive ability to vary significantly with regard to this variable.

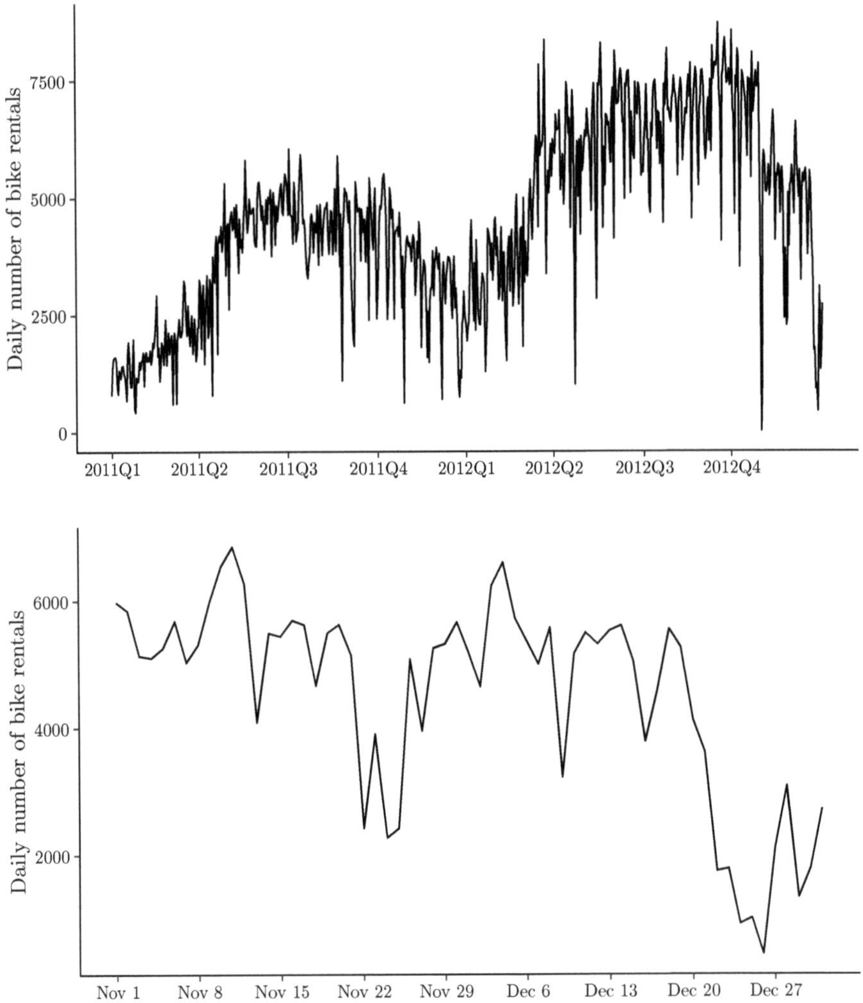

Fig. 4 Bike-sharing data. Full time period (above) and the last two months (below)

3.2 Estimating Predictive Ability

To estimate local predictive ability, we use the GP (1/3) method, as the number of data points makes the GP(χ_1^2) method inconveniently slow. More specifically, we use a GP with a squared exponential kernel with automatic relevance determination (ARD) so that each pooling variable is associated with its own length-scale hyperparameter (Rasmussen & Williams, 2006). Following the Gaussian process regression example in the Stan User's Guide, we give the length scales independent

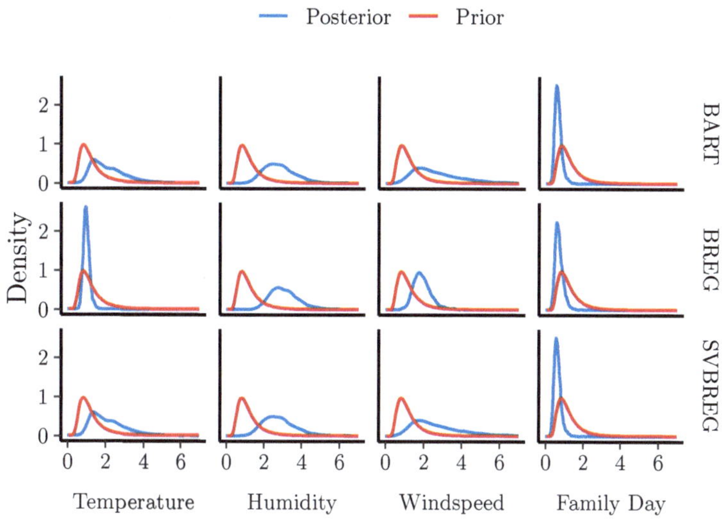

Fig. 5 Marginal posterior and prior distributions of the length scales for all three experts at the final time point, December 31, 2012

Inverse-Gamma(5, 5) priors, and the noise and signal variance are given $\mathcal{N}^+(0, 1)$ priors (Stan Development Team, 2024).

Figure 5 shows the prior and posterior distribution of the length scales for all three experts at the final time point. The length scale tells us how quickly the correlation between two points decreases, so smaller length scales are indicative of greater locality, in the sense of local predictive ability varying more rapidly over the pooling space. *Temperature* and in particular *Family Day* show the strongest evidence of locality with a posterior that is heavily concentrated on smaller length scales. There is also some variation between the experts, with the Bayesian regression model showing more evidence of local behavior.

To illustrate how local predictive ability changes over the pooling space, Fig. 6 shows the actual posterior of local predictive ability for Christmas Day 2012 together with what the posterior would have looked like if we keep all pooling variables constant except the value of *family day*. The BART model would, if this had not been a family day, been expected to perform roughly as well as the Bayesian regression model, with the stochastic volatility expected to have the poorest performance. However, for *family day* $= 1$, which is what was observed, the model expects the stochastic volatility model to perform well and the BART to have the worst local predictive ability by a wide margin.

Figure 7 shows a 95% highest posterior density (HPD) interval for the predictive distribution of a subset of the log scores. For all experts, the majority of observed log scores land within the HPD. However, around twice as many as would be expected fall outside, with BART doing the worst and SVBREG the best. This may indicate that the decision-maker could be missing an important pooling variable.

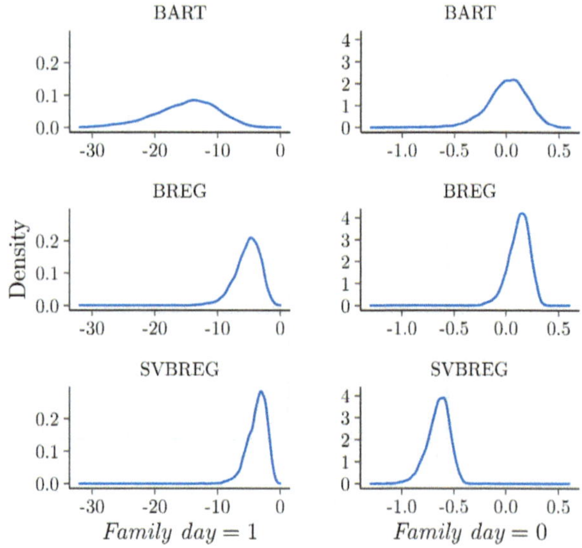

Fig. 6 The left side shows the posterior distribution of local predictive ability for Christmas Day 2012. The right side shows the same distribution but with *family day* set to 0

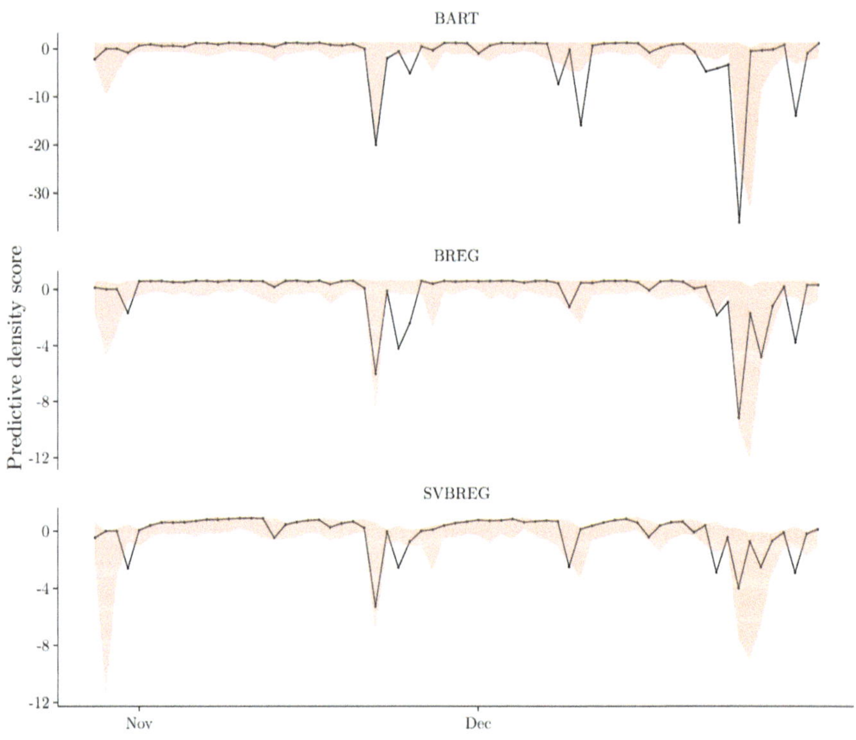

Fig. 7 Estimates of local predictive ability for the end of 2012, bike-sharing data. The black line shows the observed log scores, and the red band shows a 95% HPD interval for the posterior predictive distribution of log scores

Table 1 Mean and median log predictive score of one-step-ahead predictions of log scores, from February 6, 2012, until December 31, 2012. See (25) for a description of the models

Expert	Method	Median	Mean
BART	ℓ' random walk	−1.80	−3.14
	ℓ' cumulative mean	−1.83	−2.62
	ℓ'' random walk	−0.20	−0.58
	ℓ'' cumulative mean	−0.07	**−0.42**
	GP(1/3)	**0.02**	−2.10
BREG	ℓ' random walk	−1.08	−1.80
	ℓ' cumulative mean	−1.11	−1.54
	ℓ'' random walk	0.23	−0.03
	ℓ'' cumulative mean	0.44	**0.13**
	GP(1/3)	**0.58**	−3.67
SVBREG	ℓ' random walk	−1.38	−1.81
	ℓ' global mean	−1.40	−1.61
	ℓ'' random walk	0.19	−0.14
	ℓ'' global mean	0.58	**0.20**
	GP(1/3)	**0.68**	−2.27

Table 1 evaluates the quality of the one-step-ahead predictions of local predictive ability from the GP(1/3) model and compares it with four benchmark models:

$$\begin{aligned}
\ell' \text{ random walk}: &\quad \ell'_{ik} \mid \ell'_{i-1,k} \sim \mathcal{N}(\ell'_{i-1,k}, \sigma^2)\\
\ell' \text{ global mean}: &\quad \ell'_{ik} \sim \mathcal{N}(\mu, \sigma^2)\\
\ell'' \text{ random walk}: &\quad \ell''_{ik} \mid \ell''_{i-1,k} \sim \mathcal{N}(\ell''_{i-1,k}, \sigma^2)\\
\ell'' \text{ global mean}: &\quad \ell''_{ik} \sim \mathcal{N}(\mu, \sigma^2),
\end{aligned} \tag{25}$$

where all models are finally transformed to predict the original log scores ℓ_{ik}. The two global mean methods use as one-step-ahead prediction for time point $t + 1$ the mean of the first t log scores, while the random walk methods both use the last observed log score. Table 1 shows that the two reference methods based on the linearly transformed log scores ℓ' perform badly across the board, giving additional support to the cube root transformation. The global model based on power-transformed log scores is the best performer in terms of the mean, but GP(1/3) outperforms it when it comes to the median. This discrepancy between mean and median is due to a few extreme outliers, the worst being a log score of roughly −700.

Figure 8 shows how well the power-transformed benchmark models and the GP(1/3) model predict the observed log scores for each expert. The figure shows (kernel density estimates of) the log density of all observed log scores, $\log p(\ell_{i+1} \mid \ell_{1:i})$, in the test data, with respect to each local ELPD model. The median of each distribution is marked out with a red vertical line. We see that the

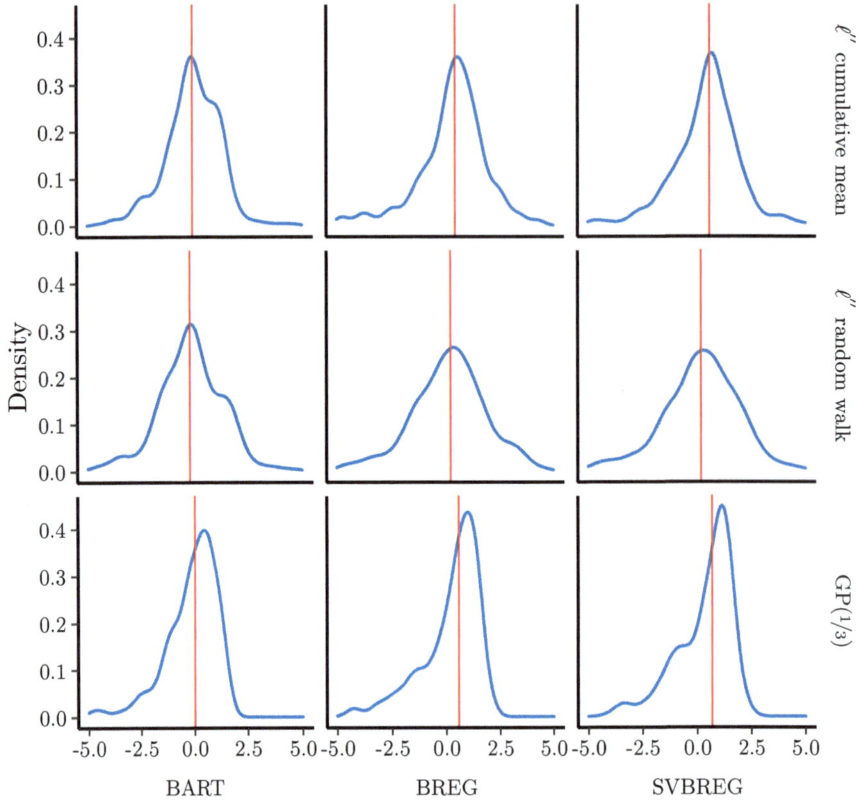

Fig. 8 Kernel density estimates of one-step-ahead log density evaluations of the predicted log scores, $\log p(\ell_{i+1} \mid \ell_{1:i})$, in the test data for the bike rental data. The rows in the figure correspond to predictions from the GP(1/3) model and two global reference methods, respectively. The median of each distribution is marked out with a red vertical line

GP(1/3) model performs better on a majority of the observations, but has a longer left tail caused by a very low predictive density for some observed log scores.

To estimate the joint predictive ability of all three experts, we use the multi-GP(1/3) model, also with a squared exponential automatic relevance determination (ARD) kernel (Rasmussen & Williams, 2006). To allow for the possibility that some pooling variables are unimportant to some experts, we use a fairly flat prior for the length scales, Cauchy$(0, 5)$, restricted to the interval $(0, 100)$. For the $\mathbf{C}$ matrix, we use a LKJ$(\eta = 3)$ prior (Lewandowski et al., 2009), and zero-truncated $\mathcal{N}^+(0, 1)$ priors for the signal and error term variances.

Figure 9 displays the bivariate joint predictive distribution of each expert pair at the last point in the dataset, with the first column showing the joint predictive distributions from single-output GPs—implicitly assuming independence—and the second column showing the joint distributions based on the multi-output model. The multi-output model captures positive correlations between all experts, especially

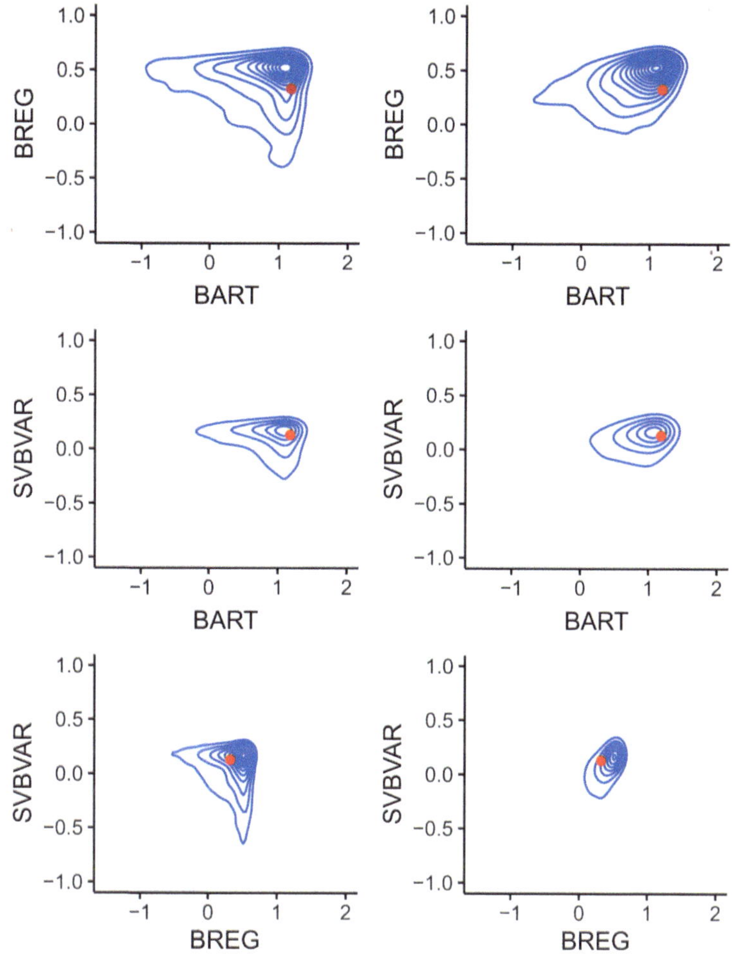

Fig. 9 Joint predictive distributions of log scores for the last data point in the bike-sharing data (December 31, 2012). The graphs show kernel density estimates for univariate (left side) and multivariate (right side) Gaussian processes. The observed log scores are marked with a red dot

between the Bayesian regression model and the stochastic volatility model. The smaller correlations with BART are not surprising, since the BART model is a quite different model with possibly highly nonlinear mean function. Note also that the predictive variances for the multi-output distributions are 10–20% lower at this time point.

3.3 Forecast Combination Based on Estimates Predictive Ability

There are many reasons to model joint predictive ability. For example, it can be done purely for inference purposes, such as when we are interested in examining how the quality of predictions for different experts covaries over a pooling space, or as a part of model evaluation. In this section, we focus on how to create a linear prediction pool (Geweke & Amisano, 2011) based on these estimates.

A linear prediction pool can be made local by letting its weights depend on $\mathbf{z}$, so that the predictive density at a new point $\tilde{\mathbf{z}}$ in the pooling space is given by

$$p\left(\tilde{y} \mid \tilde{\mathbf{z}}\right) = \sum_{k=1}^{K} w_k(\tilde{\mathbf{z}}) p_k(\tilde{y} \mid \tilde{\mathbf{z}}), \quad \sum_{k=1}^{K} w_k(\tilde{\mathbf{z}}) = 1, \quad w_k(\tilde{\mathbf{z}}) \geq 0, \tag{26}$$

where $w_k(\tilde{\mathbf{z}})$ is the weight of expert k at $\tilde{\mathbf{z}}$. To bring the predictive ability of the expert into the pooling, we let the weight $w_k(\tilde{\mathbf{z}})$ be a function of the posterior probability that expert k has the highest predictive ability at $\tilde{\mathbf{z}}$. We denote this probability by $\psi_k(\tilde{\mathbf{z}}) = p\big(\eta_k(\tilde{\mathbf{z}}) > \eta_{j \neq k}(\tilde{\mathbf{z}})\big)$ and refer to forecast combinations based directly on $\psi_k(\tilde{\mathbf{z}})$ as GP(1/3) *natural* in tables and figures.

Often, and especially when the amount of data is small, these weights will tend to not discriminate strongly between experts and be outperformed by more aggressive weighing schemes such as the optimal linear pools of Geweke and Amisano (2011). We therefore propose scaling $p\big(\eta_k(\tilde{\mathbf{z}}) > \eta_{j \neq k}(\tilde{\mathbf{z}})\big)$ by a constant c and then feeding it through a softmax transformation

$$w_k(\tilde{\mathbf{z}}) = \frac{\exp\left(c \cdot \psi_k(\tilde{\mathbf{z}})\right)}{\sum_{j=1}^{K} \exp\left(c \cdot \psi_j(\tilde{\mathbf{z}})\right)}. \tag{27}$$

The constant c, which we will refer to as the discrimination factor, allows the decision-maker to tune how aggressively to discriminate between the experts. Setting $c = 0$ will result in equal weights, and setting $c = \infty$ will assign weight one to the expert with the highest probability of having greater predictive ability than all other experts. Typically, a value of c between these extremes will yield the best results.

When the decision-maker has no strong prior opinion of c, it can be dynamically selected at each time point by selecting the value that optimizes the historical predictive ability of the pool. We refer to this method as GP(1/3) *dynamic* in tables and figures. As a special case of interest, we will include $c = \infty$ under the name GP(1/3) *model selection*, as it assigns weight 0 to all but one of the experts.

Figure 10 shows how the optimal discrimination changes over time for the bike-sharing data for the single-output method. At each time point, the historical performance of the pool for $0 \leq c \leq 20$ is evaluated, and the next prediction is made with the value of c corresponding to the historically most accurate pool. The

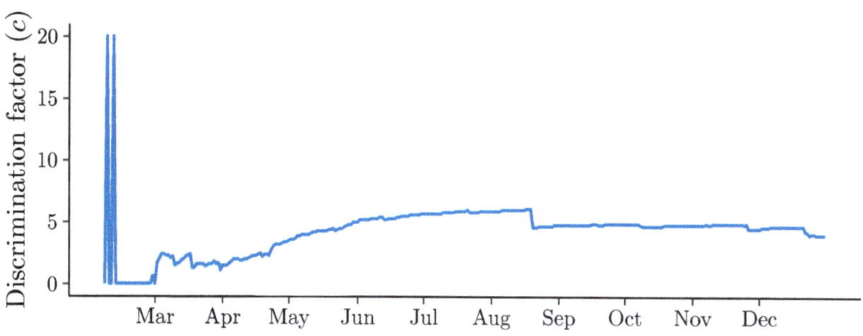

Fig. 10 Change in optimal discrimination factor over time, single output. February 6 to December 31, 2012

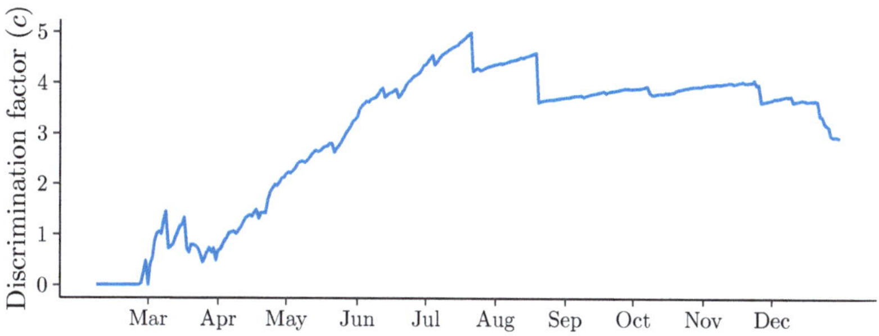

Fig. 11 Change in optimal discrimination factor over time, multi-output. February 6 to December 31, 2012

optimal discrimination c is volatile when little historical data is available, but then settles down around $c = 5$ as a compromise between the extremes of equal weights and assigning all weight to the expert with highest probability of having the best predictive ability at $\tilde{\mathbf{z}}$.

The optimal discrimination factor for the multi-output model, illustrated in Fig. 11, settles around a lower value than in the single-output version. This is due to the single-output methods failure to capture the strong positive correlation between the predictive ability of the experts, which naturally leads to a lower degree of discrimination. This is discussed further in Sect. 3.5.

All pooling methods described above will work regardless of whether the estimates of local predictive ability have been obtained with multiple single-output GPs or a single multi-output GP, but taking account of the dependence will change the weights of the experts, sometimes drastically. This can be seen in Figs. 12, 13 and 14, which compare the single- and multi-output approaches for the three aggregation methods discussed above. The multi-output model clearly outperforms its single-output counterparts when using the natural weighting scheme. When using model selection or dynamic discrimination, however, the simpler single-output GP method performs slightly better, in both cases almost entirely due to a few observations at the end of the dataset.

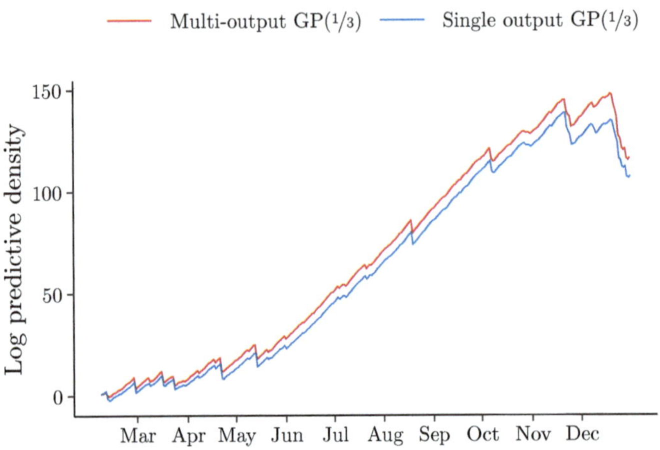

Fig. 12 Cumulative log predictive scores based on the natural aggregation scheme for single- and multi-output models. February 6 to December 31, 2012

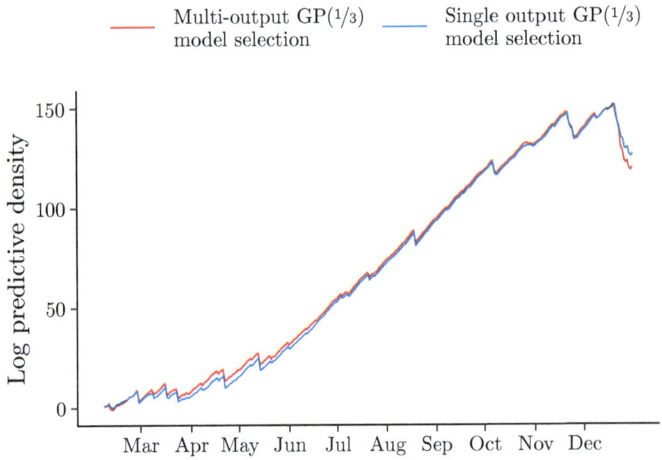

Fig. 13 Cumulative log predictive scores based on the model selection aggregation scheme for single- and multi-output models. February 6 to December 31, 2012

3.4 Comparison with Reference Methods

The sum of the log predictive scores over time for each method, together with the performance of the individual experts as well as a selection of reference methods, is shown in Table 2. The reference methods included are: the caliper method used in Oelrich et al. (2024), which is a nonparametric method that averages log scores of nearby observations in **z**-space; the global optimal pool of Geweke and Amisano (2011) which selects aggregation weights that optimize the historical performance

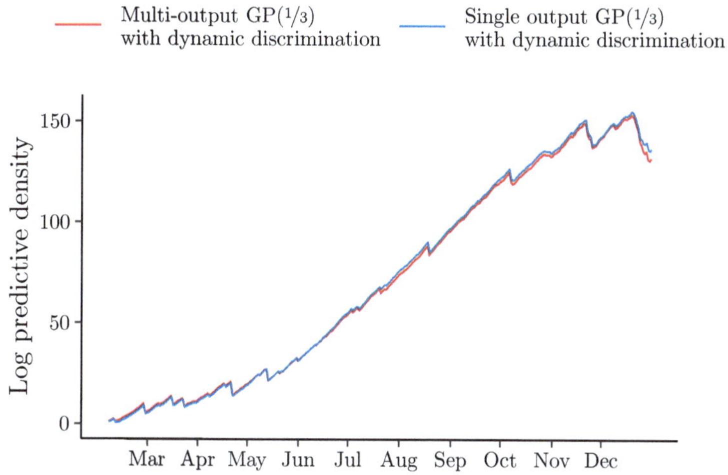

Fig. 14 Cumulative log predictive scores based on the dynamic aggregation scheme for single- and multi-output models. February 6 to December 31, 2012

Table 2 Sum of log predictive density scores over the whole test period, from February 6, 2012, until December 31, 2012

Method	Sum of log scores
SVBREG	45.1
BREG	28.8
BART	43.6
Equal weights	98.7
Caliper	133.7
Global opt.	130.2
Local opt.	145.3
GP(1/3) natural	107.1
Multi- GP(1/3) natural	115.9
GP(1/3) model sel.	126.6
Multi- GP(1/3) model sel.	119.7
GP(1/3) dynamic	136.4
Multi- GP(1/3) dynamic	131.7

of the pool in terms of log scores; the local optimal pool in Oelrich et al. (2024), which is an extension of the optimal pool of Geweke and Amisano (2011) where the optimal weights are determined on a subset of the data in the pooling space closest to the current point. For details about these reference methods, see Oelrich et al. (2024).

While the natural multi-output GP outperforms its single-output counterpart considerably, this is reversed in the dynamic version. The single-output dynamic GP(1/3) performs the best out of the methods based on local predictive ability, but is soundly beaten by the local optimal pool of Oelrich et al. (2024).

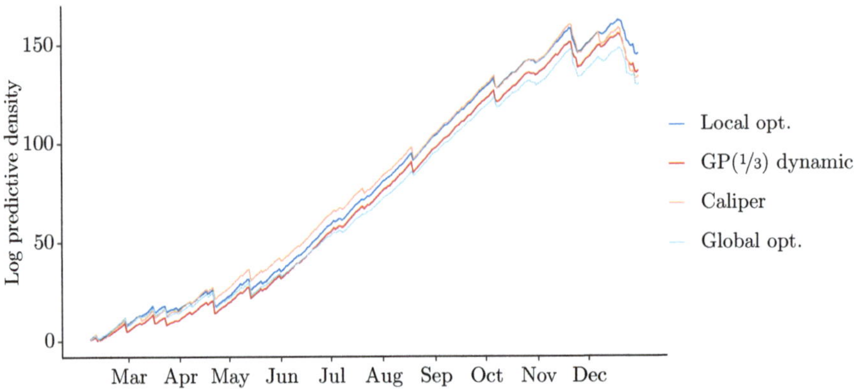

Fig. 15 Cumulative log predictive density for the four best methods. The legend is sorted in descending order of total log predictive score. February 6 to December 31, 2012

Figure 15 compares the best performing of the GP(1/3) pooling schemes with the reference methods. The GP(1/3) initially struggles when there is little training data, but then quickly improves over time, and at the end of the sample, it is only outperformed by the local optimal pool.

3.5 Correlated Experts and Degree of Discrimination

The multi-output model owes its superior performance when using natural weighting to a larger degree of discrimination, as the weak discrimination of the single-output method is in large part caused by a failure to capture correlation between the experts. A positive correlation that is unaccounted for will lead to weaker discrimination, and failure to take into account a negative correlation will lead to discrimination that is too strong. Typically, experts will be positively correlated, and the degree of discrimination will be underestimated. This effect is illustrated in Fig. 16.

Figure 16 displays contour plots of the posteriors for η_1 and η_2 for two experts in two setups. In the top row, the difference in posterior mean of predictive ability between the two experts is small ($\text{E}(\eta_2) - \text{E}(\eta_1) = 0.5$), and in the bottom row, the difference is more substantial ($\text{E}(\eta_2) - \text{E}(\eta_1) = 2$). In both setups, we fix the marginal posterior variances of the two experts at 1 and explore what happens when we let the correlation go from strong and negative ($\rho = -0.9$) to nonexistent ($\rho = 0$) to strong and positive ($\rho = 0.9$). As the correlation between the experts approaches one, the posterior probability of having the highest predictive ability will concentrate on the better model, no matter how small the difference in posterior means. Compared to uncorrelated experts, negative correlation increases the posterior probability that the expert with lower posterior mean has the higher

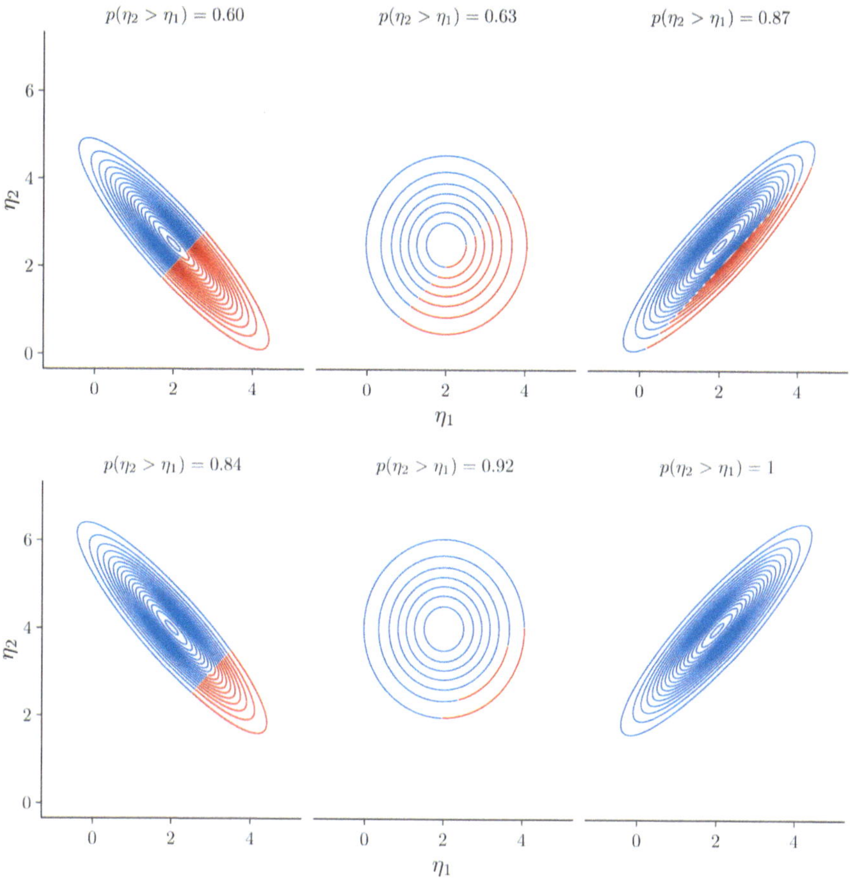

Fig. 16 Gaussian contour plot illustrating the effect of correlation in predictive ability on the posterior probability that Expert 2 has greatest predictive ability at $\tilde{\mathbf{z}}$. In the top row, the marginal difference in posterior mean of predictive ability between the two experts is small ($\mathrm{E}(\eta_2(\tilde{\mathbf{z}})) - \mathrm{E}(\eta_1(\tilde{\mathbf{z}})) = 0.5$), and in the bottom row the difference is substantial ($\mathrm{E}(\eta_2(\tilde{\mathbf{z}})) - \mathrm{E}(\eta_1(\tilde{\mathbf{z}})) = 2$). The blue part of each contour plot marks the parts of the joint density where $\eta_2(\tilde{\mathbf{z}}) > \eta_1(\tilde{\mathbf{z}})$

predictive ability, something which becomes clearer as the magnitude of the difference in mean increases.

The effect illustrated in Fig. 16 explains why the multi-output GP outperforms the single-output version when using natural weights. By capturing the correlation between experts, the multi-output GP leads to natural weights that discriminate more strongly between the experts.

4 Conclusions and Further Research

We have introduced a method for estimating the predictive ability $\eta(\mathbf{z})$ of a model or expert, in terms of ELPD, as it varies over a space of pooling variables $\mathbf{z}$. We call this conditional predictive ability *local predictive ability* and propose using a Gaussian process to model it.

Gaussian processes have the drawback of being computationally heavy, and so we suggest the use of a power transformation of log scores which allows us to use a much faster marginalized GP to generate draws from the posterior of $\eta(\mathbf{z})$. Our proposed method is shown to work well for predicting bike-sharing data (Fanaee-T & Gama, 2014), where it outperforms all compared methods except the local version of the linear prediction pool of Geweke and Amisano (2011) proposed in Oelrich et al. (2024).

Our method can be used both to model the marginal distribution of the predictive ability $\eta_k(\mathbf{z})$ for each expert and to model the predictive ability of the experts jointly using a multi-output GP that makes it possible to capture correlation between the predictive abilities of the experts as well as correlation in the noise.

While the inference and prediction from the marginalized version are much faster compared to sampling also the latent variables in the GP(1/3) model, the step from running several single-output GPs side by side to running a single multi-output GP increases the computational burden significantly, making large sample sizes time-consuming. A topic for further study is therefore methods for speeding up the calculation using sparse linear algebra, methods for large-scale homoscedastic Gaussian GPs (Liu et al., 2020), or using alternative Bayesian inference approaches than Monte Carlo sampling, such as variational inference based on optimization (Blei et al., 2017).

Another potential avenue of future research is using alternative methods to pool predictions based on local predictive ability. In the empirical application, we use three different methods, all of which have pooling weights based on ψ_k, the posterior probability that a particular expert has greater predictive ability than all other experts. Alternatively, one can form linear combinations directly on the $\eta_1(\mathbf{z}), \ldots, \eta_K(\mathbf{z})$ to obtain a posterior for the pooled local ability $\sum_{k=1}^{K} w_k \eta_k(\mathbf{z})$. Given a set of criteria for this distribution, such as maximizing the mean while keeping the variance below a certain threshold, a set of local weights can be obtained through numerical optimization.

Allowing for time-varying weights has been shown to be beneficial in macroeconomic forecasting (Del Negro et al., 2014; Billio et al., 2013; McAlinn et al., 2020). While we have not explicitly discussed modeling the persistence of local predictive ability over time, it is straightforward to include some function of time in our vector of pooling variables via an appropriate kernel function.

Acknowledgments The computations in this chapter were enabled by resources provided by the National Supercomputer Centre (NSC), funded by Linköping University, and by resources in project SNIC 2020/15-262 provided by the Swedish National Infrastructure for Computing (SNIC) at UPPMAX, partially funded by the Swedish Research Council through grant agreement no. 2018-05973.

Appendix

In this appendix, we investigate the properties of our methods on simulated data. We are particularly interested in investigating how the computationally faster GP(1/3) model compares to the computationally costly GP(χ_1^2) model.

We generate data from the model

$$y_i = x_{1i} + x_{2i} + \epsilon_i, \qquad \epsilon_i \overset{\text{iid}}{\sim} \mathcal{N}(0, 1), \tag{28}$$

where the covariates x_1 and x_2 are generated as i.i.d. standard normal variates. As our expert, we use a Bayesian regression model that is misspecified in that one of the covariates is missing, i.e.,

$$y_i = \alpha + \beta x_{1i} + \varepsilon_i, \text{ where } \varepsilon_i \overset{\text{iid}}{\sim} \mathcal{N}(0, \sigma_\varepsilon) . \tag{29}$$

To make the expert model as simple as possible for this illustration, we assume that it has been trained with so much data that the posterior of all coefficients will have variances close to zero and posterior means equal to 0 for α, 1 for β, and 2 for σ_i^2 (since ε_i absorbs both ϵ_i in the DGP and the variation from the missing x_{2i}).

We further assume that our decision-maker believes that the predictive ability of the expert varies over the variable x_2, but does not know the exact relationship between the ELDP η and x_2. We will examine how well the decision-maker will be able to estimate the true predictive ability, as summarized by the posterior distribution of η, using the two models described in Sect. 2.

The true distribution of the log scores in this setup is given by

$$\ell_i \overset{d}{=} -\frac{1}{2}\log 2\pi\sigma_i^2 - \frac{1}{2\sigma_i^2}(y_i - \mu_i)^2,$$

with each $y_i \mid x_{1i}, x_{2i} \overset{\text{indep.}}{\sim} \mathcal{N}(x_{1i} + x_{2i}, 1)$. For our expert $\mu_i = x_{1i}$ and $\sigma_i^2 = 2$, so this simplifies to

$$\ell_i \overset{d}{=} -\frac{1}{2}\log 4\pi - \frac{1}{4}(x_{2i} + \epsilon_i)^2 .$$

As $x_{2i} + \epsilon \sim \mathcal{N}(x_{2i}, 1)$, it follows that $(x_{2i} + \epsilon)^2 \sim \chi_1^2(\lambda_i = x_{2i}^2)$, and the true ELPD over x_2 is given by

$$\eta(x_2) = \mathrm{e}(\ell \mid x_2) = -\frac{1}{2}\log 4\pi - \frac{1}{4}(1 + x_2^2) .$$

Note that when $|x_2|$ is small, the distribution of ℓ is close to the central χ_1^2 distribution with a heavy tail to the right generating extreme values. As $|x_2|$ grows, the distribution approaches the normal distribution, but the variance grows large. This shows that modeling log scores is a hard problem.

To compare the GP $\left(\chi_1^2\right)$ and GP ($1/3$) models, we simulate 1000 datasets of 150 observations each from the DGP in Eq. (28). For each simulation, we then generate draws from the posterior distribution of $\eta(x_2)$ over a grid of x_2 values between -3 and 3 and compare with the true ELPD as a function of x_2. Both GPs use a squared exponential kernel (Rasmussen & Williams, 2006). We use $l \sim$ Inverse-Gamma(5, 5) priors for the length scales, and a $\alpha \sim \mathcal{N}^+(0, 1)$ prior for the signal variances. The noise variance σ_n^2 of the GP($1/3$) is given a $\mathcal{N}^+(0, 1)$ prior as well.

Figure 17 shows the posterior mean of $\eta(x_2)$ together with a 95% credible interval for three different datasets. The datasets have been selected by ranking the mean integrated squared error (MISE) of the ELPD estimates of each run—the MSE integrated over the distribution of x_2—and then selecting the 2.5:th, 50:th, and 97.5:th percentile. Both methods perform well for $x_2 \in (-2, 2)$, but poorly in the tails. This is to be expected, as two factors conspire to make estimation in the tails difficult: The DGP variance of the log scores increases with the square of x_2, and since $x_2 \sim \mathcal{N}(0, 1)$ only roughly 5% of the observations are in the tails.

Figure 18 shows the difference in log MISE of the two models and that the GP ($1/3$) model actually works marginally better in terms of squared error. However, the MISE metric only takes the posterior mean into account, so we also look at the mean integrated log score, which is calculated as the log predictive density of the true ELPD given the posterior distribution of $\eta(x_2)$, integrated over the distribution of x_2. The GP ($1/3$) model works marginally better measured by this metric as well.

Note that the uncertainty around the estimate increases quickly for both models when there are no, or few, previous predictions nearby. This will lead to the desirable behavior that when combining experts in regions where we have little to go on in terms of previous predictions, we will quickly tend toward an equal-weight solution; see Sect. 3.

The GP(χ_1^2) is prohibitively costly from a computational standpoint once the number of observations goes above a few hundred. That the more scalable GP ($1/3$) model works so well compared to the GP(χ_1^2) model is therefore an important finding for practical applications. There is also a large collection of methods for reducing computational cost in GP regression with additive homoscedastic errors (Liu et al., 2020) that can be directly used to speed up inference in the GP ($1/3$) model.

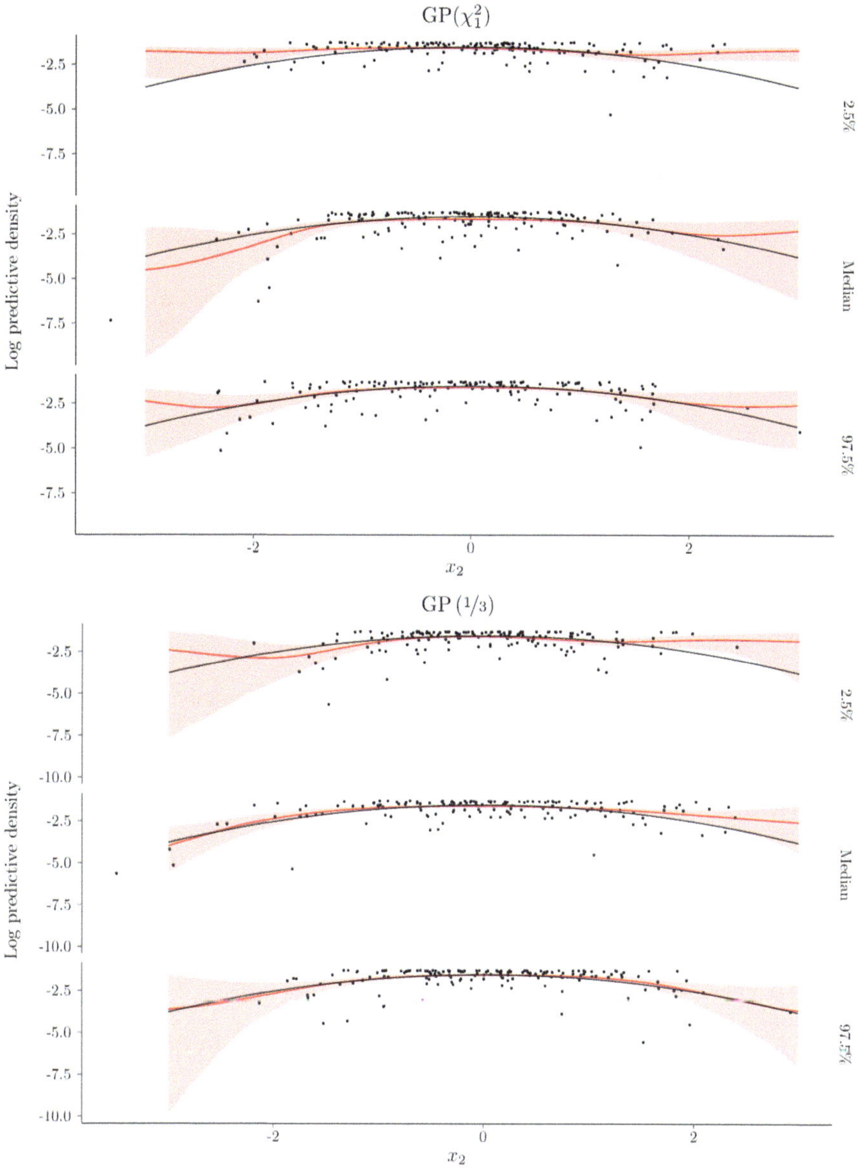

Fig. 17 Posterior distribution of $\eta(x_2)$ over a grid of x_2 values for the GP(χ_1^2) (top) and GP (1/3) (bottom) models. The black line is the true $\eta(x_2)$. The red line is the posterior median, and the red band is the 95% highest posterior density intervals. Median, 2.5%, and 97.5% indicate which sample, out of 1000, was used for each plot where the samples are ranked according to mean integrated squared error (MISE)

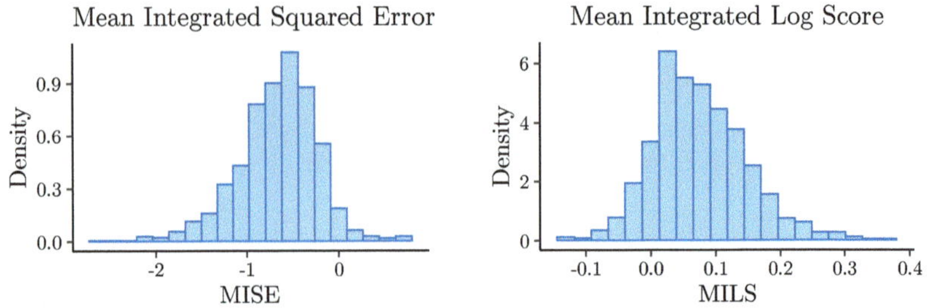

Fig. 18 The left figure shows the difference in log mean integrated squared error (MISE) of the GP (1/3) and GP(χ_1^2) model, taken over the distribution of x_2. The right figure shows the difference in mean integrated log score (MILS) of the GP(1/3) and GP(χ_1^2) model, taken over the distribution of x_2

References

Abdel-Aty, S. H. (1954). Approximate formulae for the percentage points and the probability integral of the non-central chi2 distribution. *Biometrika, 41*(3/4), 538.

AMS Council (2008). Enhancing weather information with probability forecasts. *Bulletin of the American Meteorological Society, 89*, 1049–1053.

Bates, J. M. & Granger, C. W. J. (1969). The combination of forecasts. *Journal of the Operations Research Society, 20*(4), 451–468.

Billio, M., Casarin, R., Ravazzolo, F., & van Dijk, H. K. (2013). Time-varying combinations of predictive densities using nonlinear filtering. *Journal of Econometrics, 177*(2), 213–232.

Blei, D. M., Kucukelbir, A., & McAuliffe, J. D. (2017). Variational inference: A review for statisticians. *Journal of the American Statistical Association, 112*(518), 859–877.

Croushore, D. & Stark, T. (2019). Fifty years of the survey of professional forecasters. *Economic Insight, 4*(4), 1–11.

de Vincent-Humphreys, R., Dimitrova, I., Falck, E., Henkel, L., Meyler, A., et al. (2019). Twenty years of the ECB survey of professional forecasters. *Economic Bulletin Articles, 1*, 34–58.

Del Negro, M., Hasegawa, R., & Schorfheide, F. (2014). Dynamic prediction pools: An investigation of financial frictions and forecasting performance. *Journal of Econometrics, 192*(2), 391–405.

Draper, D. (1995). Assessment and propagation of model uncertainty. *Journal of the Royal Statistical Society. Series B (Methodological), 57*(1), 45–97.

Fanaee-T, H. & Gama, J. (2014). Event labeling combining ensemble detectors and background knowledge. *Progress in Artificial Intelligence, 2*(2–3), 113–127.

Gelman, A., Hwang, J. & Vehtari, A. (2014). Understanding predictive information criteria for Bayesian models. *Statistics and Computing, 24*(6), 997–1016.

Geweke, J. & Amisano, G. (2011). Optimal prediction pools. *Journal of Econometrics, 164*(1), 130–141.

Gneiting, T. & Raftery, A. E. (2007). Strictly proper scoring rules, prediction, and estimation. *Journal of the American Statistical Association, 102*(477), 359–378.

Haldane, J. B. S. (1942). Moments of the distributions of powers and products of normal variates. *Biometrika, 32*(3/4), 226.

Hall, S. & Mitchell, J. (2007). Combining density forecasts. *International Journal of Forecasting, 23*(1), 1–13.

Hoeting, J. A., Madigan, D., Raftery, A. E. & Volinsky, C. T. (1999). Bayesian model averaging: A tutorial. *Statistical Science, 14*(4), 382–417.

Lewandowski, D., Kurowicka, D. & Joe, H. (2009). Generating random correlation matrices based on vines and extended onion method. *Journal of Multivariate Analysis, 100*(9), 1989–2001.

Li, L., Kang, Y., & Li, F. (2023). Bayesian forecast combination using time-varying features. *International Journal of Forecasting, 39*(3), 1287–1302.

Lindley, D. V., Tversky, A., & Brown, R. V. (1979). On the reconciliation of probability assessments. *Journal of the Royal Statistical Society: Series A (General), 142*(2), 146–180.

Liu, H., Ong, Y.-S., Shen, X., & Cai, J. (2020). When Gaussian process meets big data: A review of scalable GPs. *IEEE Transactions on Neural Networks and Learning Systems, 31*(11), 4405–4423.

McAlinn, K., Aastveit, K. A., Nakajima, J., & West, M. (2020). Multivariate Bayesian predictive synthesis in macroeconomic forecasting. *Journal of the American Statistical Association, 115*(531), 1092–1110.

Neal, R. M. (1996). *Bayesian learning for neural networks*. Number 118 in Lecture Notes in Statistics. Springer.

Oelrich, O., Villani, M., & Ankargren, S. (2024). Local prediction pools. *Journal of Forecasting, 43*(1), 103–117.

Rasmussen, C. E. & Williams, C. K. I. (2006). *Gaussian processes for machine learning*. Adaptive Computation and Machine Learning. MIT Press. OCLC: ocm61285753.

Sivula, T., Magnusson, M., Matamoros, A. A., & Vehtari, A. (2022a). Uncertainty in Bayesian leave-one-out cross-validation based model comparison. *arXiv:2008.10296* [stat].

Sivula, T., Magnusson, M., & Vehtari, A. (2022b). Unbiased estimator for the variance of the leave-one-out cross-validation estimator for a Bayesian normal model with fixed variance. *arXiv:2008.10859* [stat].

Stan Development Team (2024). *Stan modeling language users guide and reference manual, version 2.35*.

Teh, Y. W., Seeger, M., & Jordan, M. (2005). Semiparametric latent factor models. In *Proceedings of the Tenth International Workshop on Artificial Intelligence and Statistics*. Proceedings of Machine Learning Research (Vol. R5, pp. 333–340). PMLR.

Wang, X., Hyndman, R. J., Li, F., & Kang, Y. (2023). Forecast combinations: An over 50-year review. *International Journal of Forecasting, 39*(4), 1518–1547.

Winkler, R. L. (1981). Combining probability distributions from dependent information sources. *Management Science, 27*(4), 479–488.

Yao, Y., Pirs, G., Vehtari, A., & Gelman, A. (2021). Bayesian hierarchical stacking: Some models are (somewhere) useful. *Bayesian Analysis, 4*(17), 1043–1071.

Yao, Y., Vehtari, A., Simpson, D., & Gelman, A. (2018). Using stacking to average Bayesian predictive distributions (with discussion). *Bayesian Analysis, 13*(3), 917–1007.

Spectral Domain Likelihoods for Bayesian Inference in Time-Varying Parameter Models

Oskar Gustafsson, Mattias Villani, and Robert Kohn

Abstract Inference for locally stationary processes is often based on some local Whittle-type approximation of the likelihood function defined in the frequency domain. The main reasons for using such a likelihood approximation are that (i) it has substantially lower computational cost and better scalability to long time series compared to the time domain likelihood, particularly when used for Bayesian inference via Markov chain Monte Carlo (MCMC), (ii) convenience when the model itself is specified in the frequency domain, and (iii) it provides access to bootstrap and subsampling MCMC which exploits the asymptotic independence of Fourier transformed data. Most of the existing literature compares the asymptotic performance of the maximum likelihood estimator (MLE) from such frequency domain likelihood approximation with the exact time domain MLE. Our chapter uses three simulation studies to assess the finite-sample accuracy of several frequency domain likelihood functions when used to approximate the posterior distribution in time-varying parameter models. The methods are illustrated on an application to egg price data.

1 Introduction

Innovations in computing and storage technology allow low-cost data collection in a variety of applications. Data from sensors, cameras, and other recording devices are naturally time-stamped observations, which has motivated methodological developments in modeling, estimation, and prediction for time series. Long time series with many observations are increasingly common, collected either over long study periods or over shorter periods at high frequency.

O. Gustafsson (✉) · M. Villani
Department of Statistics, Stockholm University, Stockholm, Sweden
e-mail: oskar.gustafsson@stat.su.se; mattias.villani@stat.su.se

R. Kohn
School of Economics, University of New South Wales, Sydney, NSW, Australia
e-mail: r.kohn@unsw.edu.au

S. Mazur, P. Österholm (eds.), *Recent Developments in Bayesian Econometrics and Their Applications*, https://doi.org/10.1007/978-3-032-00110-8_8

Inference for a long time series can be computationally challenging, even under stationarity. Evaluating the time domain likelihood for a stationary Gaussian model without special structure has a computational complexity of $O(T^3)$, where T is the length of the time series, due to the inversion of the $T \times T$ covariance matrix. A computationally more attractive alternative is the Whittle likelihood (Whittle, 1953) which approximates the Gaussian likelihood in the frequency domain with a complexity of only $O(T \log T)$. The approximation is motivated by the asymptotic independence of the discrete Fourier transform of the time series, a result that has been generalized to non-Gaussian processes (Hannan, 1973; Peligrad & Wu, 2010), long-memory processes (Fox & Taqqu, 1986), and nonlinear processes (Shao et al., 2007). The Whittle likelihood requires only the spectral density of the model, which makes it particularly useful for models defined directly in the frequency domain, see, e.g., Rosen et al. (2012). The asymptotically independent data in the frequency domain has also been exploited for the bootstrap (Franke & Hardle, 1992) and for speeding up Markov chain Monte Carlo (MCMC) algorithms by subsampling periodogram data (Salomone et al., 2020).

The MLE based on the Whittle approximation is known to be efficient in finite samples for Gaussian processes, except for short time series with high autocorrelation; see Dahlhaus (1988) and Contreras-Cristán et al. (2006) and for certain non-Gaussian processes (Contreras-Cristán et al., 2006). For AR processes, the Whittle MLE reduces to the Yule-Walker estimator (Dzhaparidze & Yaglom, 1983, Example 5.1), which is known to have bias that increases with the autocorrelation of the time series (Tjøstheim & Paulsen, 1983). There are many suggestions on how to reduce the bias and to improve the finite sample efficiency of the Whittle MLE: tapering (Dahlhaus, 1988), prewhitening (Priestly, 1981), boundary correction (Subba Rao & Yang, 2021), and replacing the spectral density in the likelihood approximation with the expected periodogram (Sykulski et al., 2019).

The traditional assumption of a time-invariant stationary process is often unreasonable, particularly for long time series which typically undergo some structural changes. Much effort has therefore gone into developing more flexible models that allow the properties of the process to change over time. Markov switching models are one popular class of models that allows the process to switch between a finite set of stationary regimes or segments, see, e.g., Markov switching models in Hamilton (1989), change-point models in Chib (1998) and Rosen et al. (2012), and dynamic mixture innovations models in Gerlach et al. (2000). Another line of research develops models and theory for locally stationary process where the parameters change more smoothly over time, such that the process is locally stationary in a small interval around each time point, see Dahlhaus (2012) for a review. This framework is extended in a Gaussian setting by Dahlhaus and Polonik (2009) to allow for jumps in the parameters. Perhaps the most widely used models in applications are time-varying autoregressive (tvAR) (Prado & West, 2010) or vector autoregressive (tvVAR) models (Karlsson, 2013) where the parameters follow a stochastic process over time. The parameter process is typically taken to be a random walk or an AR(1) process with Gaussian innovations, but more recent attempts use local-global shrinkage priors (Carvalho et al., 2010) to allow for both smoothly changing parameters, periods with no change and abrupt jumps; see, e.g., Kalli and Griffin (2014), Kowal et al. (2019), Cadonna et al. (2020), and Knaus and Frühwirth-Schnatter (2023).

The Whittle likelihood has been generalized to models with time-varying parameters, often using a variant of the original Whittle likelihood over time segments. The *block Whittle likelihood* (Dahlhaus, 1997) uses time-varying periodograms over sliding and partially overlapping time windows. The block Whittle MLE is asymptotically efficient under some conditions on the parameter evolution (Dahlhaus, 2012). The *generalized Whittle likelihood* (Dahlhaus, 2000) does not use periodograms over segments but is instead based on so-called preperiodogram data (Neumann & Von Sachs, 1997) computed at every time point and frequency as the Fourier transform of a locally estimated autocovariance at lag k from a single pair of observations k steps apart. The preperiodogram is attractive since it gives maximal time and frequency resolution and is used in several studies (e.g., van Delft & Eichler, 2019) and has good asymptotic properties (Dahlhaus, 2009; Dahlhaus & Polonik, 2009) but has interfering cross-terms that can make inference more difficult (Sandsten, 2020). Everitt et al. (2013) used a modified preperiodogram which is smoothed in both time and frequency directions for efficient online learning of time-varying spectral densities. The *dynamic Whittle likelihood* in Tang et al. (2023) is based on the *moving local periodogram* introduced in Häfner and Kirch (2017). The moving local periodogram evaluates a single frequency at each point in time, instead of all Fourier frequencies, and still maintains approximate independence among most observations since neighboring time points use different frequencies. Tang et al. (2023) use their approximate likelihood for Bayesian inference, prove posterior consistency, and obtain contraction rates.

The local Whittle likelihoods for time-varying parameter models compute local periodograms in relatively short moving time windows and are therefore prone to the same small-sample problems as the Whittle likelihood in the stationary case. This chapter explores the finite sample performance of the posterior distributions from local Whittle likelihoods for time-varying parameter models through several simulation experiments. In contrast to most previous studies exploring the frequentist properties of the Whittle MLE, we focus on how well the posterior distribution for the time-varying parameters is approximated when a local Whittle likelihood is used. We compare the block Whittle likelihood in Dahlhaus (1997) and the dynamic Whittle likelihood in Tang et al. (2023). The data generating process in the experiments is taken to be a simple tvAR process with parameter evolutions following a random walk with Gaussian innovations. The tvAR can be estimated without spectral methods at a reasonable computational cost and is therefore used here as it makes it possible to easily compare the results to the gold standard posterior from the time domain likelihood. We use a tvAR process that is restricted to be stable at every time period using the parameterization in Barndorff-Nielsen and Schou (1973) and Monahan (1984), and we sample from the posterior distribution using Gibbs sampling with a particle MCMC (Lindsten et al., 2014) update of the time-varying parameters. We also investigate the potential of three proposed modifications for improving the local Whittle approximations: tapering (Dahlhaus,

1988), prewhitening (Priestly, 1981), and boundary correction (Subba Rao & Yang, 2021). The effect of the segment length and overlap is also explored. The chapter concludes by illustrating the differences in the tvAR posterior between the compared likelihoods and the modifications in an application to a widely used time series on egg prices, with the additional complication of having a heteroscedastic variance.

The Appendix contains additional results.

2 The Whittle Likelihood for Stationary Processes

This section reviews the frequency domain concepts needed for the Whittle likelihood. A simulated time-invariant AR(1) example introduces the basic measures used to assess the performance of the Whittle likelihood as an approximation to the time domain likelihood.

Let $\{X_t\}_{t=1}^{T}$ be a covariance stationary zero-mean time series with covariance function $\gamma_\tau := \mathbb{E}(X_t X_{t-\tau})$ for integer τ. The *spectral density* is the Fourier transform of $\{\gamma_\tau\}_{\tau=-\infty}^{\infty}$ (Lindgren, 2013)

$$f(\omega) := \frac{1}{2\pi} \sum_{\tau=-\infty}^{\infty} \gamma_\tau e^{-\mathrm{i}\omega\tau}, \tag{1}$$

where $\omega \in (-\pi, \pi]$ is the *angular frequency*. As an example, consider the AR(p) process

$$\phi_p(L)(X_t - \mu) = \varepsilon_t, \quad \varepsilon_t \overset{\text{iid}}{\sim} N(0, \sigma_\varepsilon^2), \tag{2}$$

where $\phi_p(L) := 1 - \phi_1 L - \cdots - \phi_p L^p$ is the autoregressive polynomial in the lag operator L, defined by $L^k X_t = X_{t-k}$. The AR(p) process in (2) has spectral density (Brockwell & Davis, 1991)

$$f(\omega) = \frac{\sigma_\varepsilon^2}{2\pi} \left| \phi_p(e^{-\mathrm{i}\omega}) \right|^{-2}. \tag{3}$$

The *discrete Fourier Transform* (DFT) of $\{X_t\}_{t=1}^{T}$ is the complex-valued series

$$J(\omega_k) := \sum_{t=1}^{T} X_t \exp(-\mathrm{i}\omega_k t), \tag{4}$$

for $\omega_k \in \Omega_T$, where Ω_T is the set of Fourier frequencies

$$\Omega_T = \{2\pi k/T, \text{ for } k = -\lceil T/2 \rceil + 1, \ldots, \lfloor T/2 \rfloor\}.$$

The Fast Fourier Transform (FFT) algorithm computes the DFT efficiently (Lindgren, 2013). The *periodogram*

$$I_T(\omega_k) := \frac{1}{2\pi T}|J(\omega_k)|^2 \tag{5}$$

is an asymptotically unbiased estimate of $f(\omega_k)$ for a stationary process with an absolutely summable covariance function (Brillinger, 2001).

The Whittle log-likelihood (Whittle, 1953) is defined as

$$\ell_T^W(\boldsymbol{\theta}) := -\frac{1}{2}\sum_{\omega_k \in \Omega_T}\left(\log f_{\boldsymbol{\theta}}(\omega_k) + \frac{I_T(\omega_k)}{f_{\boldsymbol{\theta}}(\omega_k)}\right), \tag{6}$$

where $\boldsymbol{\theta}$ is the vector of model parameters and $f_{\boldsymbol{\theta}}(\omega)$ is the model's spectral density. The term for $\omega = 0$ is removed since we assume that the process has zero mean; the term for $\omega = \pi$ is also typically removed for convenience since $I_T(\pi)$ follows a different distribution (χ_1^2) than the other periodogram data points.

The motivation for using (6) as an approximate likelihood comes from the following asymptotic result for the periodogram for a stationary Gaussian process (Whittle, 1951)

$$I_T(\omega_k) \overset{\text{ind}}{\sim} \text{Exp}\big(f(\omega_k)\big), \quad \omega_k \in \Omega_n, \tag{7}$$

as $T \to \infty$, with the exponential distribution in the scale parameterization, i.e., parameterized by its mean; the symbol $\overset{\text{ind}}{\sim}$ means independently distributed. Hence, the Whittle likelihood in (6) is obtained when the result in (7) is assumed to hold for a finite sample size, T. Whittle's original result holds under more general conditions; see Hannan (1973) for potentially non-Gaussian linear processes, Peligrad and Wu (2010) for stationary ergodic processes, Shao et al. (2007) for nonlinear processes with moment conditions, and Peligrad and Zhang (2019) for extensions to random fields.

The Whittle likelihood has the following advantages for likelihood and Bayesian inference: (i) Its computational cost is only $O(T \log T)$ compared to the $O(T^3)$ in the Gaussian likelihood without special structure of the covariance matrix, (ii) it can be conveniently used for models directly defined for the spectral density (Wahba, 1980; Rosen et al., 2012), and (iii) the asymptotic independence of the periodogram data can be exploited in the bootstrap (Franke & Hardle, 1992) and for fast subsampling MCMC (Quiroz et al., 2019) by subsampling periodogram observations (Salomone et al., 2020).

We first illustrate the efficiency and accuracy of the Whittle likelihood as an approximation to the time domain likelihood when fitting a simple AR(1) model with time-invariant parameters. We follow the setup in Contreras-Cristán et al. (2006) with the data generating process being

$$x_t = \phi x_{t-1} + \varepsilon_t, \quad \varepsilon_t \sim N(0, 1), \tag{8}$$

so that the only parameter to estimate is ϕ. We use a uniform prior and compute the posterior distribution over a fine grid of values for ϕ in the interval $[-0.999, 0.999]$. Following Contreras-Cristán et al. (2006), the time domain likelihood is the exact Gaussian likelihood with the first observation following the stationary distribution of the process.

We assess the performance of the Whittle likelihood using both a frequentist efficiency measure and a measure more adapted to Bayesian inference. The root mean square error (RMSE) of a point estimator $\hat{\boldsymbol{\phi}}$ of the AR parameters $\boldsymbol{\phi} = (\phi_1, \ldots, \phi_p)^\top$ over N_{rep} replicated time series from the data generating process is defined as

$$\text{RMSE} = \sqrt{\frac{1}{N_{\text{rep}} \cdot p} \sum_{i=1}^{N_{\text{rep}}} \|\hat{\boldsymbol{\phi}}^{(i)} - \boldsymbol{\phi}\|^2}, \tag{9}$$

where $\boldsymbol{\phi}$ is the vector of true AR parameters in the data generating process, $\hat{\boldsymbol{\phi}}^{(i)}$ is the vector of estimated AR parameters for the ith replicated time series, and $\|\cdot\|$ is the Euclidean norm. Our first performance measure is the efficiency of the Whittle MLE versus the exact time domain MLE

$$\text{Efficiency} = \frac{\text{RMSE}_{\text{time}}}{\text{RMSE}_{\text{whittle}}}. \tag{10}$$

Our second performance measure quantifies how much the Whittle posterior distribution of $\boldsymbol{\phi}$ deviates from the time domain posterior distribution, for a given dataset. Any measure of divergence between two distributions can be used, but we opt for an easily interpretable divergence between a set of quantiles. Let $\boldsymbol{q}_j^{(i)} = (q_{j1}^{(i)}, \ldots, q_{jr}^{(i)})^\top$ be r quantiles in the Whittle posterior distribution for ϕ_j based on dataset i, and let $\boldsymbol{s}_j^{(i)}$ denote the corresponding quantiles in the time domain posterior distribution. We measure how perturbed the Whittle posterior is with respect to the gold standard time domain posterior by

$$\text{Perturb} = \sqrt{\frac{1}{N_{\text{rep}} \cdot p \cdot r} \sum_{i=1}^{N_{\text{rep}}} \sum_{j=1}^{p} \|\boldsymbol{q}_j^{(i)} - \boldsymbol{s}_j^{(i)}\|^2}, \tag{11}$$

using the $r = 7$ quantiles: 2.5%, 10%, 25%, 50%, 75%, 90%, and 97.5%.

Figure 1 plots the efficiency in (10) and the degree of perturbation in (11) as a function of the true AR parameter ϕ for three different sample sizes. The left panel in Fig. 1 shows that the relative efficiency of the Whittle MLE deteriorates as $|\phi| \to 1$; this extends the results in Table 6 of Contreras-Cristán et al. (2006), who report the efficiency for $T = 50$ and $T = 200$ at $\phi = 0.2$ and $\phi = 0.8$. The loss of efficiency near-unit roots is also forcefully demonstrated by Dahlhaus

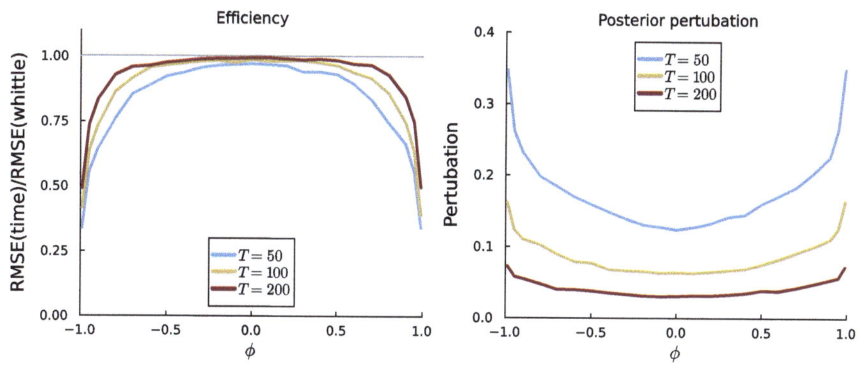

Fig. 1 Efficiency and posterior perturbation when fitting a stationary AR(1). The left panel shows the efficiency in (10) of the Whittle MLE compared to the exact time domain MLE for point estimation of ϕ in the stationary AR(1) process $x_t = \phi x_{t-1} + \varepsilon_t$, $\varepsilon_t \sim N(0, 1)$ for three different sample sizes, T. The right panel shows the degree of perturbation in (11) of the Whittle posterior from time domain posterior

(1988) in a simulation experiment with $T = 256$ observations from an AR(14) process. The right-hand panel of Fig. 1 shows that the posterior perturbation from the exact time domain posterior increases with $|\phi|$. Both the efficiency and the posterior perturbation improve with T.

Figure 2 plots the exact and Whittle posteriors for four randomly generated datasets from the AR(1) generating process with $\phi = 0.2$ for $T = 50$ (top row), $T = 100$ (middle row), and $T = 200$ (bottom row). Figure 3 is a similar plot for $\phi = 0.8$. The greater perturbation at $\phi = 0.8$ is clearly visible for $T = 50$, but for $T = 200$, the exact and Whittle posteriors are nearly indistinguishable for all four datasets.

3 Frequency Domain Likelihoods for Time-Varying Processes

Suppose that $\{X_t\}_{t=1}^{T}$ is a zero-mean process with time-varying spectral density $f_{\boldsymbol{\theta}}(\omega, t)$ modeled by a vector of parameters $\boldsymbol{\theta}$. As an example, Dahlhaus (2012) uses an AR(p) process with time-varying spectral density

$$f_{\boldsymbol{\theta}}(\omega, t) = \frac{\sigma_\varepsilon^2}{2\pi} \left|\phi_{p,t}(e^{-i\omega})\right|^{-2},$$

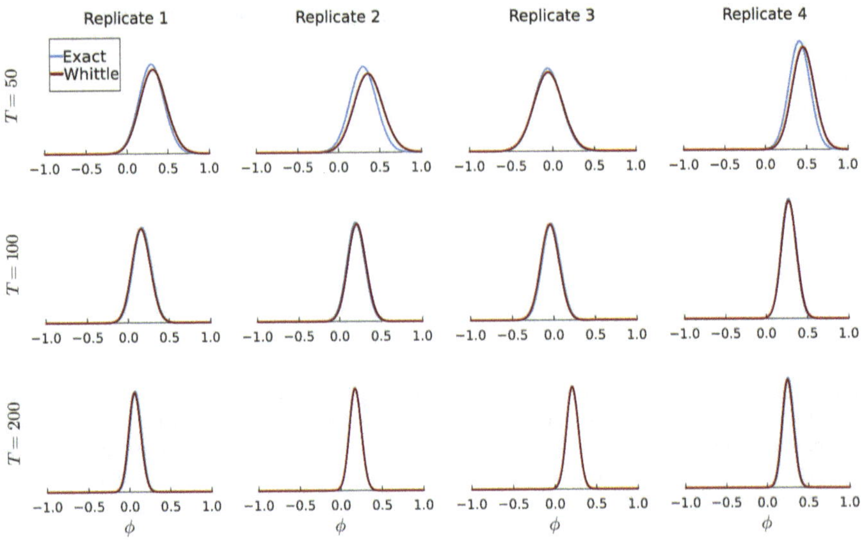

Fig. 2 Illustrating the posterior perturbation of the Whittle likelihood for ϕ in the stationary AR(1) process, $x_t = \phi x_{t-1} + \varepsilon_t, \varepsilon_t \sim N(0, 1)$, with $\phi = 0.2$. Each column corresponds to a simulated dataset with $T = 200$ observations, and the rows use the first $T = 50$, $T = 100$, or all $T = 200$ observations

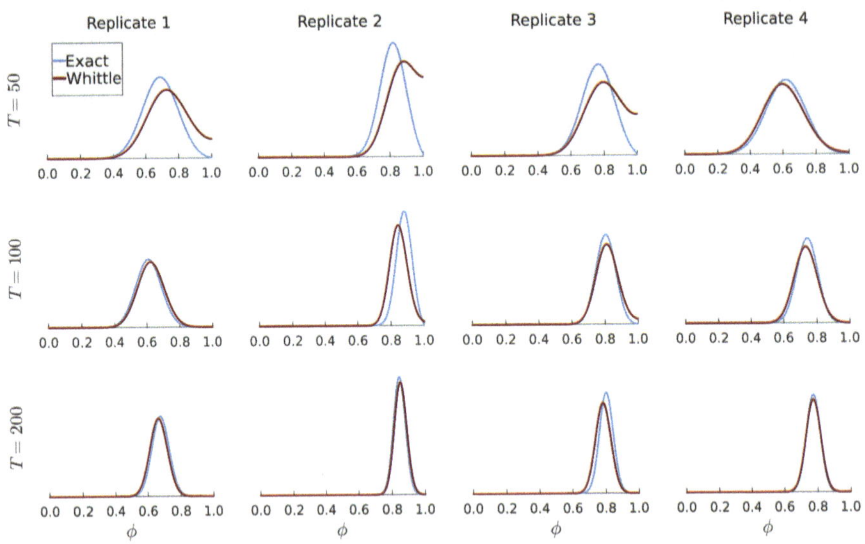

Fig. 3 Illustrating the posterior perturbation of the Whittle likelihood for ϕ in the stationary AR(1) process, $x_t = \phi x_{t-1} + \varepsilon_t, \varepsilon_t \sim N(0, 1)$, with $\phi = 0.8$. Each column corresponds to a simulated dataset with $T = 200$ observations, and the rows use the first $T = 50$, $T = 100$, or all $T = 200$ observations

where $\phi_{p,t}(L) := 1 - \phi_{1t}L - \cdots - \phi_{pt}L^p$ is the AR(p) lag polynomial at time t, and the jth AR parameter ϕ_{jt} is evolving smoothly in time as a polynomial of order r

$$\phi_{jt} = \theta_{j0} + \theta_{j1}t + \ldots, +\theta_{jr}t^r, \qquad \text{for } t = 1, \ldots, T.$$

Here $\boldsymbol{\theta} = \{\theta_{jq}\}_{j=1,\ldots,p;q=0,\ldots,r}$.

3.1 The Block Whittle Likelihood

One of the more well-known extensions of the Whittle likelihood to the time-varying case is the *block Whittle* log-likelihood function introduced by Dahlhaus (1997), based on local periodograms computed over a sliding time window of length N, where N is assumed to even for simplicity. To reduce the dependence between the periodograms computed at neighboring time points, the window slides S time steps between each computed local periodogram; see the upper part of Fig. 4 for an illustration. We will also refer to the window length N as the *segment size* and S as the *step size*. The number of segments is $M = (T - N)/S + 1$.

The block Whittle likelihood is defined as

$$\ell_T^{BW}(\boldsymbol{\theta}) := -\frac{1}{4\pi}\frac{1}{M}\sum_{j=1}^{M}\int_{-\pi}^{\pi}\left\{\log 4\pi^2 f_{\boldsymbol{\theta}}(\omega, t_j) + \frac{I_T(\omega, t_j)}{f_{\boldsymbol{\theta}}(\omega, t_j)}\right\} d\omega, \tag{12}$$

with $t_j := S(j-1) + N/2$ for $j = 1, \ldots, M$, and $I_T(\omega, t)$ is the local periodogram around t

$$I_T(\omega, t) := \frac{1}{2\pi N}\left|\sum_{s=1}^{N} X_{t-N/2+s}\exp(-i\omega s)\right|^2. \tag{13}$$

The integral over the frequencies is computed as a Riemann sum over a grid of frequencies $\omega_1, \ldots, \omega_K$. Dahlhaus (2012) defines this likelihood using a so-called taper on each segment, and we return to this in Sect. 3.3. Dahlhaus (2012) defines the block Whittle in terms of normalized time $u = t/T$ in order to use in-fill asymptotics. The block Whittle likelihood is equivalent to the model

$$I_T(\omega_k, t_j) \overset{\text{ind}}{\sim} \text{Exp}(f_{\boldsymbol{\theta}}(\omega_k, t_j)), \qquad \text{for } j = 1, \ldots, M \text{ and } k = 1, \ldots, K. \tag{14}$$

Since the segments are partially overlapping (when $S < N$), neighboring periodogram observations $I_T(\omega_k, t_j)$ share some observations and are therefore not exactly independent, making the likelihood in (12) and the corresponding equivalent model in (14) only approximate. We note that the (assumed) independence of the periodogram data, and the corresponding sum in the Block Whittle log-likelihood

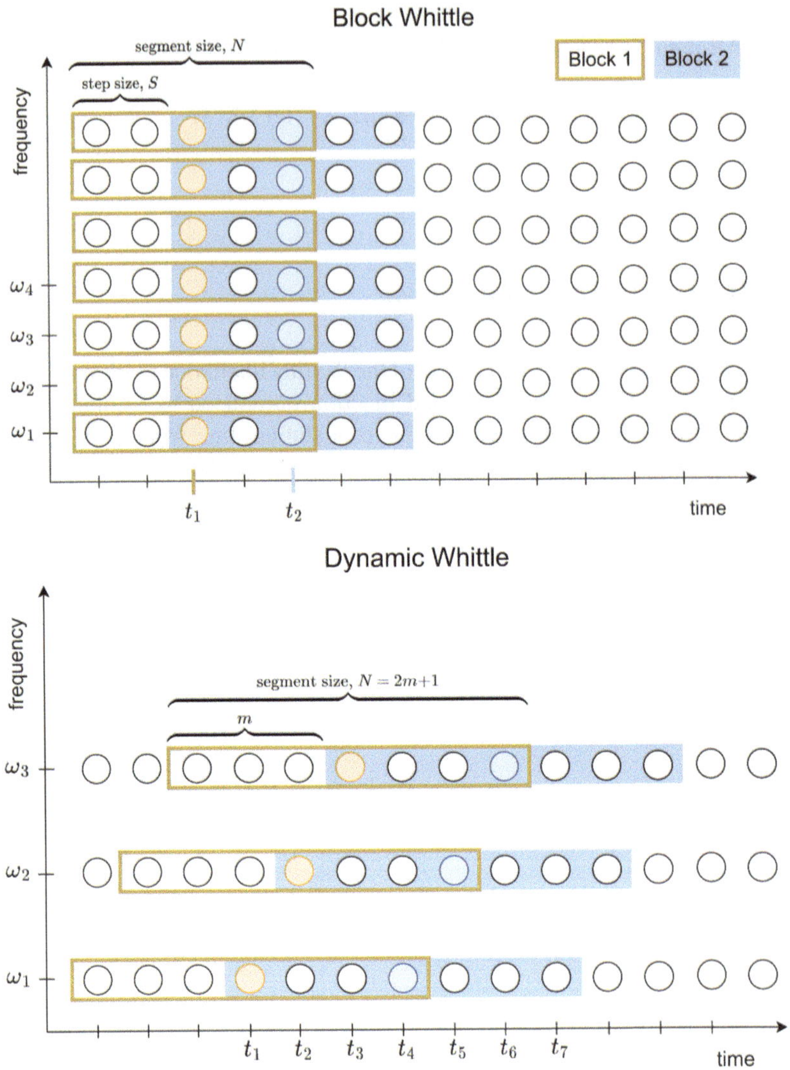

Fig. 4 Illustrating the time-frequency data structure for block Whittle (top) where the time series is split up into M overlapping segments, each of size $N = 5$ by letting the local window jump $S = 2$ observations each time. The first window is marked with an orange rectangle and the second window by a blue shaded rectangle. The dynamic Whittle (bottom) also uses segments of size $N = 2m + 1$ with $m = 3$ but step size $S = 1$ and computes the periodogram for a single frequency in each segment; the frequencies are systematically selected starting with the first frequency ω_1 in the time segment centered at time t_1. The filled circles mark out the time-frequency combinations where periodogram data is computed

approximation, makes it possible to also apply subsampling MCMC (Quiroz et al., 2019) to locally stationary time series data, as a direct extension of Salomone et al. (2020) from the stationary case.

The block Whittle likelihood in (12) can be motivated as the asymptotic Kullback-Leibler divergence between the true process with spectral density $f(\omega, t)$ and the model with spectral density $f_{\boldsymbol{\theta}}(\omega, t)$ (Dahlhaus, 1997). Dahlhaus (1997) proves consistency and asymptotic efficiency of the block Whittle MLE of $\boldsymbol{\theta}$ if $S/N \rightarrow 0$ under smoothness assumptions of the parameter evolution paths; see Dahlhaus (2012) for a large number of results on local Whittle likelihoods for a wide range of models under several different types of conditions, including time-varying ARMA models with results on the uniqueness of a time-varying spectral density.

Dahlhaus (2012) also reviews results for the *generalized Whittle log-likelihood* proposed in Dahlhaus (2000), which is defined as

$$\ell_T^{GW}(\boldsymbol{\theta}) := -\frac{1}{T}\sum_{t=1}^{T}\frac{1}{4\pi}\int_{-\pi}^{\pi}\left\{\log 4\pi^2 f_{\boldsymbol{\theta}}(\omega, t) + \frac{\check{J}_T(\omega, t)}{f_{\boldsymbol{\theta}}(\omega, t)}\right\}d\omega, \tag{15}$$

where $\check{J}_T(\omega, t)$ is the preperiodogram (Neumann & Von Sachs, 1997)

$$\check{J}_T(\omega, t) := \frac{1}{2\pi}\sum_{\substack{k \\ 1\leq\lfloor uT+0.5+k/2\rfloor,\lfloor uT+0.5-k/2\rfloor\leq T}} X_{\lfloor uT+0.5+k/2\rfloor}X_{\lfloor uT+0.5-k/2\rfloor}\exp\left(-\mathrm{i}\omega k\right). \tag{16}$$

The preperiodogram can be seen as a local version of the periodogram since

$$I_T(\omega) = \frac{1}{T}\sum_{t=1}^{T}\check{J}_T(\omega, t). \tag{17}$$

The MLE from (15) has good asymptotic properties under weak conditions of bounded variation on the parameter evolution (Dahlhaus, 2009; Dahlhaus & Polonik, 2009). However, the preperiodogram is quite noisy with complex interfering cross-terms which can make inference more difficult (Sandsten, 2020). Everitt et al. (2013) therefore use a modified preperiodogram which is smoothed in both time and frequency directions for efficient online learning of time-varying spectral densities using tvAR with similar parameter evolutions as in Sect. 5. Our preliminary experiments show that the generalized Whittle likelihood is more challenging to deal with in a Bayesian setting, and we therefore opt for comparing the block Whittle likelihood with the dynamic Whittle likelihood presented next.

3.2 The Dynamic Whittle Likelihood

Tang et al. (2023) define the dynamic Whittle (DW) likelihood as

$$\ell_T^{DW}(\boldsymbol{\theta}) := -\sum_{t=1}^{T} \left\{ \log f_{\boldsymbol{\theta}}(\omega_{\text{mod}(t)}, t) + \frac{\tilde{I}(\omega_{\text{mod}(t)}, t)}{f_{\boldsymbol{\theta}}(\omega_{\text{mod}(t)}, t)} \right\}, \tag{18}$$

where $\text{mod}(t) = 1 + ((t-1) \bmod m)$, $N = 2m + 1$ is the segment length, and $\omega_1, \ldots, \omega_m$ are the positive Fourier frequencies on the segment. The moving periodogram (Häfner & Kirch, 2017)

$$\tilde{I}(\omega_{\text{mod}(t)}, t) := \frac{1}{2\pi N} \left| \sum_{s=1}^{N} X_{t-m+s,T} \exp(-\mathrm{i}\omega_{\text{mod}(t)} s) \right|^2 \tag{19}$$

is calculated in a sliding time window of length N centered at time t, but in contrast to the block Whittle, only one frequency $\omega_{\text{mod}(t)}$ is used for each t; see Fig. 4 for a visual representation. Using different frequencies in windows at two neighboring time points makes it possible to slide the window over each time point, without introducing dependence between the periodogram data. However, when the window has completed a pass over all frequencies, it returns to the first frequency, which then partially overlaps with the previous window at the first frequency; see, for example, how the segments centered over t_1 and t_4 overlap for the same frequency ω_1 in the bottom panel of Fig. 4.

The dynamic Whittle likelihood is equivalent to the model

$$\tilde{I}_t \overset{\text{ind}}{\sim} \text{Exp}\big(f_\theta(\omega_{\text{mod}(t)}, t)\big), \quad t = 1, \ldots, T, \tag{20}$$

where $\tilde{I}_t := \tilde{I}(\omega_{\text{mod}(t)}, t)$ and T is the length of the time series, excluding the first m and last m observations which are needed to compute the moving periodograms in (19) at the boundaries. The dynamic Whittle likelihood yields a univariate observation equation with a single observation for each time point (except on the boundaries) contrasted against the block Whittle which gives roughly T/S multivariate observations observed across all frequencies.

Tang et al. (2023) prove that the posterior distribution from a bivariate extension of the Bernstein-Dirichlet process prior combined with the dynamic Whittle likelihood is consistent for the time-varying spectral density; L2-norm posterior contraction rates are also given.

Importantly, both the block and dynamic Whittle log-likelihoods are sums, due to the asymptotic independence of periodogram observations. This means that one can speed up Markov chain Monte Carlo for long time series by subsampling periodogram observations; see Salomone et al. (2020) and the extension to multivariate time series in Villani et al. (2024). This continues to hold for the three modifications of the basic block and dynamic Whittle likelihoods presented next.

3.3 Modified Frequency Likelihoods

The block Whittle and dynamic Whittle likelihoods rely on periodograms from relatively short sliding time windows and are therefore likely to inherit the small-sample problems of the original Whittle likelihood for a stationary process demonstrated in Sect. 2.

We investigate three modifications of the standard Whittle likelihood, which are applied to each segment when used for time-varying processes. The first is tapering which is well known in the literature and is recommended in Dahlhaus (1988) for small samples in the stationary case and in Dahlhaus (2012) for the block Whittle likelihood. The second approach is prewhitening (Priestly, 1981), which seems less common in the statistical literature. The last, and most recently proposed, modification is the *boundary correction* in Subba Rao and Yang (2021), which performs well both as a sampling distribution for the spectral density (Das et al., 2021) and for parametric estimation via the Whittle likelihood (Subba Rao & Yang, 2021).

The debiased Whittle (Sykulski et al., 2019) is another method for reducing the bias in the Whittle MLE. The debiased Whittle likelihood is

$$\ell_T^W(\boldsymbol{\theta}) := -\frac{1}{2} \sum_{\omega_k \in \Omega_T} \left(\log \bar{f}_{\boldsymbol{\theta}}(\omega_k) + \frac{I_T(\omega_k)}{\bar{f}_{\boldsymbol{\theta}}(\omega_k)} \right), \tag{21}$$

which debiases the Whittle MLE by replacing the spectral density with the expected periodogram $\bar{f}_{\boldsymbol{\theta}}(\omega) := \mathbb{E}(I_T(\omega))$. Sykulski et al. (2019) show how the expected periodogram can be computed by two FFTs, thereby retaining the $O(T \log T)$ cost of the original Whittle likelihood. However, for the original Whittle, this cost is incurred once when computing the periodogram data, which can then be used for subsequent evaluations of the likelihood for different $\boldsymbol{\theta}$ values, which then remains a $O(T)$ operation, or substantially lower if subsampling is used (Quiroz et al., 2019). This is especially important for Bayesian inference with MCMC where the likelihood is typically evaluated thousands of times during the course of the algorithm. The FFTs in the debiased Whittle likelihood have to be recomputed at cost $O(T \log T)$ for every new $\boldsymbol{\theta}$ value, since $\bar{f}_{\theta}(\omega)$ depends on $\boldsymbol{\theta}$; the overall cost of computing the likelihood therefore remains $O(T \log T)$, even if subsampling is used. We will therefore not consider this likelihood in our simulation study.

Tapering

Dahlhaus (1988) shows that the Whittle likelihood for the stationary case in (6) is inefficient due to bias from leakage effects in the periodogram. He suggests using a taper to correct the bias in the periodogram and defines a modified tapered Whittle likelihood. In the context of local Whittle likelihoods, this means replacing the periodogram on the segments with a tapered version

$$I_T(\omega, t) := \frac{1}{2\pi H_N} \left| \sum_{s=1}^{N} h\left(\frac{s}{N}\right) X_{t-N/2+s,T} \exp(-\mathrm{i}\omega s) \right|^2, \tag{22}$$

where $h : [0, 1] \to \mathbb{R}$ is a symmetric data taper with $h(x) = h(1 - x)$ and $H_N := \sum_{j=0}^{N-1} h^2(j/N)$ is the normalization factor. Dahlhaus (1988) shows that this modified Whittle likelihood is as efficient as the exact time domain MLE and satisfies a central limit theorem. We use a taper based on the Hanning window $h(t) = \frac{1}{2}(1 - \cos(2\pi t/N))$ applied to each segment in the local Whittle likelihood.

Prewhitening

A consequence of using the standard periodogram is that spectral power leaks to nearby frequencies, with the effect being largest at the highest peaks in the spectrum. This can hide or contaminate low-power frequencies in the spectral density, leading to distortions in the Whittle likelihood and efficiency loss for the Whittle MLE (Dahlhaus, 1988). Prewhitening (Priestly, 1981) first removes the most dominant autocorrelation in the time domain by filtering with an estimated AR(p) process (or ARMA(p,q)) so that the residual series $e_t = \hat{\phi}_p(B)x_t$ has a periodogram from a flatter spectral density. The residual periodogram is then transformed into a periodogram for the original time series, using that the spectral density of the unfiltered process is

$$f_x(\omega) = |H_{\hat{\phi}}(\omega)|^{-2} f_e(\omega),$$

where $H_{\hat{\phi}}(\omega) = \hat{\phi}_p(e^{-\mathrm{i}\omega})$ is the transfer function of the fitted AR(p) process and $f_e(\boldsymbol{\omega})$ is the spectral density of the filtered series. That is, the periodogram $I_x(\omega)$ used in the Whittle likelihood is computed as

$$I_x(\omega_k) = |H_{\hat{\phi}}(\omega_k)|^{-2} I_e(\omega_k), \qquad \text{for } k = 1, \ldots, K,$$

where $I_e(\omega)$ is the residual periodogram. Note that the AR model used for the prewhitening does not have to be very precise as the remaining lower peaks in the spectral density are captured by the residual periodogram $I_e(\omega_k)$.

Boundary Correction

Subba Rao and Yang (2021) express the Gaussian time domain likelihood in the frequency domain and prove that the approximation error of the Whittle likelihood is the result of boundary effects. The Gaussian likelihood can be viewed from this perspective to circumvent boundary effects by predicting the time series outside its range using a best linear predictor. The approximation error increases with the persistence of the time series (Subba Rao & Yang, 2021), in agreement with the empirical evidence in Sect. 2, Dahlhaus (1988) and Contreras-Cristán et al. (2006). Subba Rao and Yang (2021) suggest using the *complete DFT* that augments the standard DFT with the Fourier transform of the best linear predictor to remove the discrepancy between the Whittle likelihood and the Gaussian likelihood.

The *complete discrete Fourier transform* (DFT) of $\{X_t\}_{t=1}^T$

$$\tilde{J}(\omega; f) = J(\omega) + \hat{J}(\omega; f) \tag{23}$$

is the sum of the standard DFT $J(\omega)$ and the *predictive DFT*

$$\hat{J}(\omega; f) := \frac{1}{\sqrt{2\pi}} \sum_{t=1}^{T} X_t \sum_{\tau \leq 0} \left(\phi_{t,T}(\tau; f) e^{\mathrm{i}\omega\tau} + e^{\mathrm{i}\omega T} \phi_{T+1-t,T}(\tau; f) e^{-\mathrm{i}(\tau-1)\omega} \right),$$

where $\phi_{t,T}(\tau; f)$ are coefficients of the best linear predictor of X_τ for $\tau \leq 0$ and $\tau > T$ given $\{X_t\}_{t=1}^T$ with spectral density f. Note that $e^{iT\omega_T} = 1$ and that $\hat{J}(\omega; f)$ depends on the spectral density f of the model. Subba Rao and Yang (2021) suggest approximating the best linear predictor with a finite-order AR model determined by AIC. Note that the computation of $\hat{J}(\omega_k; f)$ for $k = 1, \ldots, K$ is a one-time cost, performed before the MCMC. Put differently, once the completed DFT in (23) is computed, it can be used as fixed data for any subsequent model fitted by MCMC or by other methods.

The example with the predictive DFT of the AR(1) model with parameter ϕ in Subba Rao and Yang (2021) is illuminating. Since the prediction to the left of the first observation X_1 for an AR(1) process is $\hat{X}_{\tau,T} = \phi^{|\tau|+1} X_1$ for $\tau \leq 0$ and to the right of the last observation X_T is $\hat{X}_{\tau,T} = \phi^{\tau-T} X_T$ for $\tau > T$, we have

$$\hat{J}(\omega; f) = \frac{\phi}{\sqrt{2\pi}} \left(\frac{1}{\phi(\omega)} X_1 + \frac{e^{\mathrm{i}(T+1)\omega}}{\overline{\phi(\omega)}} X_T \right),$$

where $\phi(\omega) = 1 - \phi \exp(-\mathrm{i}\omega)$.

The *complete periodogram* based on the complete DFT in (23) is then

$$\breve{I}_T(\omega) := \frac{1}{2\pi T} \hat{J}(\omega; f) \overline{J(\omega)}, \tag{24}$$

which is now unbiased, but not consistent, for $f(\omega)$ (Subba Rao & Yang, 2021).

The boundary-corrected Whittle likelihood proposed by Subba Rao and Yang (2021) replaces the regular periodogram with the complete periodogram in (24). In a simulation study, they find that the MLE from the boundary-corrected Whittle likelihood outperforms other spectral likelihoods including the standard and tapered Whittle likelihood and the debiased Whittle likelihood (Sykulski et al., 2019). Subba Rao and Yang (2021) recommend a hybrid using the boundary-corrected likelihood together with tapering, and this is the version used in our simulation studies.

4 Bayesian Inference

4.1 *Time-Varying Locally Stable AR Processes*

The class of time-varying AR(p) models (tvAR) is used to compare the spectral likelihood approximations in Sect. 3. This relatively simple class makes it possible to easily compare with the time domain likelihood, which is the gold standard. The AR(p) process is of the form

$$x_t = \mu_t + \sum_{j=1}^{p} \phi_{jt}(x_{t-j} - \mu_t) + \varepsilon_t, \quad \varepsilon_t \sim N(0, \sigma_{\varepsilon,t}^2), \tag{25}$$

where ϕ_{jt} is the jth AR coefficient at time t. We will for simplicity assume a zero mean μ_t and that the disturbance variance is constant over time, $\sigma_{\varepsilon,t}^2 = \sigma_\varepsilon^2$ for all t. In the application to egg price data in Sect. 6, we also allow for a time-varying variance.

The instantaneous spectral density (Kitagawa & Gersch, 1985) of the tvAR model in (25) is

$$f_{\phi_t,\sigma_\varepsilon}(\omega) = \frac{\sigma_\varepsilon^2}{2\pi} \left|\phi_{p,t}(e^{-\mathrm{i}\omega})\right|^{-2}, \tag{26}$$

where $\phi_{p,t}(L) := 1 - \phi_{1t}L - \cdots - \phi_{pt}L^p$ is the AR(p) lag polynomial at time t. The data generating processes used in Sect. 5 are tvAR processes with smooth parametric curves for the evolving AR parameters, so their instantaneous spectral density is the unique time-varying spectral density (Dahlhaus, 2012, Theorem 4 and Example 7).

Following Fagerberg et al. (2024) we enforce the stability of the AR process at every time period using the reparametrization in Barndorff-Nielsen and Schou (1973) and Monahan (1984). This transformation maps a set of unconstrained real-valued coefficients $\boldsymbol{\theta} = (\theta_1, \ldots, \theta_p)^\top$ to the set of stable parameters, i.e., parameters that determine a stable AR process. Let $\mathbb{S}^p \subset \mathbb{R}^p$ denote the region in parameter space where the AR(p) process is stable. The reparametrization is defined as follows.

Definition 1 (Stability Parameterization) An AR(p) process can be restricted to be stable by a 1:1 and onto map

$$\boldsymbol{\theta} \rightarrow \mathbf{r} \rightarrow \boldsymbol{\phi}$$

of unrestricted parameters $\boldsymbol{\theta} = (\theta_1, \ldots, \theta_p)^\top \in \mathbb{R}^p$ via the partial autocorrelations $\mathbf{r} = (r_1, \ldots, r_p)^\top \in (-1, 1)^p$ to the stable AR parameters $\boldsymbol{\phi} = (\phi_1, \ldots, \phi_p)^\top \in \mathbb{S}^p$. The mapping from $\mathbf{r}$ to $\boldsymbol{\phi}$ sets $\phi_{1,1} = r_1$ followed by the recursion

$$\phi_{k,j} = \phi_{k-1,j} + r_k \phi_{k-1,k-j}, \quad \text{for } k = 2, \ldots, p \text{ and } j = 1, \ldots, k-1,$$

and finally returning $\boldsymbol{\phi} = (\phi_{p,1}, \ldots, \phi_{p,p})^\top$, where $\phi_{p,p} = r_p$.

We follow Monahan (1984) and use the map

$$r_k = \frac{\theta_k}{\sqrt{1 + \theta_k^2}}, \qquad \text{for } k = 1, \ldots, p, \tag{27}$$

from the unrestricted parameters to the partial autocorrelations. The stability parameterization in Definition 1 is imposed at every time period in the time-varying parameter setting. We use $\boldsymbol{\phi}_t = \mathbf{g}(\boldsymbol{\theta}_t)$ to denote the composite map $\boldsymbol{\theta}_t \rightarrow \mathbf{r}_t \rightarrow \boldsymbol{\phi}_t$.

4.2 Likelihood and Parameter Evolution

The unrestricted parameters $\boldsymbol{\theta}_t$ in the fitted AR models are assumed to follow a multivariate random walk with Gaussian innovations, as in the large literature on time-varying AR and vector AR models, see, e.g., Prado and West (2010). Given the frequency domain approximations of the likelihood discussed in Sect. 3, we can write the model as a state-space model with $\boldsymbol{\theta}_t$ as state vector. In the block Whittle case

$$\begin{aligned}
I\left(\omega_k, t_j\right) &\overset{\text{ind}}{\sim} \text{Exp}\left(f_{\boldsymbol{\phi}_j, \sigma_\varepsilon}(\omega_k)\right), \qquad \text{for } k = 1, \ldots, K \text{ and } j = 1, \ldots, M, \\
\boldsymbol{\phi}_j &= \mathbf{g}(\boldsymbol{\theta}_j), \\
\boldsymbol{\theta}_j &= \boldsymbol{\theta}_{j-1} + \boldsymbol{\eta}_j, \quad \boldsymbol{\eta}_j \overset{\text{iid}}{\sim} N(\mathbf{0}, \boldsymbol{Q}), \\
\boldsymbol{\theta}_0 &\sim N(\mathbf{a}_0, \mathbf{P}_0),
\end{aligned} \tag{28}$$

where $f_{\boldsymbol{\phi}_j, \sigma_\varepsilon}(\omega)$ is the instantaneous spectral density in (26) of the assumed AR model at time t_j. In the case with the dynamic Whittle likelihood the observation equation of (28) changes to

$$\tilde{I}_t \overset{\text{ind}}{\sim} \text{Exp}\big(f_{\boldsymbol{\phi}_t,\sigma_\varepsilon}(\omega_{\text{mod}(t)})\big), \qquad \text{for } t = 1, \ldots, T, \tag{29}$$

and the parameter evolution in (28) is indexed by $t = 1, \ldots, T$.

The state evolution for $\boldsymbol{\theta}_t$ with covariance matrix $\boldsymbol{Q}$ in (28) is linear and Gaussian, but the observation equation is non-Gaussian due to the exponentially distributed periodogram data and nonlinear as a result of the stability parameterization. This prevents using the usual Forward Filtering Backward Sampling (FFBS) algorithm (Carter & Kohn, 1994; Frühwirth-Schnatter, 1994) based on the Kalman filter, and we instead rely on particle MCMC, in particular the particle Gibbs with ancestor sampling (PGAS) algorithm in Lindsten et al. (2014). The remaining parameters are inferred using standard Gibbs sampling updates.

4.3 Bayesian Inference by Particle MCMC Within Gibbs Sampling

We derive the Gibbs sampling steps for the block Whittle likelihood; the updating steps for the dynamic Whittle likelihood follow from obvious changes. The full conditional distribution for σ_ε^2 can be obtained as follows. The observation equation under the tvAR process is

$$I\left(\omega_k, t_j\right) \overset{\text{ind}}{\sim} \text{Exp}\left(f_{\boldsymbol{\phi}_j,\sigma_\varepsilon}(\omega_k)\right) = \text{Exp}\left(\frac{\sigma_\varepsilon^2}{2\pi}\left|\phi_{p,j}(e^{-\mathrm{i}\omega_k})\right|^{-2}\right), \tag{30}$$

where $\phi_{p,j}(L) := 1 - \phi_{1j}L - \cdots - \phi_{pj}L^p$ is the AR(p) lag polynomial at time t_j. Since the exponential distribution is closed under scaling, we have

$$\tilde{I}\left(\omega_k, t_j\right) \overset{\text{ind}}{\sim} \text{Exp}\left(\sigma_\varepsilon^2\right), \tag{31}$$

where $\tilde{I}(\omega, t) = 2\pi\left|\phi_{p,t}(e^{-\mathrm{i}\omega})\right|^2 I(\omega, t)$. Using the conjugate prior for the exponential likelihood, $\sigma_\varepsilon^2 \sim \text{inv-gamma}(\alpha, \beta)$, the conditional posterior distribution is

$$\sigma_\varepsilon^2|\boldsymbol{\phi}_{0:T}, I(\boldsymbol{\omega}, 1{:}M) \sim \text{inv-gamma}\left(\alpha + KM, \sum_{k=1}^{K}\sum_{j=1}^{M}\tilde{I}\left(\omega_k, t_j\right) + \beta\right), \tag{32}$$

where $I(\boldsymbol{\omega}, 1{:}M)$ is the periodogram data over all frequencies $\boldsymbol{\omega} = (\omega_1, \ldots, \omega_K)^\top$ and time segments centered at t_j, for $j = 1, \ldots, M$.

The prior for $\boldsymbol{Q}$ is inv-Wishart(ν_0, $\boldsymbol{S}_0$). The full conditional for $\boldsymbol{Q}$ is the same as in the time domain:

$$\boldsymbol{Q}|\boldsymbol{\phi}_{0:M}, \sigma_\varepsilon^2, \boldsymbol{I}(\boldsymbol{\omega}, 1{:}M) \sim \text{inv-Wishart}\left(M + \nu_0, \boldsymbol{S}_0 + \boldsymbol{\phi}_{1:M}^\top \boldsymbol{\phi}_{1:M}\right). \tag{33}$$

Since the prior for the initial state is Gaussian $\boldsymbol{\theta}_0 \sim N(\boldsymbol{a}_0, \mathbf{P}_0)$ and $\boldsymbol{\theta}_1 = \boldsymbol{\theta}_0 + \boldsymbol{\eta}_1$ from the state transition equation, standard Gaussian updating gives that the full conditional posterior for $\boldsymbol{\theta}_0$ is

$$\boldsymbol{\theta}_0|\boldsymbol{\phi}_{1:M}, \sigma_\varepsilon^2, \boldsymbol{I}(\boldsymbol{\omega}, 1{:}M) \sim N(\boldsymbol{m}, \boldsymbol{\Omega}), \tag{34}$$

with $\boldsymbol{\Omega}^{-1} = \boldsymbol{Q}^{-1} + \boldsymbol{P}_0^{-1}$ and $\boldsymbol{m} = \boldsymbol{\Omega}\left(\boldsymbol{Q}^{-1}\boldsymbol{\theta}_1 + \boldsymbol{P}_0^{-1}\boldsymbol{a}_0\right)$.

Algorithm 1 summarizes the Gibbs sampler.

Algorithm 1: Gibbs sampling with block Whittle likelihood on M segments

---

Input: segment length N
step size S
time-varying periodogram, $I(\omega, 1{:}M)$
initial value for static parameters σ_ε^2, $\boldsymbol{Q}$, $\boldsymbol{\phi}_0$.

for $j = 1$ **to** n **do**

Update $\boldsymbol{\phi}_{1:M}|\boldsymbol{Q}, \sigma_\varepsilon^2, \boldsymbol{\phi}_0$ using PGAS (Lindsten et al., 2014).
Update $\boldsymbol{Q}|\boldsymbol{\phi}_{1:M}, \sigma_\varepsilon^2, \boldsymbol{\phi}_0$ using the inverse Wishart in (33).
Update $\sigma_\varepsilon^2|\boldsymbol{Q}, \boldsymbol{\phi}_{1:M}, \boldsymbol{\phi}_0$ using the inverse gamma in (32).
Update $\boldsymbol{\phi}_0|\sigma_\varepsilon^2, \boldsymbol{Q}, \boldsymbol{\phi}_{1:M}$ using the multivariate normal in (34) and set $\boldsymbol{\phi}_0 = \mathbf{g}(\boldsymbol{\theta}_0)$.

end

Output: draws from the joint posterior for $p(\boldsymbol{\phi}_{0:M}, \sigma_\varepsilon^2, \boldsymbol{Q}|I(\boldsymbol{\omega}, 1{:}M))$.

---

5 Experiments

This section evaluates the likelihood in three simulation settings with the data generating processes as an AR process with parameters following three different deterministic paths:

- **Experiment 1**—locally stationary time-varying AR(2) from Dahlhaus (2012)
- **Experiment 2**—stationary AR(3) process with time-invariant parameters
- **Experiment 3**—locally stationary AR(2) where the parameters come close to the non-stable/explosive region several times during the observed time period

One hundred realizations of length $T = 1500$ are simulated from the data generating processes in each experiment. For each generated time series, we run the Gibbs sampler in Algorithm 1 to sample from the posterior based on the time domain likelihood, which is the target posterior. We then run the Gibbs sampler on the same datasets for both the block and dynamic Whittle approximate likelihoods and also for these likelihoods combined with one of the three improvements in Sect. 3.3: tapering, prewhitening, and boundary correction. Each Gibbs run is based on 12,000 draws, from which 2000 are discarded as burn-in. The remaining draws are thinned by a factor of 2, leaving 5000 draws for inference.

The dynamic Whittle likelihood gives rise to m missing values in the inferred parameter evolution at the beginning and end of the series; see the bottom part of Fig. 4. With the block Whittle likelihood the posterior for the parameters is only observed at every Sth time point. Sophisticated interpolation methods are available (Häfner & Kirch, 2017; Dahlhaus, 2012), but we use a simple piecewise constant extrapolation of the evolution between observed posterior estimates, see e.g., Fig. 26.

To compute the periodograms when prewhitening and boundary correction are used we need a model on each segment to filter or make predictions outside the boundary. We use an AR model with the number of lags determined by the Hannan-Quinn information criteria (Hannan & Quinn, 1979), with coefficients estimated using Burg's method.

Since we are mainly interested in the similarity to the time domain likelihood, we do not put any particular effort into finding good prior settings and use the same prior for all data generating processes. This prior is $\boldsymbol{Q} \sim IW(10, 0.005 \cdot (10 - p - 1)\boldsymbol{I}_p)$, where p is the number of lags in the AR model. For the block Whittle likelihood, we multiply the prior covariance matrix $\boldsymbol{Q}$ by the step size, S, to allow for similar parameter evolution over the whole time span as in the dynamic Whittle likelihood. The prior on the error variance is $\sigma_\varepsilon^2 \sim IG(0.01, 0.01)$. The unrestricted AR parameters at time $t = 0$ follow a multivariate normal prior with mean $\mathbf{a}_0 = \mathbf{0}_p$ and covariance matrix $\mathbf{P}_0 = 10 \cdot \boldsymbol{I}_p$.

5.1 Experiment 1: Dahlhaus Time-Varying AR(2)

The first experiment uses the AR(2) process

$$y_t = \phi_{1,t} y_{t-1} + \phi_{2,t} y_{t-2} + \varepsilon_t, \quad \varepsilon_t \sim N(0, 1), \tag{35}$$

in Dahlhaus (2012), where the first AR parameter evolves like a cosine and the second AR parameter is time invariant:

$$\begin{aligned}
\phi_{1,t} &= -1.8 \cos\left(1.5 - \cos(4\pi t_u)\right), \text{ for } t_u = (t-1)/(T-1), \\
\phi_{2,t} &= -0.9,
\end{aligned}$$

for $t = 1, \ldots, T$. The process is always relatively far from having a unit root; the third experiment explores the case where we occasionally get close to a unit root.

Figure 5 plots the efficiency in (10) for the posterior median estimate and the posterior perturbation measure in (11) for different combinations of approximate likelihood (dynamic Whittle (DW) and block Whittle (BW)) and the three modifications: tapering (TA), prewhitening (PW), and boundary correction (BC); since there is now a time-varying step, the MSE measures in (9) and (11) are also averages over

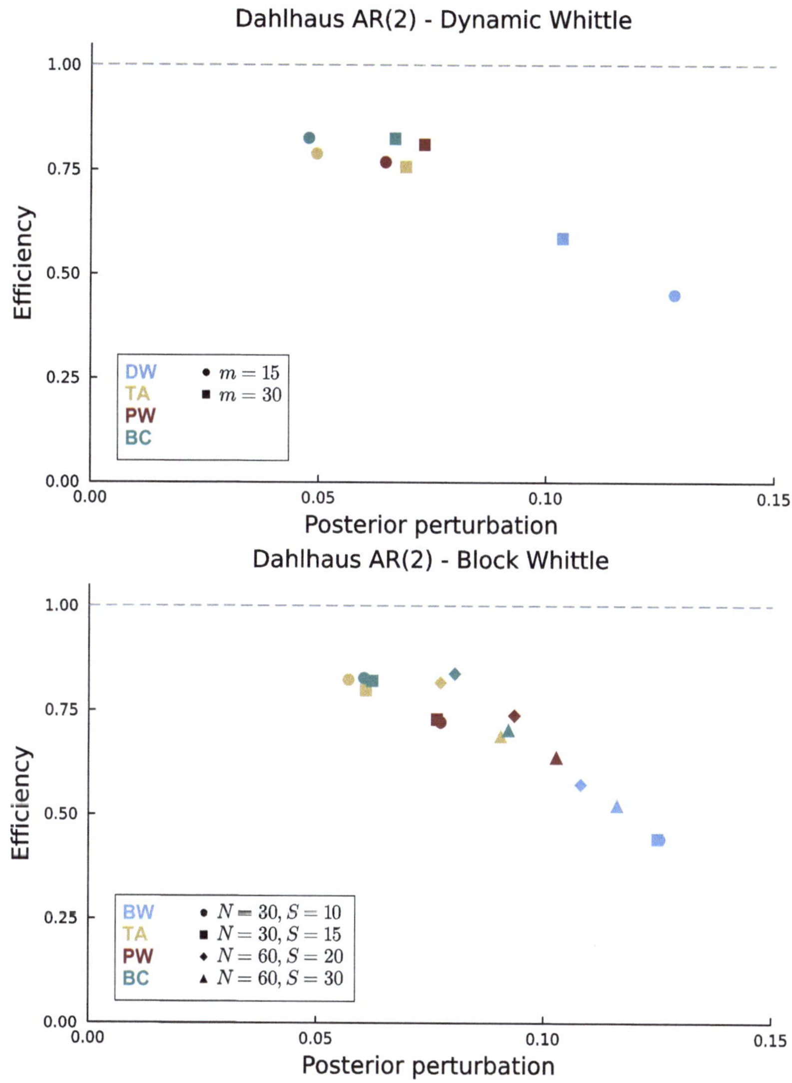

Fig. 5 Experiment 1—Dahlhaus AR(2) process. Efficiency and posterior perturbation in the posterior for the AR parameters for the dynamic Whittle likelihood (DW) in the top panel and the block Whittle likelihood (BW) in the bottom panel and with modifications including tapering (TA), prewhitening (PW), and boundary correction (BC)

time. The top panel of Fig. 5 shows that the dynamic Whittle likelihood without any modification (blue circle for $m = 15$ and blue square for $m = 30$) can be much improved on both measures by using any of the three modifications; the best modification is BC here (green markers), but tapering (yellow markers) is close second. Recall that BC is the recommended version in Subba Rao and Yang (2021) which also includes tapering. Note also the appreciable effect of segment size when no modification is used, but less so when any of the three modifications are applied. The bottom panel of Fig. 5 shows similar results for the block Whittle likelihood.

Figure 6 plots the repeated sampling distribution of the posterior median estimate of ϕ_{1t} and ϕ_{2t} over the 100 replicated time series for the time domain likelihood (top row) and the dynamic Whittle likelihood with $m = 15$ giving a segment size of $N = 2m + 1 = 31$ (second row) with each of the three modifications (bottom three

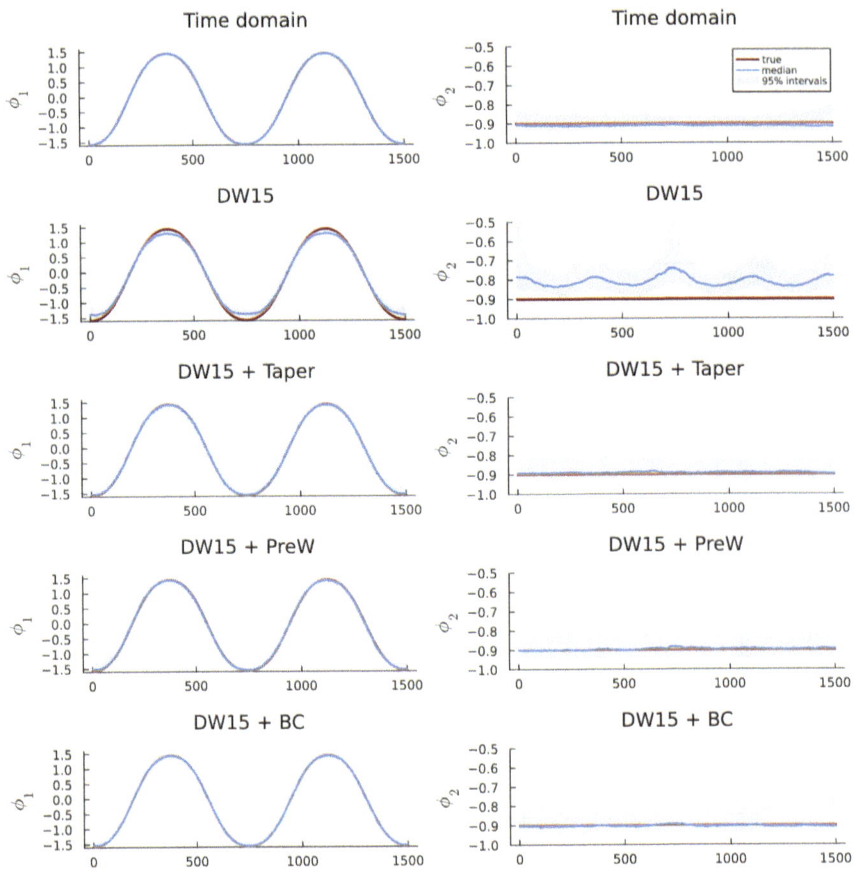

Fig. 6 Experiment 1. Sampling distribution for the posterior median estimates of the parameters using time domain likelihood, and the dynamic Whittle likelihood with $m = 15$ (DW15), and with modifications including tapering (Taper), prewhitening (PreW), and boundary correction (BC)

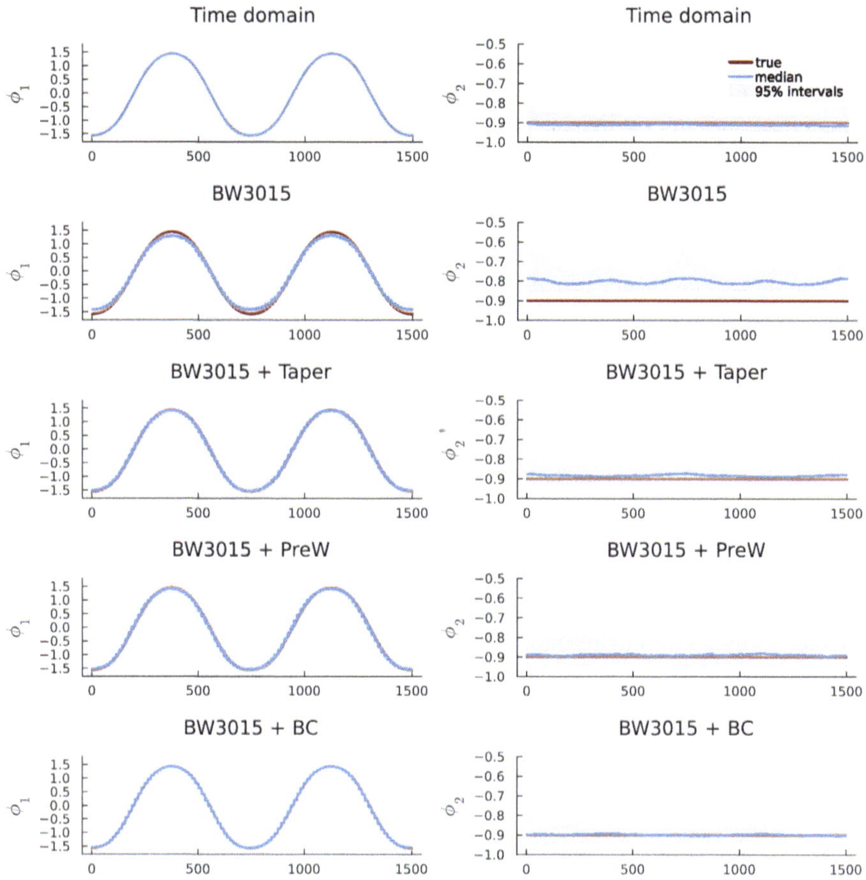

Fig. 7 Experiment 1. Sampling distribution for the posterior median estimates of the parameters using time domain likelihood, and the block Whittle likelihood with $N = 30$, $S = 15$ (BW3015), and with modifications including tapering (Taper), prewhitening (PreW), and boundary correction (BC)

rows). The bias in the dynamic Whittle posterior median estimate is clear, while the three modifications successfully remove the bias. The sampling variability in the Whittle estimates from the modifications is slightly larger than those from the time domain posterior. Figure 7 plots the sampling distributions for the block Whittle likelihood with $N = 30$, $S = 15$, which are qualitatively similar to the ones from the dynamic Whittle in Fig. 6.

Figures 18 and 19 in the Appendix show the posterior distributions for the AR parameters from the fifth worst dataset for respective method in terms of RMSE. Even when the model performs relatively poorly in a relative sense, the results are still decent.

5.2 Experiment 2: Time-Invariant AR(3) Process

The second experiment explores how well the likelihood approximations perform when time-varying AR models are fitted to an AR(3) data generating process with time-invariant parameters

$$y_t = 1.4y_{t-1} - 0.7y_{t-2} + 0.2y_{t-3} + \varepsilon_t, \quad \varepsilon_t \sim N(0, 1). \tag{36}$$

Figure 8 shows again that the raw dynamic Whittle and block Whittle can be rather inefficient and have sizable perturbations from the time domain posterior. All three modifications are substantially less perturbed and more efficient. The dynamic Whittle with modifications gives the smallest posterior perturbation, while the modified block Whittle likelihoods have the largest efficiencies, often even a bit larger than the time domain MLE. The larger segment size performs a lot better than the smaller one for the raw Whittle-type likelihoods, which is expected given the time-invariant nature of the data. The step size in the block Whittle has virtually no effect. Both the dynamic and block Whittle likelihoods become much more robust to the segment length when any of the three modifications are applied.

Figures 20 and 21 in Appendix show the sampling distributions of the posterior median estimate for the dynamic Whittle with $m = 15$ and for block Whittle with segment size $N = 30$ and step size $S = 15$. The bias for the raw Whittle likelihoods is clearly visible, while the sampling distributions for the three modifications are unbiased and similar.

5.3 Experiment 3: Time-Varying AR(2) Occasionally Near Unit Root

The final experiment investigates the case with parameters that give a near-unit root process at certain time periods. The parameter paths are determined in the unrestricted parameter space as

$$\theta_{1t} = \begin{cases} 1.5 - 4(t-1)/500, \\ -2.5 + 4.5(t-501)/500, \\ 2 - 4(t-1001)/500, \end{cases} \quad \theta_{2t} = \begin{cases} 1 - 2(t-1)/500, & \text{for } t = 1, \dots, 500 \\ -1 + (t-501)/500, & \text{for } t = 501, \dots, 1000 \\ -0.5(t-1001)/500, & \text{for } t = 1001, \dots, 1500 \end{cases}$$

and then transformed to the stability region via a composition of the transformations in Definition 1 and (27). Figure 9 plots the time evolution of the AR parameters. The leftmost panel shows that there are four periods where the parameters are close to the non-stable region.

Figure 10 displays the efficiency and perturbation of the approximate posterior from the local Whittle likelihoods from the time domain posterior. The three modifications again lead to increased efficiency and lower perturbation when applied to the raw dynamic and block Whittle likelihoods. The dynamic Whittle

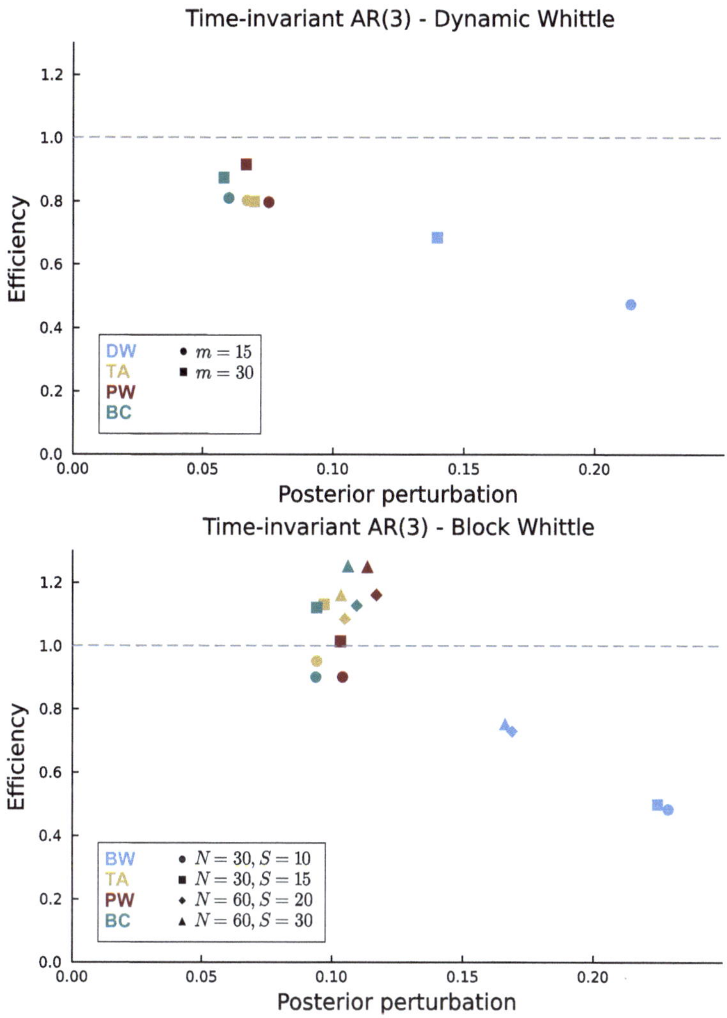

Fig. 8 Experiment 2—time-variant AR(3) process. Efficiency and posterior perturbation in the posterior for the AR parameters for the dynamic Whittle likelihood (DW) in the top panel and the block Whittle likelihood (BW) in the bottom panel and with modifications including tapering (TA), prewhitening (PW), and boundary correction (BC)

performs slightly better than the block Whittle with lower posterior perturbation than the block Whittle variants. Surprisingly, the effect of the modifications is small here, even though the process is close to unit roots on several occasions, possibly because the time spent in those regions is too short to cause problems.

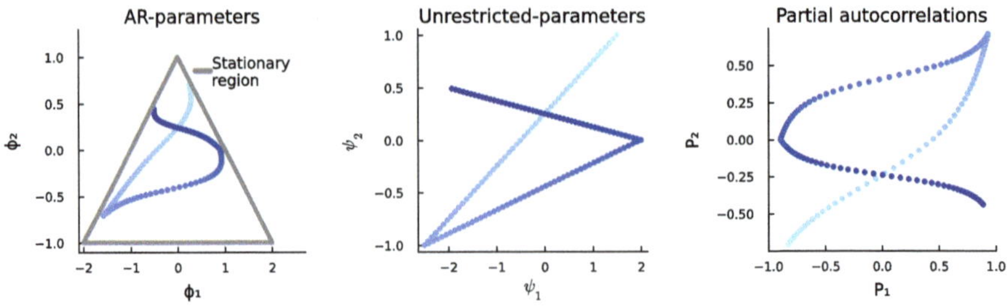

Fig. 9 Parameter evolution over time where the parameters start with light blue scatters and move in steps of 10 until it ends at the dark blue color at time 1500

6 Application to Egg Price Data

We analyze a dataset from Fan and Yao (2008), with 1201 time series observations on weekly egg prices (per egg in Deutsche Mark/100 paid to producers) in the German market from April 1967 to May 1990. Following Paparoditis (2010) and Tang et al. (2023), the first difference of the series is analyzed as locally stationary. Figure 11 shows that the change in egg prices has a decreasing variance over time. We therefore extend the tvAR model in (25) to have a stochastic volatility component

$$\log \sigma^2_{\varepsilon,t} = \log \sigma^2_{\varepsilon,t-1} + \zeta_t, \qquad \zeta_t \overset{\text{iid}}{\sim} N(0, \sigma^2_\zeta)$$

and add $\log \sigma^2_{\varepsilon,t}$ as the last element of the state vector.

Our aim here is to study the differences in approximating the time domain smoothing posterior of the latent AR parameters, rather than to obtain the optimal model for the data. To keep it simple, we split the time series into different non-overlapping segments and use the `auto.arima` function in R to find the optimal number of lags. The `auto.arima` function is restricted to search for models without differencing, without moving average lags, and to only to consider stationary models. Segment lengths of either 60 or 100 observations are used, and in most cases, either 1 or 2 lags are chosen. A tvAR(2) model with stochastic volatility is therefore used.

The prior is set to $\boldsymbol{Q} \sim \text{Inv-Wishart}\left(10, \text{Diag}(10^{-4}, 10^{-4}, 10^{-3})\right)$ and $\boldsymbol{\theta}_0 \sim N\left(\mathbf{0}_{p+1}, 10 \cdot \boldsymbol{I}_{p+1}\right)$ for both the time domain and dynamic Whittle approaches. For the block Whittle the state variances are multiplied by the step size, S.

We use a segment size $N = 100$ (+1 for dynamic Whittle since $N = 2m + 1$), following Tang et al. (2023), and step size $S = 50$ for the block Whittle. The last section of the Appendix also presents results for a shorter segment size of $N = 50$ combined with a step size of $S = 25$.

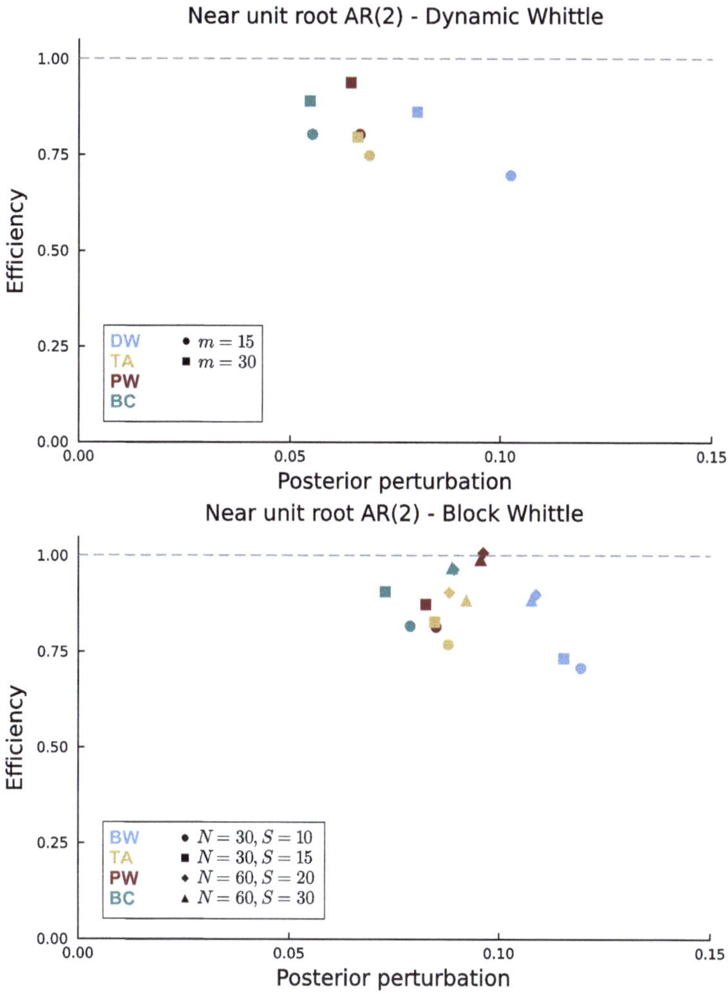

Fig. 10 Experiment 3—occasionally near-unit root AR(2) process. Efficiency and posterior perturbation in the posterior for the AR parameters for the dynamic Whittle likelihood (DW) in the top panel and the block Whittle likelihood (BW) in the bottom panel and with modifications including tapering (TA), prewhitening (PW), and boundary correction (BC)

Figures 12 and 13 show that the raw Whittle likelihood and the prewhitened Whittle likelihood (the red line for posterior median and shaded light gray regions for intervals) do a fairly good job when used to approximate the posterior distribution for the AR parameters from the time domain likelihood (black lines). However, the variance evolution is over-smoothed with too narrow probability bands, which is probably an effect of the fairly long data segments. Figures 23 and 24 in the Appendix show that the smoothing posterior of the variance can be better captured when using shorter segments.

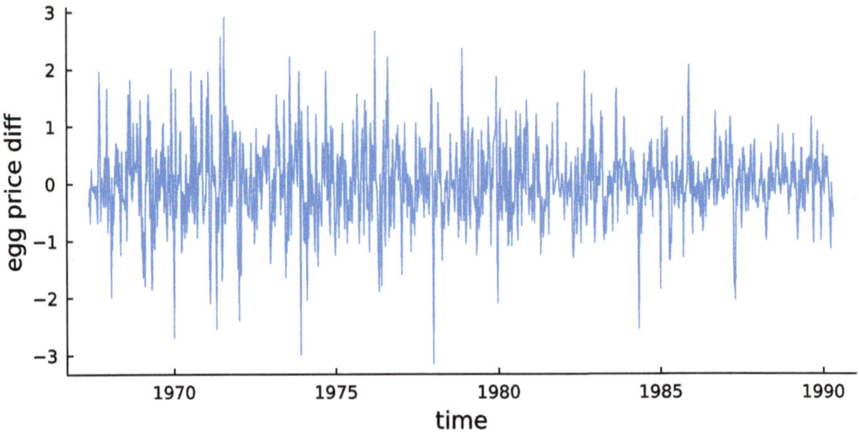

Fig. 11 Weekly changes in egg prices paid to producers in Deutsche Marks/100

The tapered likelihood produces excessively wiggly posteriors compared to their time domain counterpart. Tapering seems to exaggerate the movement of the variance. The normalized Hanning taper used here ensures that the limiting distribution of $I_T(\omega_k)$ is the same whether or not tapering is used (Brillinger, 2001, p.128). Tapering can however distort the variance on a short finite segment, and the overlap between segments can then trick the model into wrongly inferring a volatility cluster. To mitigate this problem we rescale each tapered segment in the time domain to have the same empirical variance as the non-tapered segment data. This rescaling gives a much less perturbed posterior for $\sigma^2_{\varepsilon,t}$, see Fig. 14, although the posterior for ϕ_{1t} is still too wiggly.

The boundary corrected approach uses tapering, so Figs. 12 and 24 show the same problem for this modification. In addition, as noted in Das et al. (2021), the boundary correction induces an inflated variance through the predictive DFT. One can consider a similar rescaling as for the tapered versions, but the rescaling would have to be done on the different periodogram data and is not attempted here. We instead use boundary correction without tapering which performs very well for the AR paths and better on the log variance than BC with tapering; see Figs. 15 and 26.

Figure 16 plots the posterior median for the log-spectral density. To isolate the time variation in the AR parameter from the time variation in the variance, Fig. 17 plots the posterior median of the log of the squared transfer function, $\log|\phi_p(e^{-i\omega})|^2$. The raw and prewhitened Whittle likelihoods approximate the time domain median of the log of the squared transfer function fairly well, while the log spectrogram is over-smoothed.

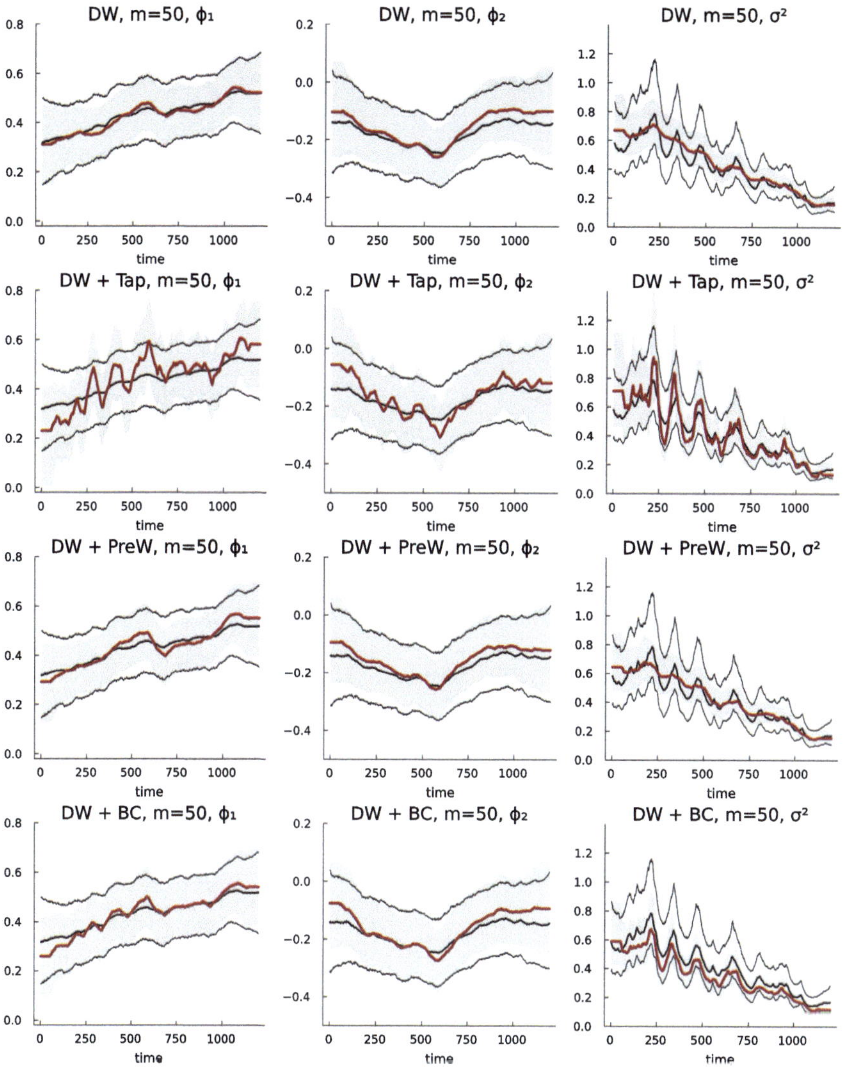

Fig. 12 Egg price data. Posterior median (red) and 95% probability bands (light-shaded gray) for the AR and variance parameter evolutions using the dynamic Whittle likelihood (DW) with tapering (Tap), prewhitening (PreW), and boundary correction (BC), compared to the corresponding measures for the time domain likelihood (black lines)

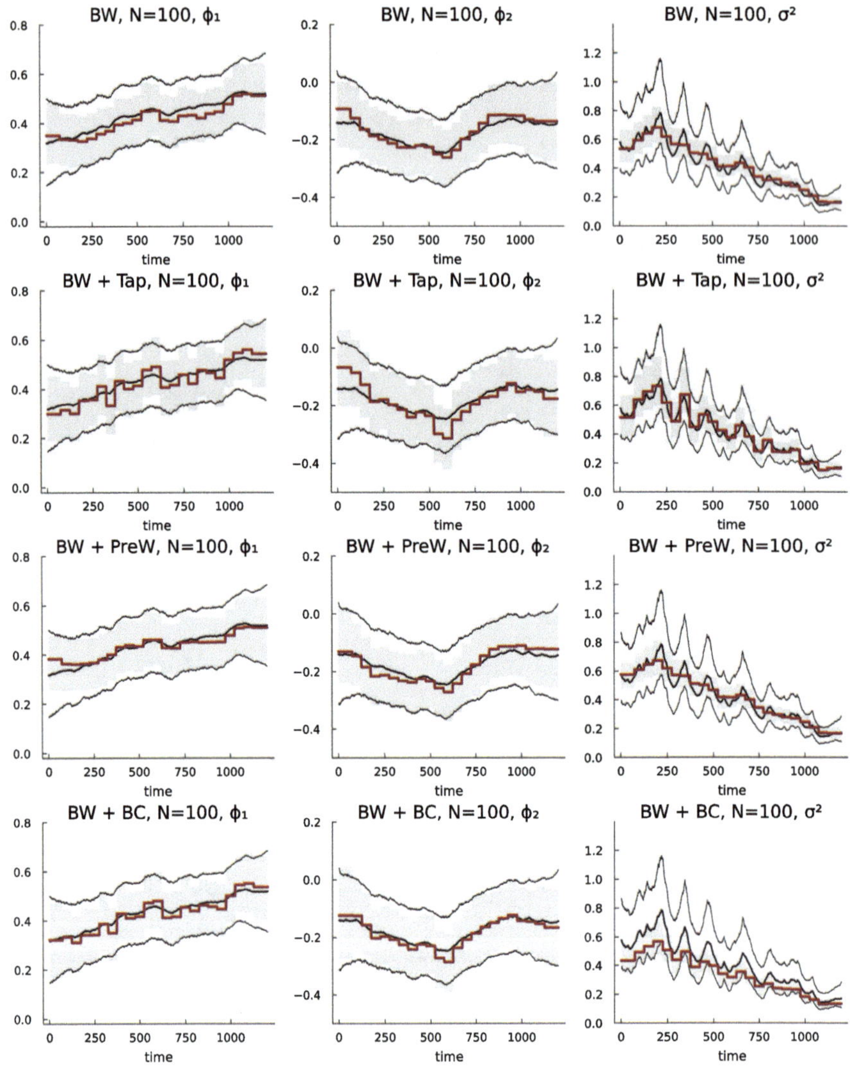

Fig. 13 Egg price data. Posterior median (red) and 95% probability bands (light-shaded gray) for the AR and variance parameter evolutions using the block Whittle likelihood (BW), with tapering (Tap), prewhitening (PreW), and boundary correction (BC), compared to the corresponding measures for the time domain likelihood (black lines)

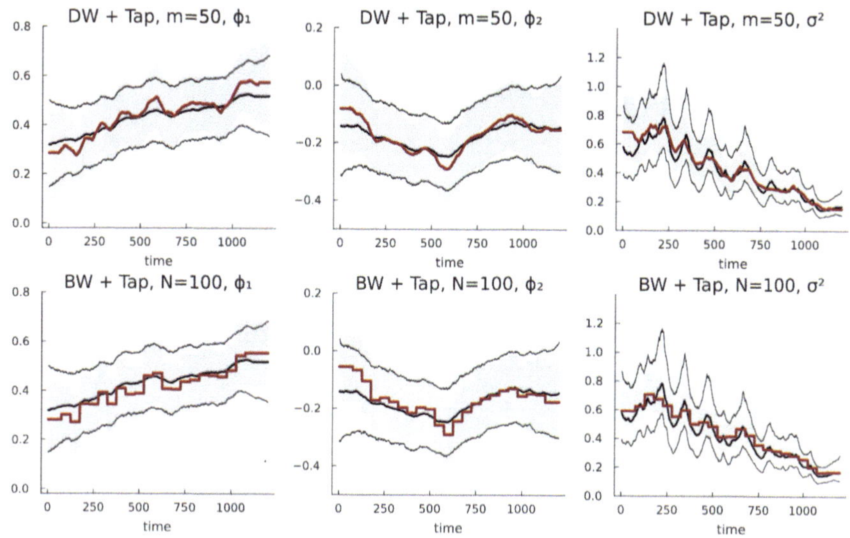

Fig. 14 Egg price data. Posterior median (red) and 95% probability bands (light-shaded gray) for the AR and variance parameter evolutions using the tapered dynamic (DW + Tap) and block Whittle (BW + Tap) likelihood, *with rescaling* on each segment, compared to the corresponding measures for the time domain likelihood (black lines)

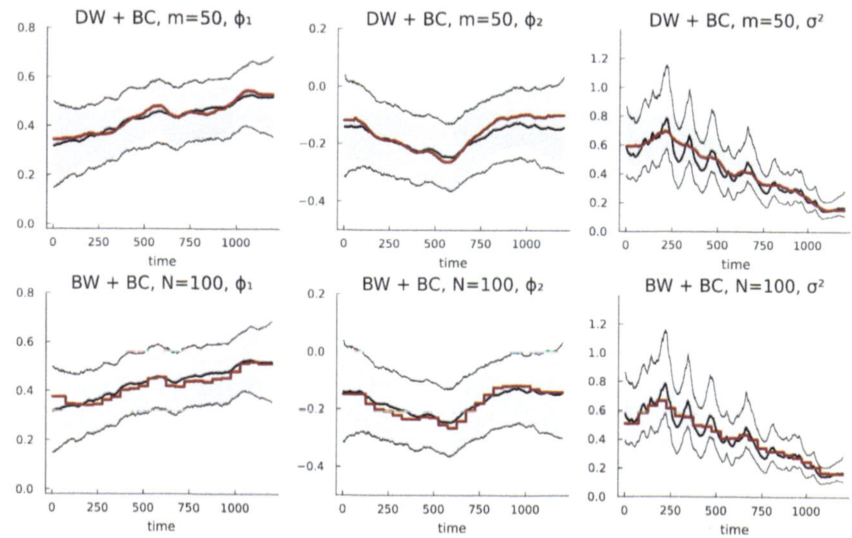

Fig. 15 Egg price data. Posterior median (red) and 95% probability bands (light-shaded gray) for the AR and variance parameter evolutions using the boundary corrected dynamic (DW+BC) and block Whittle (BW+BC) likelihood, *without tapering*. Results are compared to the corresponding measures for the time domain likelihood (black lines)

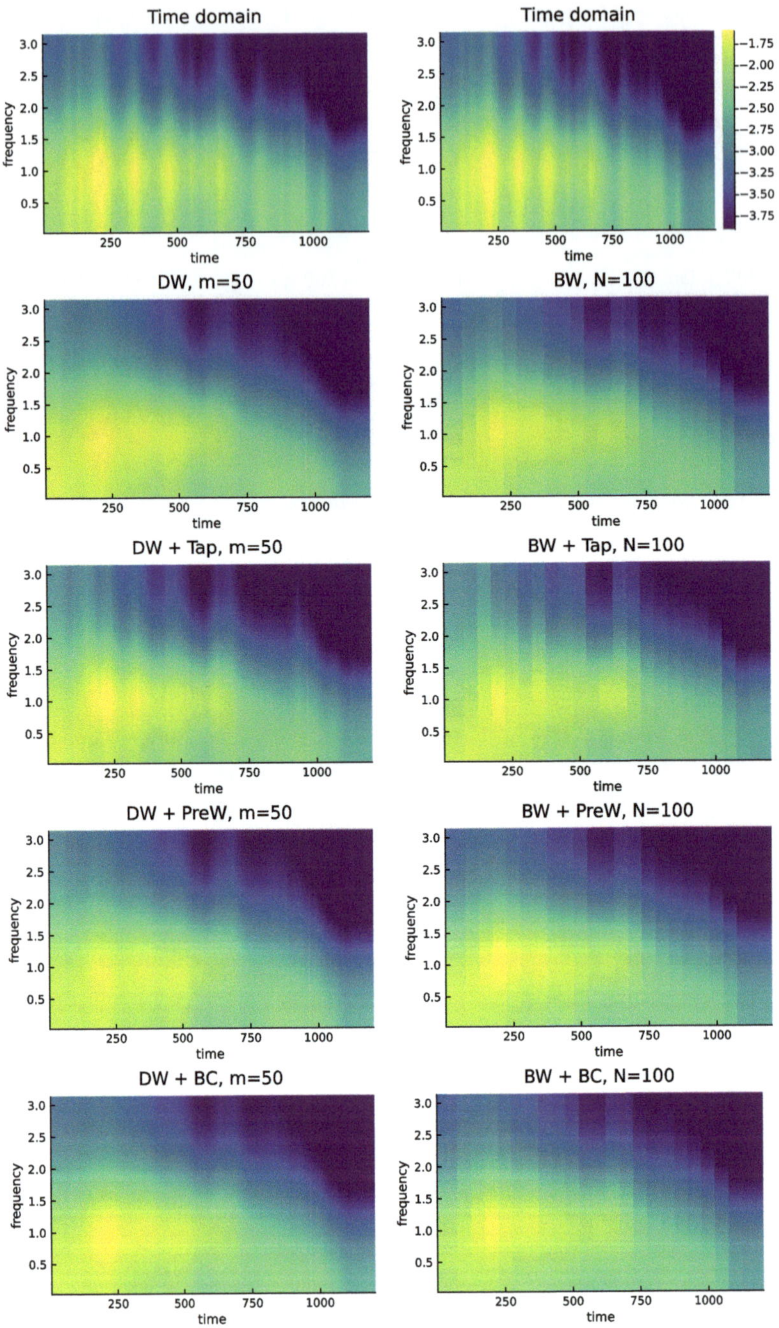

Fig. 16 Egg price data. Posterior median of the log spectrogram for the dynamic (DW) and block (BW) Whittle likelihoods with tapering (Tap), prewhitening (PreW), and boundary correction (BC), using the longer segment length. The tapered segment data is rescaled to match the non-tapered variance, and boundary correction is used without tapering

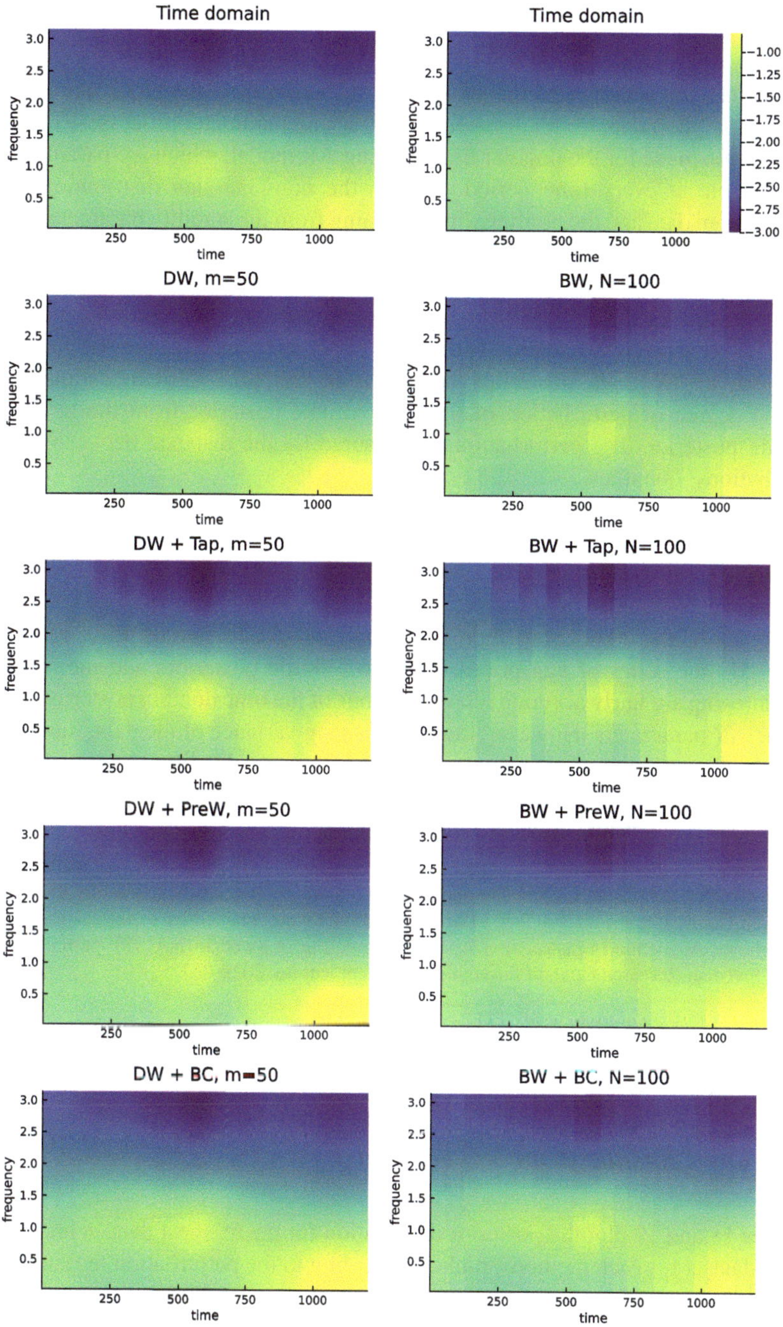

Fig. 17 Egg price data. Posterior median of the log of the squared transfer function for the dynamic (DW) and block (BW) Whittle likelihoods with tapering (Tap), prewhitening (PreW), and boundary correction (BC), using the longer segment length. The tapered segment data is rescaled to match the non-tapered variance, and boundary correction is used without tapering

7 Conclusions

We evaluate the accuracy of the posterior distribution from the block and dynamic Whittle likelihood approximations for time-varying parameter processes. The comparisons are made for the popular time-varying AR model, which are parameterized to be stable at every time period. Due to the non-Gaussian distribution of the periodogram data and the nonlinearity stemming from the stability parameterization, the posterior distribution is sampled by Gibbs sampling with a particle MCMC step for the parameter evolutions.

The quality of the approximations is compared in three simulation experiments along two criteria (i) the (relative) efficiency of point estimates of the AR parameters from the approximate posteriors compared to the time domain estimates and (ii) the degree of perturbation of the approximate posterior from the target time domain posterior. The perturbation is measured by the distance between the two distributions' quantiles.

Both the dynamic and block Whittle likelihoods perform relatively well in all experiments, with an edge for the dynamic Whittle. Both likelihood approximations are much improved by any of the three modifications: tapering, prewhitening, and boundary correction.

We apply the likelihood approximations and modifications to a previously analyzed time series of egg prices. The raw Whittle posteriors and the one from prewhitening are fairly accurate representations of the time domain posterior, at least for the AR parameters. However, the time-varying variance of the data, modeled by a stochastic volatility model, causes some problems for likelihood approximations based on tapering, which needs to be rescaled to give a more correct representation of the posterior for the variance evolution.

Acknowledgments Oskar Gustafsson and Mattias Villani were partially funded by the Swedish Research Council, grant 2020-02846. The computations were enabled by resources provided by the National Academic Infrastructure for Supercomputing in Sweden (NAISS), partially funded by the Swedish Research Council through grant agreement no. 2022-06725.

Appendix

Additional Results from Experiment 1

Figures 18 and 19 plot the posterior distribution for the 5% best dataset and the 5% worst dataset for each methods, ranked according to the perturbation measure. Note that the posteriors for different methods are therefore typically for different datasets.

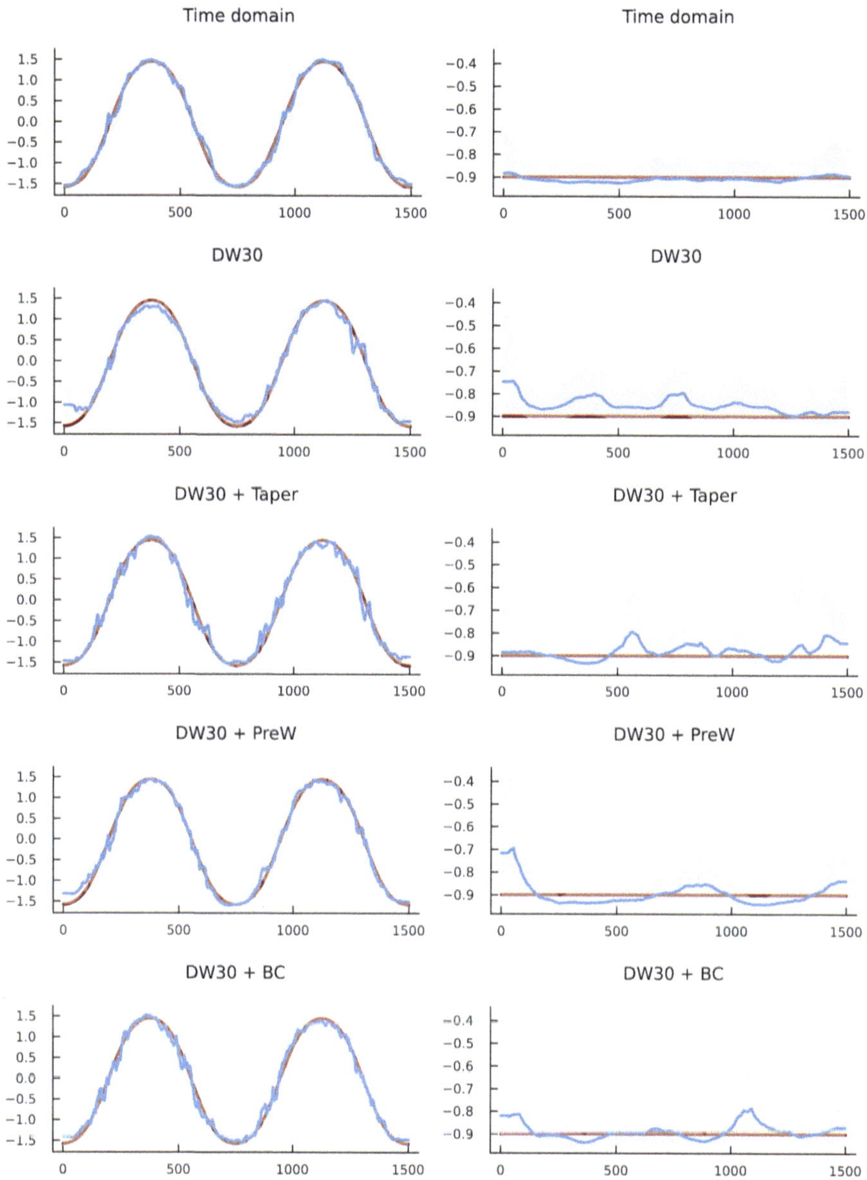

Fig. 18 Experiment 1. Posterior distribution based on the fifth worst dataset for each method ranked according to the perturbation measure. Results are shown for the dynamic Whittle likelihood with $m = 30$ (DW30) and with tapering (Taper), prewhitening (PreW), and boundary correction (BC)

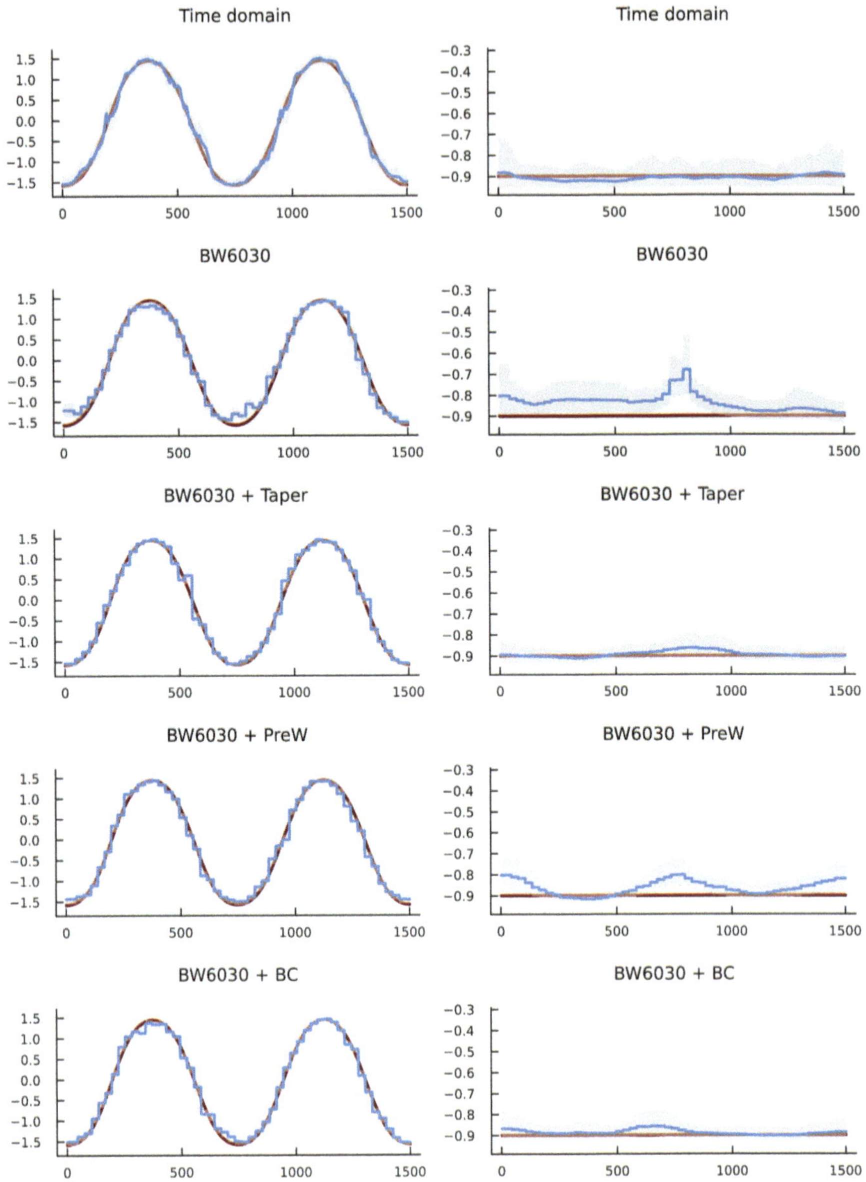

Fig. 19 Experiment 1. Posterior distribution based on the fifth worst dataset for each method ranked according to the perturbation measure. Results are shown for the block Whittle likelihood with $N = 60$ and $S = 30$ (BW6030) and with tapering (Taper), prewhitening (PreW), and boundary correction (BC)

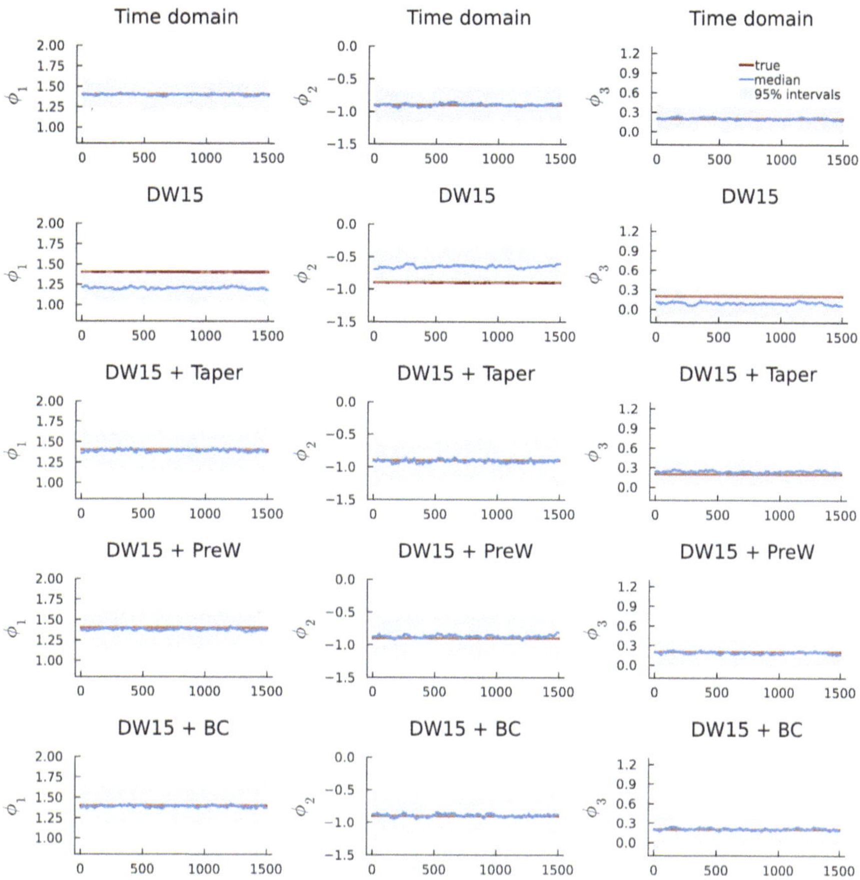

Fig. 20 Experiment 2. Sampling distribution for the posterior median estimates of the parameters from the dynamic Whittle with segment length $m = 15$ (DW15) and with tapering (Taper), prewhitening (PreW), and boundary correction (BC)

Additional Results from Experiment 2

Figures 20 and 21 plot the sampling distributions for the posterior median estimates from the dynamic and block Whittle posteriors for experiment 2.

Additional Results from Experiment 3

Figure 22 plots the sampling distributions for the posterior median estimates from the dynamic Whittle posterior for experiment 3; the results for the block Whittle posterior are similar and not shown.

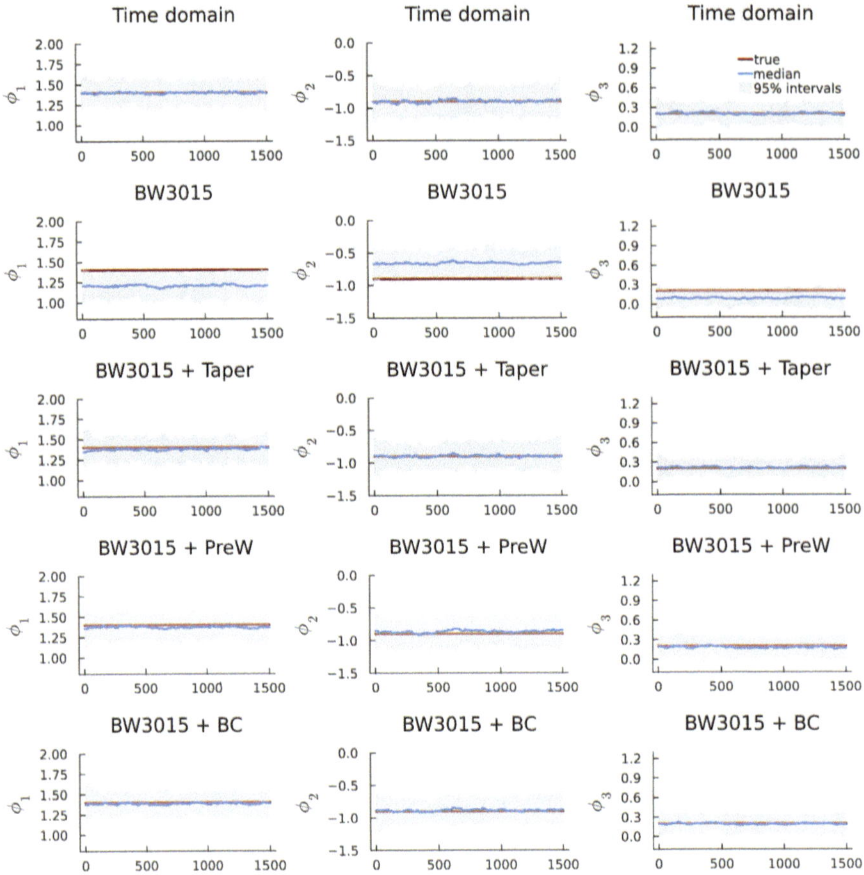

Fig. 21 Experiment 2. Sampling distribution for the posterior median estimates of the parameters from the block Whittle with $N = 30$ and $S = 15$ (BW3015) and with tapering (Taper), prewhitening (PreW), and boundary correction (BC)

Additional Results from the Application to Egg Price Data

Figures 23, 24, 25, and 26 display the result for the egg prices data in Sect. 6 using the smaller segment size.

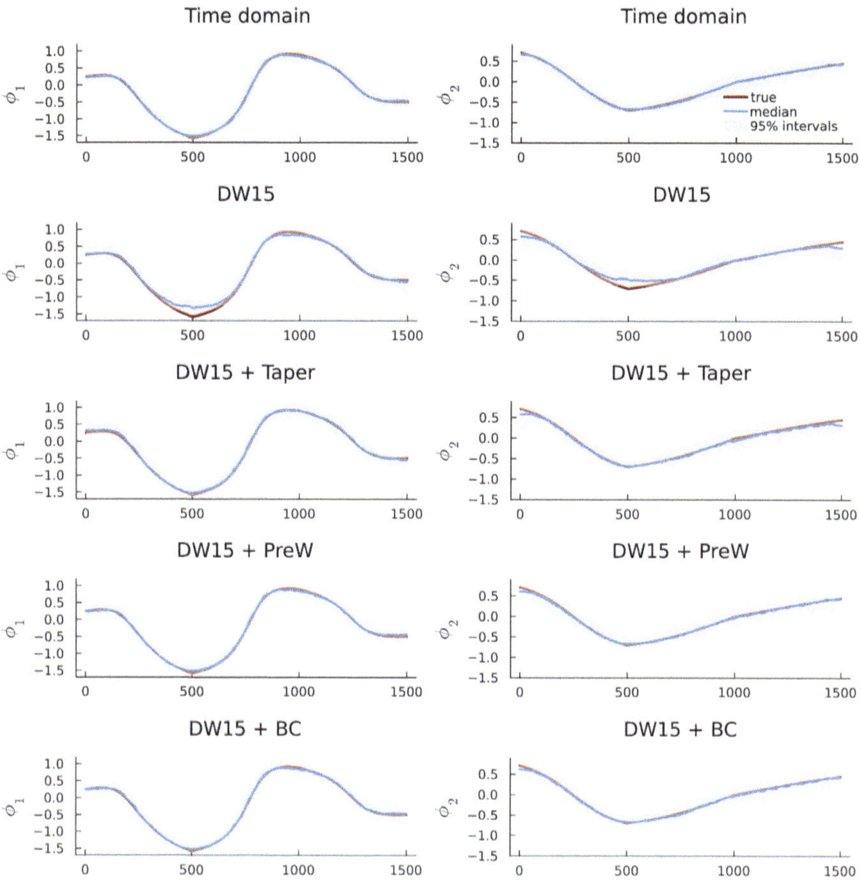

Fig. 22 Experiment 3. Sampling distribution for the posterior median estimates of the parameters from the dynamic Whittle with $m = 15$ (DW15) and with tapering (Taper), prewhitening (PreW), and boundary correction (BC)

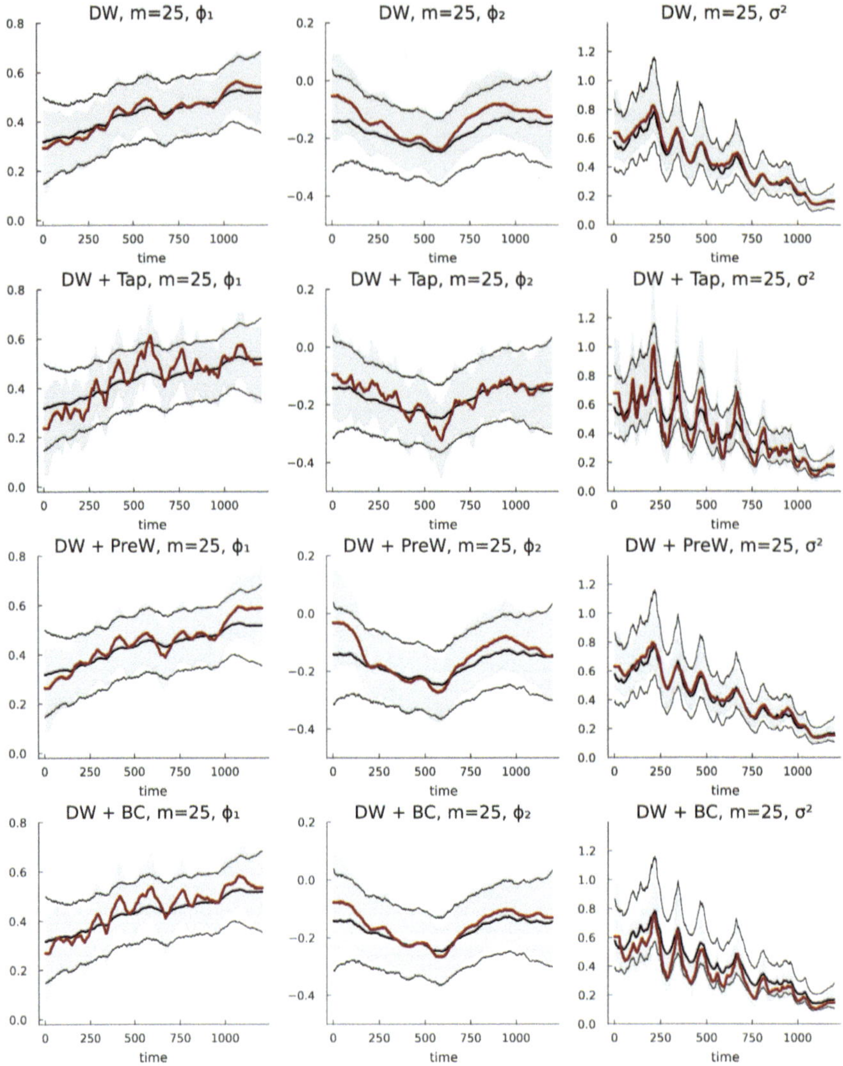

Fig. 23 Egg price data. Posterior median (red) and 95% probability bands (light-shaded gray) for the variance and AR parameter evolutions estimated from the dynamic Whittle likelihood (DW) and with tapering (Tap), prewhitening (PreW), and boundary correction (BC). The results are compared to the corresponding measures for the time domain likelihood (black lines)

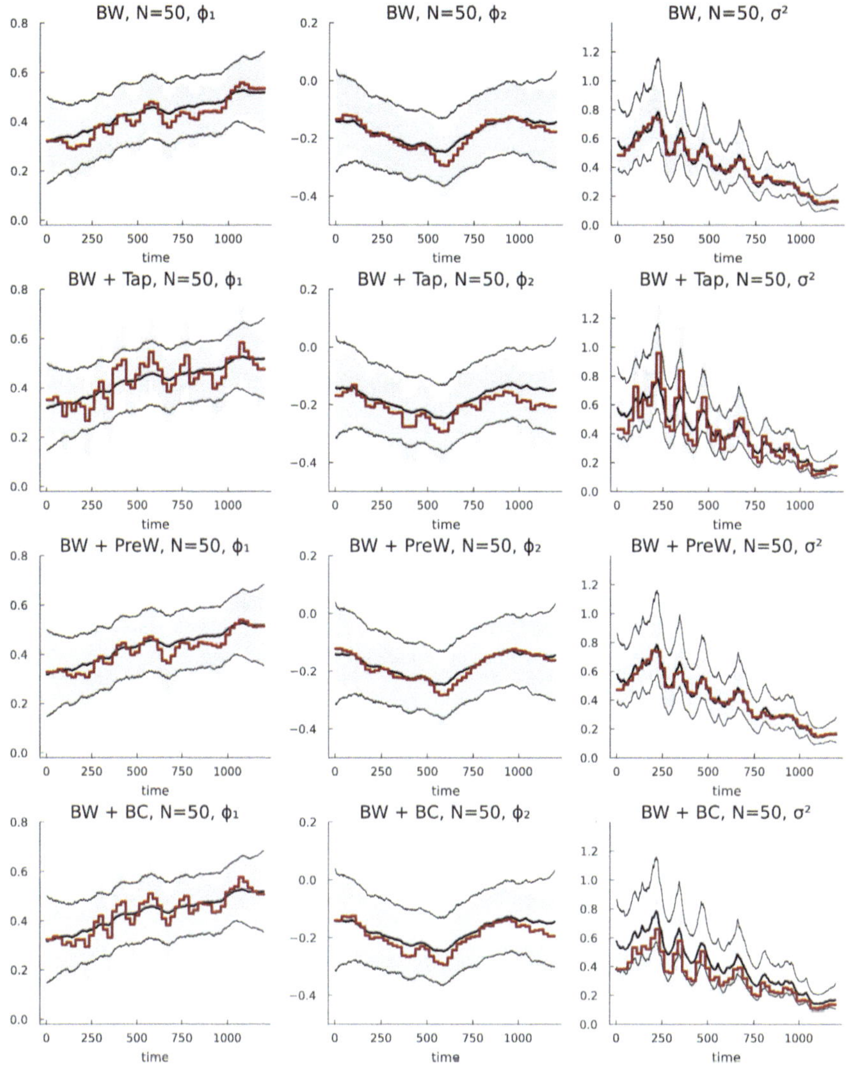

Fig. 24 Egg price data. Posterior median (red) and 95% probability bands (light-shaded gray) for the variance and AR parameter evolutions estimated from the block Whittle likelihood (BW) and with tapering (Tap), prewhitening (PreW), and boundary correction (BC). The results are compared to the corresponding measures for the time domain likelihood (black lines)

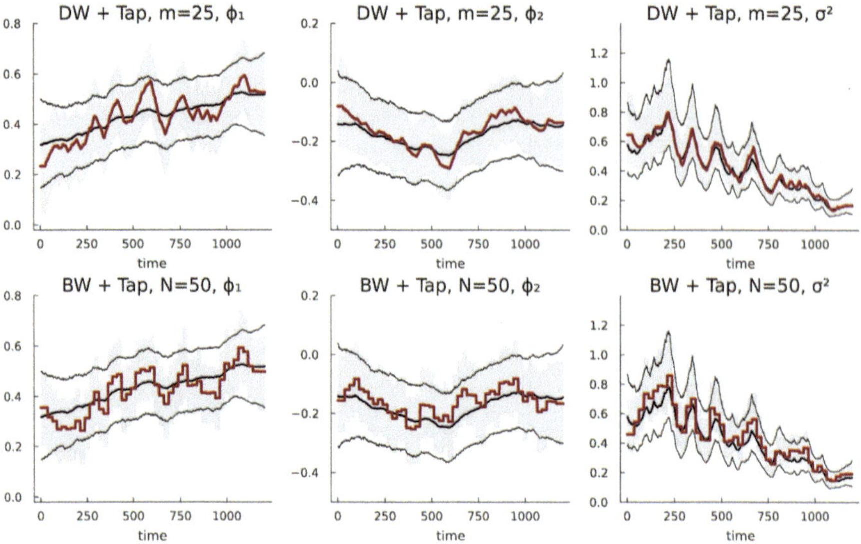

Fig. 25 Egg price data. Posterior median (red) and 95% probability bands (light-shaded gray) for the variance and AR parameter evolutions estimated from the tapered dynamic (DW + Tap) and block Whittle (BW + Tap) likelihoods, with rescaling on each segment. The results are compared to the corresponding measures for the time domain likelihood (black lines)

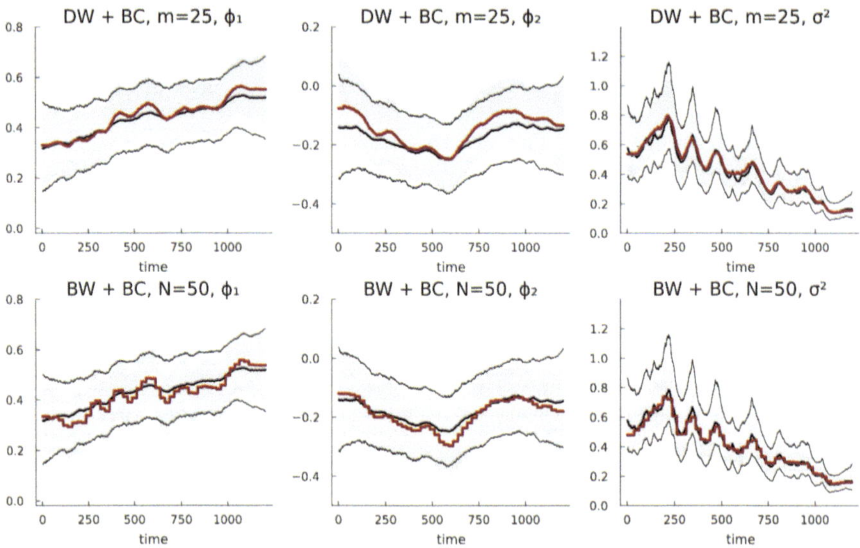

Fig. 26 Egg price data. Posterior median (red) and 95% probability bands (light-shaded gray) for the variance and AR parameter evolutions estimated from the boundary corrected dynamic (DW + BC) and block Whittle (BW + BC) likelihood, without tapering. The results are compared to the corresponding measures for the time domain likelihood (black lines)

References

Barndorff-Nielsen, O. & Schou, G. (1973). On the parametrization of autoregressive models by partial autocorrelations. *Journal of Multivariate Analysis, 3*(4), 408–419.
Brillinger, D. (2001). *Time series: Data analysis and theory*. Classics in Applied Mathematics. Society for Industrial and Applied Mathematics.
Brockwell, P. J. & Davis, R. A. (1991). *Time series: Theory and methods*. Springer Science & Business Media.
Cadonna, A., Frühwirth-Schnatter, S., & Knaus, P. (2020). Triple the gamma–a unifying shrinkage prior for variance and variable selection in sparse state space and TVP models. *Econometrics, 8*(2), 20.
Carter, C. K. & Kohn, R. (1994). On Gibbs sampling for state space models. *Biometrika, 81*(3), 541–553.
Carvalho, C. M., Polson, N. G., & Scott, J. G. (2010). The horseshoe estimator for sparse signals. *Biometrika, 97*(2), 465–480.
Chib, S. (1998). Estimation and comparison of multiple change-point models. *Journal of Econometrics, 86*(2), 221–241.
Contreras-Cristán, A., Gutiérrez-Peña, E., & Walker, S. G. (2006). A note on Whittle's likelihood. *Communications in Statistics-Simulation and Computation, 35*(4), 857–875.
Dahlhaus, R. (1988). Small sample effects in time series analysis: A new asymptotic theory and a new estimate. *The Annals of Statistics, 16*, 808–841.
Dahlhaus, R. (1997). Fitting time series models to nonstationary processes. *The Annals of Statistics, 25*(1), 1–37.
Dahlhaus, R. (2000). A likelihood approximation for locally stationary processes. *The Annals of Statistics, 28*(6), 1762–1794.
Dahlhaus, R. (2009). Local inference for locally stationary time series based on the empirical spectral measure. *Journal of Econometrics, 151*(2), 101–112.
Dahlhaus, R. (2012). Locally stationary processes. In *Handbook of statistics* (Vol. 30, pp. 351–413). Elsevier.
Dahlhaus, R. & Polonik, W. (2009). Empirical spectral processes for locally stationary time series. *Bernoulli, 15*(1), 1–39.
Das, S., Subba Rao, S., & Yang, J. (2021). Spectral methods for small sample time series: A complete periodogram approach. *Journal of Time Series Analysis, 42*(5–6), 597–621.
Dzhaparidze, K. & Yaglom, A. (1983). Spectrum parameter estimation in time series analysis. In *Developments in statistics* (Vol. 4, pp. 1–96). Elsevier.
Everitt, R. G., Andrieu, C., & Davy, M. (2013). Online Bayesian inference in some time-frequency representations of non-stationary processes. *IEEE Transactions on Signal Processing, 61*(22), 5755–5766.
Fagerberg, G., Villani, M., & Kohn, R. (2024). Time-varying multi-seasonal AR models. *arXiv preprint arXiv:2409.18640*.
Fan, J. & Yao, Q. (2008). *Nonlinear time series: Nonparametric and parametric methods*. Springer.
Fox, R. & Taqqu, M. S. (1986). Large-sample properties of parameter estimates for strongly dependent stationary Gaussian time series. *The Annals of Statistics, 14*(2), 517–532.
Franke, J. & Hardle, W. (1992). On bootstrapping kernel spectral estimates. *The Annals of Statistics, 20*(1), 121–145.
Frühwirth-Schnatter, S. (1994). Data augmentation and dynamic linear models. *Journal of Time Series Analysis, 15*(2), 183–202.
Gerlach, R., Carter, C., & Kohn, R. (2000). Efficient Bayesian inference for dynamic mixture models. *Journal of the American Statistical Association, 95*(451), 819–828.
Häfner, F. & Kirch, C. (2017). Moving Fourier analysis for locally stationary processes with the bootstrap in view. *Journal of Time Series Analysis, 38*(6), 895–922.
Hamilton, J. D. (1989). A new approach to the economic analysis of nonstationary time series and the business cycle. *Econometrica, 57*(2), 357–384.

Hannan, E. J. (1973). The asymptotic theory of linear time-series models. *Journal of Applied Probability, 10*(1), 130–145.

Hannan, E. J. & Quinn, B. G. (1979). The determination of the order of an autoregression. *Journal of the Royal Statistical Society: Series B, 41*(2), 190–195.

Kalli, M. & Griffin, J. E. (2014). Time-varying sparsity in dynamic regression models. *Journal of Econometrics, 178*(2), 779–793.

Karlsson, S. (2013). Forecasting with Bayesian vector autoregression. *Handbook of Economic Forecasting, 2*, 791–897.

Kitagawa, G. & Gersch, W. (1985). A smoothness priors time-varying AR coefficient modeling of nonstationary covariance time series. *IEEE Transactions on Automatic Control, 30*(1), 48–56.

Knaus, P. & Frühwirth-Schnatter, S. (2023). The dynamic triple gamma prior as a shrinkage process prior for time-varying parameter models. *arXiv preprint arXiv:2312.10487.*

Kowal, D. R., Matteson, D. S., & Ruppert, D. (2019). Dynamic shrinkage processes. *Journal of the Royal Statistical Society: Series B, 81*(4), 781–804.

Lindgren, G. (2013). *Stationary stochastic processes.* CRC Press.

Lindsten, F., Jordan, M. I., & Schön, T. B. (2014). Particle Gibbs with ancestor sampling. *Journal of Machine Learning Research, 15*, 2145–2184.

Monahan, J. F. (1984). A note on enforcing stationarity in autoregressive-moving average models. *Biometrika, 71*(2), 403–404.

Neumann, M. H. & Von Sachs, R. (1997). Wavelet thresholding in anisotropic function classes and application to adaptive estimation of evolutionary spectra. *The Annals of Statistics, 25*(1), 38–76.

Paparoditis, E. (2010). Validating stationarity assumptions in time series analysis by rolling local periodograms. *Journal of the American Statistical Association, 105*(490), 839–851.

Peligrad, M. & Wu, W. B. (2010). Central limit theorem for Fourier transforms of stationary processes. *The Annals of Probability, 38*, 2009–2022.

Peligrad, M. & Zhang, N. (2019). Central limit theorem for Fourier transform and periodogram of random fields. *Bernoulli, 25*(2), 499–520.

Prado, R. & West, M. (2010). *Time series: Modeling, computation, and inference.* Chapman and Hall/CRC.

Priestly, M. B. (1981). *Spectral analysis and time series* (Vol. 1). Academic.

Quiroz, M., Kohn, R., Villani, M., & Tran, M.-N. (2019). Speeding up MCMC by efficient data subsampling. *Journal of the American Statistical Association, 114*(526), 831–843.

Rosen, O., Wood, S., & Stoffer, D. S. (2012). AdaptSPEC: Adaptive spectral estimation for nonstationary time series. *Journal of the American Statistical Association, 107*(500), 1575–1589.

Salomone, R., Quiroz, M., Kohn, R., Villani, M., & Tran, M.-N. (2020). Spectral subsampling MCMC for stationary time series. In *International Conference on Machine Learning* (pp. 8449–8458). PMLR.

Sandsten, M. (2020). Time-frequency analysis of time-varying signals and non-stationary processes. *Lund University.*

Shao, X., Wu, W. B., et al. (2007). Asymptotic spectral theory for nonlinear time series. *The Annals of Statistics, 35*(4), 1773–1801.

Subba Rao, S. & Yang, J. (2021). Reconciling the Gaussian and Whittle likelihood with an application to estimation in the frequency domain. *The Annals of Statistics, 49*(5), 2774–2802.

Sykulski, A. M., Olhede, S. C., Guillaumin, A. P., Lilly, J. M., & Early, J. J. (2019). The debiased Whittle likelihood. *Biometrika, 106*(2), 251–266.

Tang, Y., Kirch, C., Lee, J. E., & Meyer, R. (2023). Bayesian nonparametric spectral analysis of locally stationary processes. *arXiv preprint arXiv:2303.11561.*

Tjøstheim, D. & Paulsen, J. (1983). Bias of some commonly-used time series estimates. *Biometrika, 70*(2), 389–399.

van Delft, A. & Eichler, M. (2019). Data-adaptive estimation of time-varying spectral densities. *Journal of Computational and Graphical Statistics, 28*(2), 244–255.

Villani, M., Quiroz, M., Kohn, R., & Salomone, R. (2024). Spectral subsampling MCMC for stationary multivariate time series with applications to vector ARTFIMA processes. *Econometrics and Statistics, 32*, 98–121.
Wahba, G. (1980). Automatic smoothing of the log periodogram. *Journal of the American Statistical Association, 75*(369), 122–132.
Whittle, P. (1951). *Hypothesis testing in time series analysis.* Almqvist & Wiksell.
Whittle, P. (1953). Estimation and information in stationary time series. *Arkiv för matematik, 2*(5), 423–434.

Bayesian Regularization of the Tangency Portfolio

Olha Bodnar, Taras Bodnar, and Vilhelm Niklasson

Abstract In the chapter, two new priors, designed directly for the tangency portfolio weights, are developed. While the first approach is based on the extension of the Laplace prior, the second procedure generalizes the spike and slab prior. The posterior distributions of the tangency portfolio weights under both priors are characterized in terms of stochastic representations. These findings are used to establish exact sampling schemes for drawing samples of tangency portfolio weights from the corresponding posterior distributions, from which both the Bayesian point and interval estimators of the tangency portfolio weights are constructed.

1 Introduction

Mean-variance portfolio theory is a well-established tool in finance starting with the seminal contribution of Harry Markowitz (see Markowitz, 1952). The idea behind the approach is to select a portfolio which has the largest expected return for the given level of risk, measured by the portfolio variance, or alternatively to choose the portfolio with the smallest risk for a preselected portfolio return.

Although the mean-variance portfolio theory provides investors with the closed-form formulas of optimal portfolio weights, the expressions cannot be directly implemented in practice since they depend on the mean vector and covariance matrix of the asset returns, which are unknown quantities in practice and have to be estimated by using historical observations of the asset return vectors. The

O. Bodnar
Division of Statistics, School of Business, Örebro University, Örebro, Sweden
e-mail: olha.bodnar@oru.se

T. Bodnar (✉)
Department of Management and Engineering, Linköping University, Linköping, Sweden
e-mail: taras.bodnar@liu.se

V. Niklasson
Department of Mathematics, Stockholm University, Stockholm, Sweden
e-mail: niklasson@math.su.se

S. Mazur, P. Österholm (eds.), *Recent Developments in Bayesian Econometrics and Their Applications*, https://doi.org/10.1007/978-3-032-00110-8_9

problem of incorporating the parameter uncertainty into the decision process has been investigated in a number of literature studies, from the viewpoints of both frequentist statistics (cf., Jobson & Korkie, 1980; Gibbons et al., 1989; Okhrin & Schmid, 2006; Bodnar & Bodnar, 2010; Gupta et al., 2013; Bodnar et al., 2018; Bodnar et al., 2023a) and Bayesian statistics (cf., Zellner & Chetty, 1965; Winkler & Barry, 1975; Bawa et al., 1979; Stambaugh, 1997; Aguilar & West, 2000; Tu & Zhou, 2010; Bauder et al., 2020; Bodnar et al., 2022c).

Recently, Bayesian portfolio selection has increased its popularity in both statistical and financial literature (see Tu & Zhou, 2010; Bodnar et al., 2017; Bauder et al., 2018; Bauder et al., 2021; Shapiro et al., 2023; Bodnar et al., 2024d). The Bayesian approaches possess several advantages in comparison to the methods of frequentist statistics. First, the application of the Bayesian procedures leads to the posterior distribution of optimal portfolio weights, while the frequentist approaches result in their point estimates. Second, the methods of Bayesian statistics allow to incorporate the challenging problem of parameter uncertainty directly into the optimization tasks, while the population portfolios are constructed and then estimated when the conventional methods are employed. Third, the Bayesian procedures, in contrast to the frequentist ones, implement different sources of information and expert knowledge into the decision process. It is remarkable that the Bayesian methods not only provide an effective way to deal with the problem of parameter uncertainty but are also based on rigorous probability theory.

When there is no possibility to invest into a risk-free asset, the set of optimal portfolio is a parabola in the mean-variance space and a hyperbola in the mean-standard deviation space (cf., Merton, 1972; Kan & Smith, 2008; Bodnar & Schmid, 2009). The situation simplifies when there is a possibility to invest into a risk-free asset. In the latter situation, the efficient frontier is a line which is determined as the tangent line drawn to the efficient frontier, constructed in the case without a risk-free asset, from the location of the risk-free asset in the mean-variance (mean-standard deviation) space. The intersection point between the two efficient frontiers is the tangency portfolio, which attract a special attention in financial literature (see Britten-Jones, 1999; Kan & Zhou, 2007; Bauder et al., 2018; Lassance et al., 2023). Recently, Bodnar et al. (2024b) developed a new Bayesian approach to estimate the tangency portfolio by using the information available in asset returns taken at different frequencies and in other indicators of the financial market, while Bodnar et al. (2024a) extended these approaches by incorporating the non-short selling constraints into the optimization problem.

In this chapter, we extend the results of Bodnar et al. (2024b) and Bodnar et al. (2024a) by introducing regularization approaches to Bayesian portfolio selection and providing a more efficient way to deal with parameter uncertainty problems in practice. Historically, the first Bayesian approach for model selection was based on the Bayes factor in the case of proper priors (cf., Kass & Raftery, 1995) and the intrinsic Bayes factor in the case of improper prior (see, e.g., Berger & Pericchi, 1996, 2001). Recently, Bayesian lasso (see Park & Casella, 2008) and Bayesian elastic net (see Li & Lin, 2010) were proposed in the literature as further methods. Another widely spread approach for performing model selection from the Bayesian perspective employs a spike and slab prior as proposed in Ishwaran and Rao (2005). The latter approach, which performs both the selection of stocks to be included in a portfolio and the estimation of portfolio structure, together with the Bayesian lasso method will be extended to derive the Bayesian inference procedures for the tangency portfolio weights in the case without short selling.

Our results will shed light on the quantitative understanding of the trade-off between portfolio diversification and parameter uncertainty. Namely, our findings provide the investor with information about the structure of the optimal portfolio which will be in line with Makowitz's principle "Don't put all your eggs in one basket" and the sparse structure of optimal portfolios which suggests to avoid investing in stocks whose weights are not statistically significant to reduce the estimation risk. The aim is achieved by reparameterizing the stochastic model of asset returns in such a way that the vector of the weights of the tangency portfolio is a parameter in the new parameterization. This result allows the investor to determine priors directly for the weights of the tangency portfolio. Therefore, Bayesian methods of model selection can be employed in the new parameterization, yielding the desired regularized Bayesian tangency portfolio. The derived stochastic representations of the tangency portfolio weights present an important advantage of the suggested approach by allowing exact approaches to draw samples from the posterior distribution of the tangency portfolio weights. In contrast to widely used Markov chain Monte Carlo (MCMC) methods, the application of stochastic representations leads to numerical methods that are exact and time-efficient.

The rest of the chapter is structured as follows. In Sect. 2, we provide the mathematical definition of the tangency portfolio and briefly review the existing approaches to estimate it. Methods of the frequentist statistics are described in Sect. 2.1, while Sects. 2.2 and 2.3 present the Bayesian approaches in both cases when short selling is allowed or not, respectively. The main results of the chapter are given in Sect. 3, where the regularized Bayesian tangency portfolio is derived. Concluding remarks are summarized in Sect. 4.

2 Tangency Portfolio and Its Estimation

Let $\tilde{\boldsymbol{x}}_1, \ldots, \tilde{\boldsymbol{x}}_n$ be the p-dimensional vectors of the asset returns on p assets, and let r_f denote the return of the risk-free asset. The vectors of the excess returns are defined by $\boldsymbol{x}_1 = \tilde{\boldsymbol{x}}_1 - r_f\mathbf{1}, \ldots, \boldsymbol{x}_n = \tilde{\boldsymbol{x}}_n - r_f\mathbf{1}$, where $\mathbf{1}$ is the p-dimensional vector of ones. Further, let $\boldsymbol{x}_{1:n} = (\boldsymbol{x}_1, \ldots, \boldsymbol{x}_n)$ be the data matrix which consists on n realizations of the p-dimensional vectors of the excess asset returns.

In the following the mean vector and the covariance matrix of the p-dimensional vector of the excess asset returns are denoted by $\boldsymbol{\mu}$ and $\boldsymbol{\Sigma}$, respectively. Following the Bayesian setup, it is assumed that the vectors of the excess returns are independent and multivariate normally distributed conditionally on $\boldsymbol{\mu}$ and $\boldsymbol{\Sigma}$. It is important to note that unconditionally the assets returns are no longer neither independent nor normally distributed (see, e.g., Bernardo & Smith, 2000; Bodnar et al., 2022c).

Under the assumptions imposed on the asset returns, we get that given $\boldsymbol{\mu}$ and $\boldsymbol{\Sigma}$ the joint distribution of the asset returns belongs to the exponential family (see Sundberg, 2019) with the likelihood function expressed as

$$\begin{aligned} f(\boldsymbol{x}_{1:n}|\boldsymbol{\mu}, \boldsymbol{\Sigma}) &\propto \det(\boldsymbol{\Sigma})^{-n/2} \exp\left(-\frac{1}{2}\sum_{i=1}^{n}(\boldsymbol{x}_i - \boldsymbol{\mu})^\top \boldsymbol{\Sigma}^{-1}(\boldsymbol{x}_i - \boldsymbol{\mu})\right) \\ &= \det(\boldsymbol{\Omega})^{n/2} \exp\left(-\frac{n}{2}\boldsymbol{\nu}^\top \boldsymbol{\Omega}^{-1}\boldsymbol{\nu}\right) \exp\left(\left(\sum_{i=1}^{n}\boldsymbol{x}_i\right)^\top \boldsymbol{\nu} + \operatorname{tr}\left(\boldsymbol{\Omega}\left(-\frac{1}{2}\sum_{i=1}^{n}\boldsymbol{x}_i\boldsymbol{x}_i^\top\right)\right)\right), \end{aligned} \tag{1}$$

where

$$\boldsymbol{\nu} = \boldsymbol{\Sigma}^{-1}\boldsymbol{\mu} \quad \text{and} \quad \boldsymbol{\Omega} = \boldsymbol{\Sigma}^{-1} \tag{2}$$

are the canonical parameters, while the canonical statistic $(\boldsymbol{t}, -\boldsymbol{T}/2)$ is given by

$$\boldsymbol{t} = \sum_{i=1}^{n}\boldsymbol{x}_i \quad \text{and} \quad \boldsymbol{T} = \sum_{i=1}^{n}\boldsymbol{x}_i\boldsymbol{x}_i^\top. \tag{3}$$

On the other side, the weights of the tangency portfolio are expressed as (see, e.g., Ingersoll, 1987; Britten-Jones, 1999)

$$\boldsymbol{w} = \gamma^{-1}\boldsymbol{\Sigma}^{-1}\boldsymbol{\mu} = \gamma^{-1}\boldsymbol{\nu}, \tag{4}$$

where $\gamma > 0$ is the risk-aversion coefficient determining the investor attitude toward risk. As a result, the canonical representation of the likelihood function allows us to assess the weights of the tangency portfolio directly. In particular, the investor beliefs about the weights of the tangency portfolio and other sources of information can be incorporated in the definition of the prior (see Bodnar et al., 2024b and Bodnar et al., 2024a), and the Bayesian inference procedures can directly be constructed on the weights.

In the next subsections, we review the methods of both frequentist and Bayesian statistics to estimate the weights of the tangency portfolio, while the new regularized Bayesian tangency portfolio will be introduced in Sect. 3.

2.1 Frequentist Methods

The conventional frequentist approach to estimate the weights of the tangency portfolio, which is widely spread in practice, utilizes the sample estimators of the mean vector and the covariance matrix of the asset returns expressed as

$$\bar{\boldsymbol{x}}_n = \frac{1}{n}\sum_{i=1}^{n} \boldsymbol{x}_i \tag{5}$$

and

$$\boldsymbol{S}_n = \frac{1}{n-1}\sum_{i=1}^{n}(\boldsymbol{x}_i - \bar{\boldsymbol{x}})(\boldsymbol{x}_i - \bar{\boldsymbol{x}})^\top, \tag{6}$$

respectively. Then, the estimated weights of the tangency portfolio are given by

$$\hat{\boldsymbol{w}}_S = \gamma^{-1}\boldsymbol{S}_n^{-1}\bar{\boldsymbol{x}}_n. \tag{7}$$

If the portfolio dimension p is considerably smaller than the size of the sample used to estimate the mean vector and the covariance matrix of the asset returns, then the estimator (7) is consistent and asymptotically normally distributed as soon as the estimators $\bar{\boldsymbol{x}}_n$ and $\boldsymbol{S}_n$ are consistent and asymptotically normally distributed following the multivariate versions of the fundamental theorems of classical asymptotic statistics, like the continuous mapping theorem and the multivariate delta methods (see, e.g., DasGupta, 2008).

The situation becomes more complicated when the portfolio dimension is comparable to the sample size, i.e., under the high-dimensional setting. In this case, the theorems of classical asymptotic statistics provide a poor approximation of the sampling distribution of the tangency portfolio weights, and, consequently, they cannot longer be used. To solve the challenging issue, the methods of random matrix theory come in the game which are derived under the high-dimensional asymptotic setup (see, e.g., Bai & Silverstein, 2010). One of the most used high-dimensional approaches is based on the shrinkage methods, which drastically reduces the variability of the sample estimators in the price of a bias (cf., Stein, 1956). The situation becomes even more difficult when $p > n$. In this case, the sample covariance matrix is singular, and its generalized inverse, like the Moore-Penrose inverse, should be used (see Bodnar et al., 2016b; Bodnar & Parolya, 2024).

The improved shrinkage estimators of the weights of the tangency portfolio are constructed by using the improved estimators of the mean vector $\boldsymbol{\mu}$ and the covariance matrix $\boldsymbol{\Sigma}$. In general, a shrinkage estimator is expressed as

$$\hat{\boldsymbol{w}}_{n,Sh} = \gamma^{-1}\hat{\boldsymbol{\Sigma}}_{n,Sh}^{-1}\hat{\boldsymbol{\mu}}_{n,Sh}, \tag{8}$$

where $\hat{\boldsymbol{\mu}}_{n,Sh}$ and $\hat{\boldsymbol{\Sigma}}_{n,Sh}$ are some shrinkage estimators of the mean vector and covariance matrix of the asset returns.

Several shrinkage estimators of the mean vector exist in the literature with the mostly spread approaches presented by:

- **James-Stein shrinkage estimator (Jorion, 1986)**:

$$\hat{\boldsymbol{\mu}}_{n,JS} = (1 - \hat{\alpha}_{JS})\bar{\boldsymbol{x}}_n + \hat{\alpha}_{JS} x_0 \mathbf{1}, \tag{9}$$

where x_0 is the shrinkage target and

$$\hat{\alpha}_{JS} = \min\left\{1, \frac{(p-2)/n}{(\bar{\boldsymbol{x}}_n - x_0\mathbf{1})^\top \mathbf{S}_n^{-1}(\bar{\boldsymbol{x}}_n - x_0\mathbf{1})}\right\}. \tag{10}$$

- **Bayes-Stein shrinkage estimator (Jorion, 1986)**:

$$\hat{\boldsymbol{\mu}}_{n,BS} = (1 - \hat{\alpha}_{BS})\bar{\boldsymbol{x}}_n + \hat{\alpha}_{BS} \hat{x}_0 \mathbf{1}, \tag{11}$$

where

$$\hat{\alpha}_{BS} = \frac{p+2}{p+2+n(\bar{\boldsymbol{x}}_n - \hat{x}_0\mathbf{1})^\top \mathbf{S}_n^{-1}(\bar{\boldsymbol{x}}_n - \hat{x}_0\mathbf{1})} \quad \text{and} \quad \hat{x}_0 = \frac{\bar{\boldsymbol{x}}_n^\top \mathbf{S}_n^{-1}\mathbf{1}}{\mathbf{1}^\top \mathbf{S}_n^{-1}\mathbf{1}}. \tag{12}$$

Moreover, the application of $\frac{n-1}{n-p-2}\mathbf{S}_n$ instead of $\mathbf{S}_n$ in the above formulas is suggested in Jorion (1986).

- **Optimal linear shrinkage estimator (Bodnar et al., 2019c)**:

$$\hat{\boldsymbol{\mu}}_{n,BOP} = \hat{\alpha}_{BOP}\bar{\boldsymbol{x}}_n + \hat{\beta}_{BOP}\boldsymbol{\mu}_0, \tag{13}$$

where $\boldsymbol{\mu}_0$ is the target mean vector,

$$\hat{\alpha}_{BOP} = \frac{\left(\bar{\boldsymbol{x}}_n^\top \mathbf{S}_n^{-1}\bar{\boldsymbol{x}}_n - \frac{p/n}{1-p/n}\right)\boldsymbol{\mu}_0^\top \mathbf{S}_n^{-1}\boldsymbol{\mu}_0 - (\bar{\boldsymbol{x}}_n^\top \mathbf{S}_n^{-1}\boldsymbol{\mu}_0)^2}{\bar{\boldsymbol{x}}_n^\top \mathbf{S}_n^{-1}\bar{\boldsymbol{x}}_n\boldsymbol{\mu}_0^\top \mathbf{S}_n^{-1}\boldsymbol{\mu}_0 - (\bar{\boldsymbol{x}}_n^\top \mathbf{S}_n^{-1}\boldsymbol{\mu}_0)^2}, \tag{14}$$

and

$$\hat{\beta}_{BOP} = (1 - \hat{\alpha}_{BOP})\frac{\bar{\boldsymbol{x}}_n^\top \mathbf{S}_n^{-1}\boldsymbol{\mu}_0}{\boldsymbol{\mu}_0^\top \mathbf{S}_n^{-1}\boldsymbol{\mu}_0}. \tag{15}$$

The two most popular estimators for the covariance matrix are the linear shrinkage estimator and the nonlinear shrinkage estimator given by:

- **Linear shrinkage estimator (Bodnar et al., 2014):**

$$\hat{\mathbf{\Sigma}}_{n;BGP} = \hat{\alpha}_{cov}\mathbf{S}_n + \hat{\beta}_{cov}\mathbf{\Sigma}_0, \tag{16}$$

with

$$\hat{\alpha}_{cov} = 1 - \frac{\frac{1}{n}||\mathbf{S}_n||_{tr}^2||\mathbf{\Sigma}_0||_F^2}{||\mathbf{S}_n||_F^2||\mathbf{\Sigma}_0||_F^2 - \left(\mathrm{tr}(\mathbf{S}_n\mathbf{\Sigma}_0)\right)^2} \tag{17}$$

and

$$\hat{\beta}_{cov} = \frac{\mathrm{tr}(\mathbf{S}_n\mathbf{\Sigma}_0)}{||\mathbf{\Sigma}_0||_F^2}\left(1 - \hat{\alpha}_{cov}\right), \tag{18}$$

where $||\mathbf{A}||_{tr} = \mathrm{tr}\left[\left(\mathbf{A}\mathbf{A}^\top\right)^{1/2}\right]$ is the trace norm and the shrinkage target $\mathbf{\Sigma}_0$ has the bounded trace norm.
- **Nonlinear shrinkage estimator (Ledoit & Wolf, 2017, 2020):**

$$\hat{\mathbf{\Sigma}}_{n,LWnonlin} = \mathbf{H}\mathrm{diag}(d_1^{or}, \ldots, d_p^{or})\mathbf{H}^\top,$$

$$d_i^{or} = \begin{cases} \frac{d_i}{|1-p/n-p/nd_i\breve{m}_F(d_i)|^2}, & \text{if } d_i > 0, \\ \frac{1}{(p/n-1)\breve{m}_{\underline{F}}(0)}, & \text{if } d_i = 0, \end{cases} \quad \text{for } i \in \{1, \ldots, p\}, \tag{19}$$

where $\breve{m}_F(x) = \lim_{z \to x} m_F(z)$ and $\breve{m}_{\underline{F}}(x) = \lim_{z \to x} m_{\underline{F}}(z) = \lim_{z \to x} \frac{p/n-1}{z} + \frac{p}{n}m_F(z)$ with $m_F(z)$ the limiting Stieltjes transform of the sample covariance matrix, which, for a distribution function $G : \mathbb{R} \to \mathbb{R}$, is defined by

$$m_G(z) = \int_{-\infty}^{+\infty} \frac{1}{\lambda - z} dG(\lambda); \quad z \in \mathbb{C}^+ \equiv \{z \in \mathbb{C} : \Im z > 0\}.$$

A numerical approach to estimate $\breve{m}_F(x)$ is provided in Ledoit and Wolf (2020), and it is implemented in the R-package *HDShOP* (see, Bodnar et al., 2024c).

Finally, a linear shrinkage estimator for the inverse covariance matrix (precision matrix) is derived in Bodnar et al. (2016a), and it is given by

$$\hat{\mathbf{\Pi}}_{n,BGP} = \hat{\alpha}_{prec}\mathbf{S}_n^{-1} + \hat{\beta}_{prec}\mathbf{\Pi}_0, \tag{20}$$

with $\mathbf{\Pi}_0$ the target precision matrix

$$\hat{\alpha}_{prec} = 1 - p/n - \frac{\frac{1}{n}||\mathbf{S}_n^{-1}||_{tr}^2||\mathbf{\Pi}_0||_F^2}{||\mathbf{S}_n^{-1}||_F^2||\mathbf{\Pi}_0||_F^2 - \left(\mathrm{tr}(\mathbf{S}_n^{-1}\mathbf{\Pi}_0)\right)^2}, \tag{21}$$

and

$$\hat{\beta}_{prec} = \frac{\text{tr}(\boldsymbol{S}_n^{-1}\boldsymbol{\Pi}_0)}{||\boldsymbol{\Pi}_0||_F^2}\left(1 - p/n - \hat{\alpha}_{prec}\right), \tag{22}$$

where $1/p\boldsymbol{\Pi}_0$ is assumed to have the bounded trace norm.

Other types of shrinkage estimators for the mean vector, covariance matrix, and precision matrix can be found in the review paper of Bodnar et al. (2022a), while the practical implementation of the described shrinkage approaches is available in the R package *HDShOP* (High-Dimensional Shrinkage Optimal Portfolio, see Bodnar et al., 2024c). Shrinkage estimators constructed directly for the weights of optimal portfolio were derived in Bodnar et al. (2018), Bodnar et al. (2023a), Bodnar et al. (2023b), and Bodnar et al. (2024e), while shrinkage-based test theory was explored in Bodnar et al. (2019a) and Bodnar et al. (2021). Finally, Bodnar et al. (2016b), Muhinyuza et al. (2020), Karlsson et al. (2021), Alfelt and Mazur (2022), Bodnar et al. (2022b), Bodnar et al. (2019b), Drin et al. (2024), and Javed et al. (2024) presented several distributional results of the estimated portfolio weights with singular and nonsingular covariance matrix, while Bodnar (2009) established sequential surveillance procedures for monitoring the structure of the tangency portfolio.

2.2 Bayesian Approaches

Several Bayesian estimators of the tangency portfolio weights exist in the literature. The first approaches in constructing Bayesian tangency portfolio utilize the fact that its weights $\boldsymbol{w}$ (see (4)) depend solely on the mean vector $\boldsymbol{\mu}$ and covariance matrix $\boldsymbol{\Sigma}$ of the asset returns. Several methods to specify a prior on $\boldsymbol{\mu}$ and $\boldsymbol{\Sigma}$ have been developed in the literature (see Winkler, 1973; Winkler & Barry, 1975; Bawa et al., 1979; Frost & Savarino, 1986; Greyserman et al., 2006; Rachev et al., 2008; Avramov & Zhou, 2010; Tu & Zhou, 2010; Bodnar et al., 2017; Bauder et al., 2018; and Bauder et al., 2019), with the diffuse prior and conjugate prior being the most popular procedures.

The diffuse prior, also known as the Jeffreys prior, is a noninformative prior (see Held & Bové, 2014) which under the assumption of independent and normally distributed asset return vectors given $\boldsymbol{\mu}$ and $\boldsymbol{\Sigma}$ is expressed as

$$\pi_d(\boldsymbol{\mu}, \boldsymbol{\Sigma}) \propto |\boldsymbol{\Sigma}|^{-(p+1)/2}. \tag{23}$$

The conjugate prior is an informative prior that assigns a multivariate normal prior for $\boldsymbol{\mu}$ conditionally on $\boldsymbol{\Sigma}$ and an inverse Wishart prior for $\boldsymbol{\Sigma}$. It is given by

$$\pi_c(\boldsymbol{\mu}|\boldsymbol{\Sigma}) \propto |\boldsymbol{\Sigma}|^{-1/2} \exp\left\{-\frac{\kappa_c}{2}(\boldsymbol{\mu} - \boldsymbol{\mu}_c)^\top \boldsymbol{\Sigma}^{-1}(\boldsymbol{\mu} - \boldsymbol{\mu}_c)\right\}, \tag{24}$$

$$\pi_c(\mathbf{\Sigma}) \propto |\mathbf{\Sigma}|^{-\nu_c/2} \exp\left\{-\frac{1}{2}\mathrm{tr}[\mathbf{V}_c\mathbf{\Sigma}^{-1}]\right\}, \tag{25}$$

where $\boldsymbol{\mu}_c$ is the prior mean, κ_c reflects the precision of the prior mean $\boldsymbol{\mu}_c$, while ν_c is the corresponding precision parameter in the prior for on $\mathbf{\Sigma}$, and $\mathbf{V}_c$ is a prior matrix for $\mathbf{\Sigma}$. Under (24) and (25), the joint prior for $\boldsymbol{\mu}$ and $\mathbf{\Sigma}$ is given by

$$\pi_c(\boldsymbol{\mu}, \mathbf{\Sigma}) \propto |\mathbf{\Sigma}|^{-(\nu_c+1)/2} \exp\left\{-\frac{\kappa_c}{2}(\boldsymbol{\mu}-\boldsymbol{\mu}_c)^\top\mathbf{\Sigma}^{-1}(\boldsymbol{\mu}-\boldsymbol{\mu}_c) - \frac{1}{2}\mathrm{tr}[\mathbf{V}_c\mathbf{\Sigma}^{-1}]\right\}. \tag{26}$$

To characterize the posterior distribution of linear combinations of tangency portfolio weights under the Jeffreys and the conjugate prior, Bauder et al. (2018) deduced very useful stochastic representations. Let $\boldsymbol{\theta} = \mathbf{L}\mathbf{w}$, where $\mathbf{L}$ is an $m \times p$ matrix of constants of full rank $m < p$, and let the symbol $\overset{d}{=}$ denote the equality in distribution. Then:

- **Under the diffuse prior** (23)

$$\boldsymbol{\theta}_d \overset{d}{=} \frac{\eta_d}{\gamma}\mathbf{L}\boldsymbol{\zeta}_d + \frac{\sqrt{\eta_d}}{\gamma}\left(\epsilon_d \cdot \mathbf{L}\boldsymbol{\Upsilon}_d\mathbf{L}^\top - \mathbf{L}\boldsymbol{\zeta}_d\boldsymbol{\zeta}_d^\top\mathbf{L}^\top\right)^{1/2}\mathbf{t}_0, \tag{27}$$

with

$$\boldsymbol{\Upsilon}_d = \boldsymbol{\Upsilon}_d(Q_d, \mathbf{U}) = \frac{1}{n-1}\mathbf{S}_n^{-1} - \frac{\frac{p}{n-p}Q_d}{1+\frac{p}{n-p}Q_d}\frac{1}{n-1}\mathbf{S}_n^{-1/2}\mathbf{U}\mathbf{U}^\top\mathbf{S}_n^{-1/2},$$

$$\boldsymbol{\zeta}_d = \boldsymbol{\zeta}_d(Q_d, \mathbf{U}) = \boldsymbol{\Upsilon}_d(Q_d, \mathbf{U})\bar{\mathbf{x}}_n + \frac{\sqrt{\frac{p}{n(n-p)}Q_d}}{1+\frac{p}{n-p}Q_d}\frac{1}{\sqrt{n-1}}\mathbf{S}_n^{-1/2}\mathbf{U},$$

$$\epsilon_d = \epsilon_d(Q_d, \mathbf{U}) - \bar{\mathbf{x}}_n^\top\boldsymbol{\Upsilon}_d(Q_d, \mathbf{U})\bar{\mathbf{x}}_n + \frac{2}{\sqrt{n-1}}\frac{\sqrt{\frac{p}{n(n-p)}Q_d}}{1+\frac{p}{n-p}Q_d}\bar{\mathbf{x}}_n^\top\mathbf{S}_n^{-1/2}\mathbf{U} + \frac{\frac{p}{n(n-p)}Q_d}{1+\frac{p}{n-p}Q_d},$$

where η_d, $\mathbf{t}_0$, Q_d, and $\mathbf{U}$ are mutually independent with $\eta_d \sim \chi^2_n$, $\mathbf{t}_0 \sim \mathcal{N}_m(\mathbf{0}, \mathbf{I}_m)$, $Q_d \sim \mathcal{F}(p, n-p)$, and $\mathbf{U}$ is uniformly distributed on the unit sphere in $\mathbb{R}^p$.

- **Under the conjugate prior** (26)

$$\boldsymbol{\theta}_c \overset{d}{=} \frac{\eta_c}{\gamma}\cdot\mathbf{L}\boldsymbol{\zeta}_c + \frac{\sqrt{\eta_c}}{\gamma}\left(\epsilon_c \cdot \mathbf{L}\boldsymbol{\Upsilon}_c\mathbf{L}^\top - \mathbf{L}\boldsymbol{\zeta}_c\boldsymbol{\zeta}_c^\top\mathbf{L}^\top\right)^{1/2}\mathbf{t}_0, \tag{28}$$

with

$$\boldsymbol{\Upsilon}_c = \boldsymbol{\Upsilon}_c(Q_c, \boldsymbol{U}) = \tilde{\boldsymbol{S}}_n^{-1} - \frac{\frac{p}{\nu_c+n-2p} Q_c}{1 + \frac{p}{\nu_c+n-2p} Q_c} \tilde{\boldsymbol{S}}_n^{-1/2} \boldsymbol{U}\boldsymbol{U}^\top \tilde{\boldsymbol{S}}_n^{-1/2},$$

$$\boldsymbol{\zeta}_c = \boldsymbol{\zeta}_c(Q_c, \boldsymbol{U}) = \boldsymbol{\Upsilon}_c(Q_c, \boldsymbol{U})\breve{\boldsymbol{x}}_n + \frac{\sqrt{\frac{p}{(\kappa_c+n)(\nu_c+n-2p)} Q_c}}{1 + \frac{p}{\nu_c+n-2p} Q_c} \tilde{\boldsymbol{S}}_n^{-1/2} \boldsymbol{U},$$

$$\epsilon_c = \epsilon_c(Q_c, \boldsymbol{U}) = \breve{\boldsymbol{x}}_n^\top \boldsymbol{\Upsilon}_c(Q_c, \boldsymbol{U})\breve{\boldsymbol{x}}_n + 2\frac{\sqrt{\frac{p}{(\kappa_c+n)(\nu_c+n-2p)} Q_c}}{1 + \frac{p}{\nu_c+n-2p} Q_c} \breve{\boldsymbol{x}}_n^\top \tilde{\boldsymbol{S}}_n^{-1/2} \boldsymbol{U}$$

$$+ \frac{\frac{p}{(\kappa_c+n)(\nu_c+n-2p)} Q_c}{1 + \frac{p}{\nu_c+n-2p} Q_c},$$

where η_c, $\boldsymbol{t}_0$, Q_c, and $\boldsymbol{U}$ are mutually independent with $\eta_c \sim \chi^2_{\nu_c+n-p}$, $\boldsymbol{t}_0 \sim \mathcal{N}_m(\boldsymbol{0}, \boldsymbol{I}_m)$, $Q_c \sim \mathcal{F}(k, \nu_c + n - 2p)$, and $\boldsymbol{U}$ is uniformly distributed on the unit sphere in $\mathbb{R}^p$, $\breve{\boldsymbol{x}}_n = \left(n\bar{\boldsymbol{x}}_n + \kappa_c\boldsymbol{\mu}_c\right)/(n + \kappa_c)$, and

$$\tilde{\boldsymbol{S}}_n = (n-1)\boldsymbol{S}_n + \boldsymbol{V}_c - (n+\kappa_c)\breve{\boldsymbol{x}}_n\breve{\boldsymbol{x}}_n^\top + n\bar{\boldsymbol{x}}_n\bar{\boldsymbol{x}}_n^\top + \kappa_c\boldsymbol{\mu}_c\boldsymbol{\mu}_c^\top.$$

The two derived stochastic representations (27) and (28) provide an efficient approach to generate multivariate samples from the posterior distribution of $\boldsymbol{\theta}$ derived under the diffuse and conjugate priors. Furthermore, in contrast to the MCMC, the mostly used approach in Bayesian statistics to draw samples from the posterior distribution (see, e.g., Held & Bové, 2014), the application of the stochastic representations (27) and (28) yields samples from the exact distributions which are obtained by generating random objects from standard probabilistic distributions. We summarize this approach in the following algorithm:

Algorithm 1: Generating samples from the posterior distribution of $\boldsymbol{\theta}$ under the diffuse and conjugate prior

---

(1) Generate independently

   (a) Diffuse prior: $\eta_d \sim \chi^2_n$; conjugate prior: $\eta_c \sim \chi^2_{\nu_c+n-p}$
   (b) $\boldsymbol{t}_0 \sim \mathcal{N}_m(\boldsymbol{0}, \boldsymbol{I}_m)$
   (c) Diffuse prior: $Q_d \sim \mathcal{F}(p, n-p)$, conjugate prior: $Q_c \sim \mathcal{F}(p, \nu_c + n - 2p)$
   (d) $\boldsymbol{W} \sim \mathcal{N}_p(\boldsymbol{0}, \boldsymbol{I}_p)$ and compute $\boldsymbol{U} = \boldsymbol{W}/\sqrt{\boldsymbol{W}^\top\boldsymbol{W}}$

(2) Compute the vector of weights using (27) for the diffuse prior and (28) for the conjugate prior using true parameters.
(3) Repeat steps (1) and (2) B times.

The stochastic representations (27) and (28) are also used to derive the Bayesian estimators of the tangency portfolio weights and to assess their uncertainty. Following Theorem 2.2 in Bauder et al. (2018), the Bayesian estimators of $\boldsymbol{w}$ are computed as posterior mean vectors, and they are given by

$$\hat{\boldsymbol{w}}_{TP;d} = \frac{1}{\gamma}\mathbf{S}_n^{-1}\bar{\boldsymbol{x}}_n \tag{29}$$

and

$$\hat{\boldsymbol{w}}_{TP;c} = \frac{\nu_c + n - p - 1}{\gamma}\tilde{\mathbf{S}}_n^{-1}\breve{\boldsymbol{x}}_n, \tag{30}$$

under the diffuse and conjugate prior, respectively. Furthermore, their posterior covariance matrices are (see Theorem 2.3 of Bauder et al., 2018)

$$\begin{aligned} Var(\boldsymbol{w}|\boldsymbol{x}_{1:n}) &= \frac{1}{n-1}\gamma^{-2}\mathbf{S}_n^{-1}\bar{\boldsymbol{x}}_n\bar{\boldsymbol{x}}_n^\top\mathbf{S}_n^{-1} \\ &+ \gamma^{-2}\left[\frac{n^2+p-2}{(n-1)n(n+2)} + \frac{n}{(n-1)^2}\frac{p-1}{p}b_d\right]\mathbf{S}_n^{-1} \end{aligned}$$

with $b_d = \bar{\boldsymbol{x}}_n^\top\mathbf{S}_n^{-1}\bar{\boldsymbol{x}}_n$ for the diffuse prior (23) and

$$\begin{aligned} Var(\boldsymbol{w}|\boldsymbol{x}_{1:n}) &= (\nu_c + n - p - 1)\gamma^{-2}\tilde{\mathbf{S}}_n^{-1}\breve{\boldsymbol{x}}_n\breve{\boldsymbol{x}}_n^\top\tilde{\mathbf{S}}_n^{-1} \\ &+ \gamma^{-2}\left[\frac{(\nu_c+n-p)^2+p-2}{(n+\kappa_c)(\nu_c+n-p+2)} + (\nu_c+n-p)\frac{p-1}{p}b_c\right]\tilde{\mathbf{S}}_n^{-1} \end{aligned}$$

with $b_c = \breve{\boldsymbol{x}}_n^\top\tilde{\mathbf{S}}_n^{-1}\breve{\boldsymbol{x}}_n$ for the conjugate prior (26).

Utilizing the $(\boldsymbol{\nu}, \boldsymbol{\Omega})$ parameterization of the likelihood function, two further priors were derived in Bodnar et al. (2024b). The conjugate prior for $\boldsymbol{\nu}$ and $\boldsymbol{\Omega}$ is given by

$$\boldsymbol{\nu}|\boldsymbol{\Omega} \sim \mathcal{N}_p\left(c_0 n_0^{-1}\boldsymbol{\Omega}\mathbf{S}_0\boldsymbol{w}_0, n_0^{-1}\boldsymbol{\Omega}\right), \tag{31}$$

$$\boldsymbol{\Omega} \sim \mathcal{W}_p\left(n_0 + p + 2, \mathbf{S}_0^{-1} + \frac{\boldsymbol{w}_0\boldsymbol{w}_0^\top}{c_0^{-2}n_0 - \boldsymbol{w}_0^\top\mathbf{S}_0\boldsymbol{w}_0}\right), \tag{32}$$

where

$$c_0 = \frac{-(n_0+p+2) + \sqrt{(n_0+p+2)^2 + 4n_0\boldsymbol{w}_0^\top\mathbf{S}_0\boldsymbol{w}_0}}{2\boldsymbol{w}_0^\top\mathbf{S}_0\boldsymbol{w}_0} \tag{33}$$

to ensure that the unconditional prior mean vector of $\boldsymbol{\nu}$ is $\boldsymbol{w}_0$. This leads to an important advantage of the conjugate prior derived in the $\boldsymbol{\nu}$ and $\boldsymbol{\Omega}$ parameterization, namely an investor can determine the belief about the tangency portfolio weights in the prior specification by setting the hyperparameter $\boldsymbol{w}_0$. If no prior information is available for $\boldsymbol{w}_0$, then one can opt for using the value-weighted portfolio (see Black & Litterman, 1992) or the equally weighted portfolio (see DeMiguel et al., 2009) as a proxy. Bodnar et al. (2024b) suggested the application of the realized covariance matrix computed by using 5-min, 15-min, or 60-min high-frequency stock returns for determining the hyperparameter $\boldsymbol{S}_0$, while a time-indexed market condition metric (MCM) is employed in the definition of n_0. The Economic Policy Uncertainty (EPU) index and the VIX index, derived from the implied volatility of S&P 500 options, were used in Bodnar et al. (2024b) as an MCM, while Bodnar et al. (2024d) suggested the application of the estimated portfolio variances based on the short and long terms in the specification of n_0.

The Jeffreys prior in the $\boldsymbol{\nu}$ and $\boldsymbol{\Omega}$ parameterization was derived in the supplement of Bodnar et al. (2024b), and it is given by

$$\pi(\boldsymbol{\nu}, \boldsymbol{\Omega}) \propto \det(\boldsymbol{\Omega})^{-(p+2)/2}. \tag{34}$$

Under both priors assigned to $\boldsymbol{\nu}$ and $\boldsymbol{\Omega}$, Bodnar et al. (2024b) derived stochastic representations for $\boldsymbol{\nu}$ from the corresponding marginal posterior distributions, which are given by:

- **Under the Jeffreys prior** (34)

$$\boldsymbol{\nu} \stackrel{d}{=} \boldsymbol{\Omega}\boldsymbol{a}_J + n^{-1/2}\boldsymbol{\Omega}^{1/2}\boldsymbol{\eta}, \tag{35}$$

where $\boldsymbol{\eta} \sim \mathcal{N}_p(\mathbf{0}, \boldsymbol{I})$ independent of $\boldsymbol{\Omega}$ and $\boldsymbol{\Omega}|\boldsymbol{x}_{1:n} \sim \mathcal{W}_p(n, \boldsymbol{V}_J)$ with

$$\boldsymbol{a}_J = n^{-1}\boldsymbol{t} \text{ and } \boldsymbol{V}_J = (\boldsymbol{T} - n^{-1}\boldsymbol{t}\boldsymbol{t}^\top)^{-1}.$$

- **Under the conjugate prior** (31) **and** (32)

$$\boldsymbol{\nu} \stackrel{d}{=} \boldsymbol{\Omega}\boldsymbol{a}_c + n_c^{-1/2}\boldsymbol{\Omega}^{1/2}\boldsymbol{\eta}, \tag{36}$$

where $\boldsymbol{\eta} \sim \mathcal{N}_p(\mathbf{0}, \boldsymbol{I})$ and $\boldsymbol{\Omega} \sim \boldsymbol{\Omega}|\boldsymbol{x}_{1:n} \sim \mathcal{W}_p(n_c + p + 2, \boldsymbol{V}_c)$ with

$$\begin{aligned}
n_c &= n + n_0,\\
\boldsymbol{S}_c &= \boldsymbol{S}_0 + \boldsymbol{T},\\
\boldsymbol{w}_c &= \boldsymbol{S}_c^{-1}(c_0\boldsymbol{S}_0\boldsymbol{w}_0 + \boldsymbol{t}),\\
\boldsymbol{a}_c &= n_c^{-1}(c_0\boldsymbol{S}_0\boldsymbol{w}_0 + \boldsymbol{t}),\\
\boldsymbol{V}_c &= = \boldsymbol{S}_c^{-1} + \frac{\boldsymbol{w}_c\boldsymbol{w}_c^\top}{n_c - \boldsymbol{w}_c^\top\boldsymbol{S}_c\boldsymbol{w}_c}.
\end{aligned}$$

Since the vector of the tangency portfolio weights is proportional to the vector $\boldsymbol{v}$, namely $\boldsymbol{w} = \gamma^{-1}\boldsymbol{v}$, these deduced stochastic representations can be used to draw samples from the posterior distribution of the tangency portfolio weights. Considerable simplifications of the stochastic representations are obtained in the case of one linear combination of the tangency portfolio weights $\boldsymbol{m}^\top\boldsymbol{v}$. They are given by:

- **Under the Jeffreys prior** (34)

$$\boldsymbol{m}^\top\boldsymbol{v} \stackrel{d}{=} \boldsymbol{m}^\top\boldsymbol{V}_J\boldsymbol{a}_J\xi + \sqrt{(\boldsymbol{a}_J^\top\boldsymbol{V}_J\boldsymbol{a}_J + n^{-1})\boldsymbol{m}^\top\boldsymbol{V}_J\boldsymbol{m} - (\boldsymbol{a}_J^\top\boldsymbol{V}_J\boldsymbol{m})^2}\xi^{1/2}\eta, \tag{37}$$

  where η and ξ are independent with $\eta \sim \mathcal{N}(0, 1)$ and $\xi \sim \chi^2_n$.
- **Under the conjugate prior** (31)–(32)

$$\boldsymbol{m}^\top\boldsymbol{v} \stackrel{d}{=} \boldsymbol{m}^\top\boldsymbol{V}_c\boldsymbol{a}_c\xi + \sqrt{(\boldsymbol{a}_c^\top\boldsymbol{V}_c\boldsymbol{a}_c + n_c^{-1})\boldsymbol{m}^\top\boldsymbol{V}_c\boldsymbol{m} - (\boldsymbol{a}_c^\top\boldsymbol{V}_c\boldsymbol{m})^2}\xi^{1/2}\eta, \tag{38}$$

  where η and ξ are independent with $\eta \sim \mathcal{N}(0, 1)$ and $\xi \sim \chi^2_{n_c+p+2}$.

Using the above stochastic representation, Bodnar et al. (2024b) derived the Bayesian estimator for the tangency portfolio weights expressed as

$$\mathbb{E}(\boldsymbol{v}|\boldsymbol{x}_{1:n}) = (\boldsymbol{T} - n^{-1}\boldsymbol{t}\boldsymbol{t}^\top)^{-1}\boldsymbol{t} \tag{39}$$

and

$$\mathbb{E}(\boldsymbol{v}|\boldsymbol{x}_{1:n}) = (n_c + p + 2)\frac{\boldsymbol{w}_c}{n_c - \boldsymbol{w}_c^\top\boldsymbol{S}_c\boldsymbol{w}_c}, \tag{40}$$

under the Jeffreys and conjugate priors, respectively.

The corresponding covariance matrices, deduced from the posterior distributions, are given by

$$\begin{aligned}\mathbb{V}ar(\boldsymbol{v}|\boldsymbol{x}_{1:n}) &= n^{-1}(\boldsymbol{T} - n^{-1}\boldsymbol{t}\boldsymbol{t}^\top)^{-1}\boldsymbol{t}\boldsymbol{t}^\top(\boldsymbol{T} - n^{-1}\boldsymbol{t}\boldsymbol{t}^\top)^{-1}\\ &+ (n^{-1}\boldsymbol{t}^\top(\boldsymbol{T} - n^{-1}\boldsymbol{t}\boldsymbol{t}^\top)^{-1}\boldsymbol{t} + 1)(\boldsymbol{T} - n^{-1}\boldsymbol{t}\boldsymbol{t}^\top)^{-1}\end{aligned} \tag{41}$$

and

$$\mathbb{V}ar(\boldsymbol{v}|\boldsymbol{x}_{1:n}) = (n_c + p + 2)\left(2\frac{\boldsymbol{w}_c\boldsymbol{w}_c^\top}{(n_c - \boldsymbol{w}_c^\top\boldsymbol{S}_c\boldsymbol{w}_c)^2} + \frac{1}{n_c - \boldsymbol{w}_c^\top\boldsymbol{S}_c\boldsymbol{w}_c}\boldsymbol{S}_c^{-1}\right), \tag{42}$$

under the Jeffreys and conjugate priors, respectively.

Finally, to specify the hyperparameters n_0 and $\boldsymbol{S}_0$ of the conjugate prior (31) and (32), it was suggested in Bodnar et al. (2024b) to use high-frequency data and the market condition metric (MCM), like the VIX index and the Economic Policy Uncertainty (EPU) index, developed by Baker et al. (2016).

2.3 *Bayesian Approaches Without Short Selling*

Both frequentist and Bayesian approaches, presented in Sects. 2.1 and 2.2, do not take any structural assumptions, possibly imposed on the weights of the population tangency portfolio, into account. In particular, the resulting frequentist and Bayesian estimators of the tangency portfolio weights may consist of both positive and negative weights. On some capital markets, negative weights, i.e., the possibility of short selling, are not allowed. As a result, the approaches of Sects. 2.1 and 2.2 cannot directly be implemented in practice without their modifications.

Methods of Bayesian statistics provide the investor with technical tools to ensure the validity of constraints imposed on the structure of the tangency portfolio weights. Recently, Bodnar et al. (2024a) proposed a prior distribution for the tangency portfolio weights whose support consists of only positive values. As a result, this leads to the posterior distribution of the tangency portfolio weights which is concentrated on the positive values only. This approach ensures that the resulting Bayesian estimator of the tangency portfolio has only positive weights. Finally, we note that other structural restrictions, like leverage constraints, imposed on the tangency portfolio weights can be handled similarly by defining a prior such that the required conditions are fulfilled.

The prior proposed in Bodnar et al. (2024a) is given by

$$\log(\boldsymbol{v}) \sim \mathcal{N}_p(\log(\boldsymbol{w}_0), \kappa \boldsymbol{I}), \tag{43}$$

$$\sigma_i \sim C^+(\beta) \quad \text{for } i = 1, 2, \ldots, p, \tag{44}$$

$$\boldsymbol{R} \sim \mathcal{LKJ}(\eta), \tag{45}$$

$$\boldsymbol{\Omega} = \text{diag}(\boldsymbol{\sigma})\boldsymbol{R}\text{diag}(\boldsymbol{\sigma}), \tag{46}$$

where $\boldsymbol{I}$ is the identity matrix, κ, β, and η are constants, controlling the shapes and scales of the prior distributions, and $\boldsymbol{w}_0$ represents the prior belief about the optimal portfolio weights. The symbol LKJ stands for the Lewandowski-Kurowicka-Joe distribution, while C^+ denotes the half-Cauchy distribution. In the prior specification (43)–(46), the hyperparameter κ determines the level of trust in the prior belief about $\boldsymbol{w}_0$ with larger values indicating a diminished reliance on the prior.

The posterior distribution, derived by employing the prior (43)–(46), has a complicated structure. As a result, the Bayesian inference procedures for the tangency portfolio weights under the prior (43)–(46) are obtained by using the MCMC approach to draw samples from the posterior distribution. The elements of the generated samples are then used to obtain the Bayesian estimator of the tangency portfolio weights by computing the sample means or medians and to construct credible regions for the tangency portfolio weights (see Bodnar et al., 2024a for details).

3 Regularized Bayesian Tangency Portfolio

Due to the definition, the application of the priors presented in Sect. 2 would lead to the portfolio strategy that will invest in all of the considered stocks. The posterior probability that at least one weight is zero will tend to be zero. As a result, the proposed procedures do not provide an automatic answer to the question of which stocks from the available set of financial instruments should be chosen by an investor. We deal with the problem in this section by adopting Bayesian variable selection procedures to construct the tangency portfolio. The results of this section are provided in the case of the tangency portfolio with short selling, while we note that similar approaches can be employed also in the case when short selling is not allowed.

The first regularized Bayesian approach extends the double Laplace prior suggested in Park and Casella (2008). In the $\boldsymbol{\nu}$ and $\boldsymbol{\Omega}$ parameterization, it is obtained by replacing the conditional multivariate normal prior in (31) by a conditional multivariate Laplace prior, and it is given by

$$\boldsymbol{\nu}|\boldsymbol{\Omega} \sim \mathcal{L}_p\left(c_0 n_0^{-1}\boldsymbol{\Omega}\mathbf{S}_0\boldsymbol{w}_0, n_0^{-1}\boldsymbol{\Omega}\right), \tag{47}$$

$$\boldsymbol{\Omega} \sim \mathcal{W}_p\left(n_0 + p + 2, \mathbf{S}_0^{-1} + \frac{\boldsymbol{w}_0\boldsymbol{w}_0^\top}{c_0^{-2}n_0 - \boldsymbol{w}_0^\top\mathbf{S}_0\boldsymbol{w}_0}\right). \tag{48}$$

Using the properties of the multivariate Laplace distribution, the prior (47) and (48) can be rewritten as

$$\boldsymbol{\nu}|\boldsymbol{\Omega}, \delta \sim \mathcal{N}_p\left(c_0 n_0^{-1}\boldsymbol{\Omega}\mathbf{S}_0\boldsymbol{w}_0, n_0^{-1}\delta\boldsymbol{\Omega}\right), \tag{49}$$

$$\boldsymbol{\Omega} \sim \mathcal{W}_p\left(n_0 + p + 2, \mathbf{S}_0^{-1} + \frac{\boldsymbol{w}_0\boldsymbol{w}_0^\top}{c_0{}^{2}n_0 - \boldsymbol{w}_0^\top\mathbf{S}_0\boldsymbol{w}_0}\right), \tag{50}$$

$$\delta \sim Exp(1), \tag{51}$$

where the symbol Exp stands for the standard univariate exponential distribution. Moreover, δ and $\boldsymbol{\Omega}$ are independent.

The second approach extends the spike and slab prior in Ishwaran and Rao (2005), and it is given by

$$\boldsymbol{\nu}|\boldsymbol{\Omega}, \boldsymbol{\Lambda} \sim \mathcal{N}_p\left(c_0 n_0^{-1}\boldsymbol{\Omega}\mathbf{S}_0\boldsymbol{w}_0, n_0^{-1}\boldsymbol{\Omega}\right), \tag{52}$$

$$\boldsymbol{\Omega}|\boldsymbol{\Lambda}, \sim \mathcal{W}_p\left(n_0+p+2, \boldsymbol{\Lambda}^{-1/2}\left(\mathbf{S}_0^{-1}+\frac{\boldsymbol{w}_0\boldsymbol{w}_0^\top}{c_0^{-2}n_0-\boldsymbol{w}_0^\top\mathbf{S}_0\boldsymbol{w}_0}\right)\boldsymbol{\Lambda}^{-1/2}\right), \quad (53)$$

$$\boldsymbol{\Lambda} = \mathrm{diag}(d_{0,1}\delta_1+(1-\delta_1),\ldots,d_{0,p}\delta_p+(1-\delta_p)),$$

$$\delta_i|v_i \sim Be(v_i) \quad \text{and} \quad v_i \sim Unif(0,1), \quad (54)$$

where the symbol $Be(v)$ denotes the Bernoulli distribution with the probability of success v and the symbol $Unif(0,1)$ stands for the uniform distribution on the interval $[0,1]$. Finally, $\boldsymbol{\Omega}$ is independent of $\boldsymbol{\Lambda}$, and (v_i,δ_i), $i=1,\ldots,p$, are mutually independent. The quantities $d_{0,i}$, $i=1,\ldots,p$, are some small positive numbers.

In Theorem 1 we derive a stochastic representation of one linear combination of the tangency portfolio weights $\boldsymbol{m}^\top\boldsymbol{v}$ under the Laplace prior (49)–(51), while the results under the spike and slab prior (52)–(54) are presented in Theorem 2.

Theorem 1 *Under the Laplace prior* (49)–(51)*, the stochastic representation of* $\boldsymbol{m}^\top\boldsymbol{v}$ *is given by*

$$\boldsymbol{m}^\top\boldsymbol{v} \overset{d}{=} \boldsymbol{m}^\top\boldsymbol{V}_c\boldsymbol{a}_c\xi+\sqrt{(\boldsymbol{a}_c^\top\boldsymbol{V}_c\boldsymbol{a}_c+n_c^{-1})\boldsymbol{m}^\top\boldsymbol{V}_c\boldsymbol{m}-(\boldsymbol{a}_c^\top\boldsymbol{V}_c\boldsymbol{m})^2}\,\xi^{1/2}\eta, \quad (55)$$

where η *and* ξ *are independent with* $\eta\sim\mathcal{N}(0,1)$ *and* $\xi\sim\chi^2_{n+n_0+p+2}$ *with*

$$n_c = n_c(\delta) = n+n_0/\delta,$$

$$\mathbf{S}_c = \mathbf{S}_c(\delta) = \mathbf{S}_0+\left(\frac{1}{\delta}-1\right)\frac{c_0^2}{n_0}\mathbf{S}_0\boldsymbol{w}_0\boldsymbol{w}_0^\top\mathbf{S}_0+\boldsymbol{T}$$

$$\boldsymbol{w}_c = \boldsymbol{w}_c(\delta) = \mathbf{S}_c^{-1}\left(\frac{c_0}{\delta}\mathbf{S}_0\boldsymbol{w}_0+\boldsymbol{t}\right),$$

$$\boldsymbol{a}_c = \boldsymbol{a}_c(\delta) = n_c^{-1}\mathbf{S}_c\boldsymbol{w}_c = n_c^{-1}\left(\frac{c_0}{\delta}\mathbf{S}_0\boldsymbol{w}_0+\boldsymbol{t}\right),$$

$$\boldsymbol{V}_c = \boldsymbol{V}_c(\delta) = \left(\mathbf{S}_c-n_c^{-1}\mathbf{S}_c\boldsymbol{w}_c\boldsymbol{w}_c^\top\mathbf{S}_c\right)^{-1} = \mathbf{S}_c^{-1}+\frac{\boldsymbol{w}_c\boldsymbol{w}_c^\top}{n_c-\boldsymbol{w}_c^\top\mathbf{S}_c\boldsymbol{w}_c}.$$

Proof of Theorem 1 Under the Laplace prior (49)–(51), we get

$$\pi(\boldsymbol{v},\boldsymbol{\Omega}|\delta,\boldsymbol{x}_{1:n}) \propto \delta^{-1/2}\det(\boldsymbol{\Omega})^{(n_0+n)/2}\exp\left(-\frac{n_0/\delta+n}{2}\boldsymbol{v}^\top\boldsymbol{\Omega}^{-1}\boldsymbol{v}\right)$$

$$\times\exp\left(\left(\frac{c_0}{\delta}\mathbf{S}_0\boldsymbol{w}_0+\boldsymbol{t}\right)^\top\boldsymbol{v}\right.$$

$$\left.+\,\mathrm{tr}\left(\boldsymbol{\Omega}\left(-\frac{1}{2}\left(\mathbf{S}_0+\left(\frac{1}{\delta}-1\right)\frac{c_0^2}{n_0}\mathbf{S}_0\boldsymbol{w}_0\boldsymbol{w}_0^\top\mathbf{S}_0+\boldsymbol{T}\right)\right)\right)\right)$$

$$= \delta^{-1/2}\det(\boldsymbol{\Omega})^{(n_0+n)/2}\exp\left(-\frac{1}{2}\mathrm{tr}\left(\boldsymbol{\Omega}\left(\boldsymbol{S}_c - n_c^{-1}\boldsymbol{S}_c\boldsymbol{w}_c\boldsymbol{w}_c^\top\boldsymbol{S}_c\right)\right)\right)$$

$$\times \exp\left(-\frac{n_c}{2}(\boldsymbol{\nu} - n_c^{-1}\boldsymbol{\Omega}\boldsymbol{S}_c\boldsymbol{w}_c)^\top\boldsymbol{\Omega}^{-1}(\boldsymbol{\nu} - n_c^{-1}\boldsymbol{\Omega}\boldsymbol{S}_c\boldsymbol{w}_c)\right), \tag{56}$$

and, hence,

$$\boldsymbol{\nu}|\boldsymbol{\Omega}, \delta, \boldsymbol{x}_{1:n} \sim \mathcal{N}_p\left(\boldsymbol{\Omega}\boldsymbol{a}_c, n_c^{-1}\boldsymbol{\Omega}\right), \tag{57}$$

$$\boldsymbol{\Omega}|\delta, \boldsymbol{x}_{1:n} \sim \mathcal{W}_p\left(n + n_0 + p + 2, \boldsymbol{V}_c\right), \tag{58}$$

with n_c, $\boldsymbol{S}_c$, $\boldsymbol{w}_c$, $\boldsymbol{a}_c$, and $\boldsymbol{V}_c$ given in the statement of the theorem.

Using the properties of normal distribution and (57), we get

$$\boldsymbol{m}^\top\boldsymbol{\nu}|\boldsymbol{\Omega}, \delta, \boldsymbol{x}_{1:n} \sim \mathcal{N}\left(\boldsymbol{m}^\top\boldsymbol{\Omega}\boldsymbol{a}_c, n_c^{-1}\boldsymbol{m}^\top\boldsymbol{\Omega}\boldsymbol{m}\right), \tag{59}$$

while the application of (58) and Theorem 3.2.5 in Muirhead (1982) leads to

$$\left.\begin{pmatrix}\boldsymbol{a}_c^\top\boldsymbol{\Omega}\boldsymbol{a}_c & \boldsymbol{a}_c^\top\boldsymbol{\Omega}\boldsymbol{m}\\ \boldsymbol{m}^\top\boldsymbol{\Omega}\boldsymbol{a}_c & \boldsymbol{m}^\top\boldsymbol{\Omega}\boldsymbol{m}\end{pmatrix}\right|\delta, \boldsymbol{x}_{1:n} \sim \mathcal{W}_2\left(n + n_0 + p + 2, \begin{pmatrix}\boldsymbol{a}_c^\top\boldsymbol{V}_c\boldsymbol{a}_c & \boldsymbol{a}_c^\top\boldsymbol{V}_c\boldsymbol{m}\\ \boldsymbol{m}^\top\boldsymbol{V}_c\boldsymbol{a}_c & \boldsymbol{m}^\top\boldsymbol{V}_c\boldsymbol{m}\end{pmatrix}\right). \tag{60}$$

From (59) the stochastic representation of $\boldsymbol{m}^\top\boldsymbol{\nu}$ is deduced as

$$\boldsymbol{m}^\top\boldsymbol{\nu} \overset{d}{=} \boldsymbol{m}^\top\boldsymbol{\Omega}\boldsymbol{a}_c + n_c^{-1/2}\sqrt{\boldsymbol{m}^\top\boldsymbol{\Omega}\boldsymbol{m}}\eta_1, \tag{61}$$

where $\eta_1 \sim \mathcal{N}(0, 1)$ and is independent of $\boldsymbol{m}^\top\boldsymbol{\Omega}\boldsymbol{a}_c$ and $\boldsymbol{m}^\top\boldsymbol{\Omega}\boldsymbol{m}$.

The application of Theorems 3.2.8 and 3.2.10 of Muirhead (1982) yields

$$\boldsymbol{m}^\top\boldsymbol{\Omega}\boldsymbol{a}_c|\boldsymbol{m}^\top\boldsymbol{\Omega}\boldsymbol{m} \sim \mathcal{N}\left(\boldsymbol{m}^\top\boldsymbol{V}_c\boldsymbol{a}_c\frac{\boldsymbol{m}^\top\boldsymbol{\Omega}\boldsymbol{m}}{\boldsymbol{m}^\top\boldsymbol{V}_c\boldsymbol{m}}, (\boldsymbol{a}_c^\top\boldsymbol{V}_c\boldsymbol{a}_c\boldsymbol{m}^\top\boldsymbol{V}_c\boldsymbol{m} - (\boldsymbol{a}_c^\top\boldsymbol{V}_c\boldsymbol{m})^2)\right.$$
$$\left.\times\frac{\boldsymbol{m}^\top\boldsymbol{\Omega}\boldsymbol{m}}{\boldsymbol{m}^\top\boldsymbol{V}_c\boldsymbol{m}}\right) \tag{62}$$

and

$$\frac{\boldsymbol{m}^\top\boldsymbol{\Omega}\boldsymbol{m}}{\boldsymbol{m}^\top\boldsymbol{V}_c\boldsymbol{m}} \sim \chi^2_{n+n_0+p+2}. \tag{63}$$

As a result, the joint stochastic representation of $\boldsymbol{m}^\top \boldsymbol{\Omega} \boldsymbol{a}_c$ and $(\boldsymbol{m}^\top \boldsymbol{\Omega} \boldsymbol{m})/(\boldsymbol{m}^\top \boldsymbol{V}_c \boldsymbol{m})$ is given by

$$\frac{\boldsymbol{m}^\top \boldsymbol{\Omega} \boldsymbol{m}}{\boldsymbol{m}^\top \boldsymbol{V}_c \boldsymbol{m}} \overset{d}{=} \xi, \ \boldsymbol{m}^\top \boldsymbol{\Omega} \boldsymbol{a}_c \overset{d}{=} \boldsymbol{m}^\top \boldsymbol{V}_c \boldsymbol{a}_c \xi + \sqrt{\boldsymbol{a}_c^\top \boldsymbol{V}_c \boldsymbol{a}_c \boldsymbol{m}^\top \boldsymbol{V}_c \boldsymbol{m} - (\boldsymbol{a}_c^\top \boldsymbol{V}_c \boldsymbol{m})^2} \xi^{1/2} \eta_2, \tag{64}$$

where ξ and η_2 are independent with $\xi \sim \chi^2_{n+n_0+p+2}$ and $\eta_2 \sim \mathcal{N}(0, 1)$.

Hence, the stochastic representation of $\boldsymbol{m}^\top \boldsymbol{v}$ is expressed as

$$\begin{aligned}\boldsymbol{m}^\top \boldsymbol{v} \overset{d}{=} & \ \boldsymbol{m}^\top \boldsymbol{V}_c \boldsymbol{a}_c \xi + \sqrt{\boldsymbol{a}_c^\top \boldsymbol{V}_c \boldsymbol{a}_c \boldsymbol{m}^\top \boldsymbol{V}_c \boldsymbol{m} - (\boldsymbol{a}_c^\top \boldsymbol{V}_c \boldsymbol{m})^2} \xi^{1/2} \eta_2 \\ & + n_c^{-1/2} \sqrt{\boldsymbol{m}^\top \boldsymbol{V}_c \boldsymbol{m}} \xi^{1/2} \eta_1,\end{aligned} \tag{65}$$

where η_1, η_2, and ξ are mutually independent. Finally, using that η_1 and η_2 are independent and standard normally distributed, the statement of the theorem follows. □

Theorem 2 *Under the spike and slab prior* (52)–(54), *the stochastic representation of* $\boldsymbol{m}^\top \boldsymbol{v}$ *is given by*

$$\boldsymbol{m}^\top \boldsymbol{v} \overset{d}{=} \boldsymbol{m}^\top \boldsymbol{V}_c \boldsymbol{a}_c \xi + \sqrt{(\boldsymbol{a}_c^\top \boldsymbol{V}_c \boldsymbol{a}_c + n_c^{-1}) \boldsymbol{m}^\top \boldsymbol{V}_c \boldsymbol{m} - (\boldsymbol{a}_c^\top \boldsymbol{V}_c \boldsymbol{m})^2} \xi^{1/2} \eta, \tag{66}$$

where η *and* ξ *are independent with* $\eta \sim \mathcal{N}(0, 1)$ *and* $\xi \sim \chi^2_{n_c+p+2}$ *with*

$$\begin{aligned} n_c &= n + n_0, \\ \boldsymbol{S}_c &= \boldsymbol{S}_c(\boldsymbol{\Lambda}) = \boldsymbol{\Lambda}^{1/2} \boldsymbol{S}_0 \boldsymbol{\Lambda}^{1/2} - \frac{c_0^2}{n_0} \boldsymbol{\Lambda}^{1/2} \boldsymbol{S}_0 \boldsymbol{w}_0 \boldsymbol{w}_0^\top \boldsymbol{S}_0 \boldsymbol{\Lambda}^{1/2} + \frac{c_0^2}{n_0} \boldsymbol{S}_0 \boldsymbol{w}_0 \boldsymbol{w}_0^\top \boldsymbol{S}_0 + \boldsymbol{T} \\ \boldsymbol{w}_c &= \boldsymbol{w}_c(\boldsymbol{\Lambda}) = \boldsymbol{S}_c^{-1} \left(c_0 \boldsymbol{S}_0 \boldsymbol{w}_0 + \boldsymbol{t} \right), \\ \boldsymbol{a}_c &= n_c^{-1} \boldsymbol{S}_c \boldsymbol{w}_c = n_c^{-1} \left(c_0 \boldsymbol{S}_0 \boldsymbol{w}_0 + \boldsymbol{t} \right), \\ \boldsymbol{V}_c &= \boldsymbol{V}_c(\boldsymbol{\Lambda}) = \left(\boldsymbol{S}_c - n_c^{-1} \boldsymbol{S}_c \boldsymbol{w}_c \boldsymbol{w}_c^\top \boldsymbol{S}_c \right)^{-1} = \boldsymbol{S}_c^{-1} + \frac{\boldsymbol{w}_c \boldsymbol{w}_c^\top}{n_c - \boldsymbol{w}_c^\top \boldsymbol{S}_c \boldsymbol{w}_c}. \end{aligned}$$

Proof of Theorem 2 Under the spike and slab prior (52)–(54), we get

$$\begin{aligned}\pi(\boldsymbol{v}, \boldsymbol{\Omega} | \boldsymbol{\Lambda}, \boldsymbol{x}_{1:n}) \propto & \det(\boldsymbol{\Omega})^{(n_0+n)/2} \exp\left(-\frac{n_0 + n}{2} \boldsymbol{v}^\top \boldsymbol{\Omega}^{-1} \boldsymbol{v} \right) \\ & \exp\left((c_0 \boldsymbol{S}_0 \boldsymbol{w}_0 + \boldsymbol{t})^\top \boldsymbol{v} \right)\end{aligned}$$

$$\times \exp\left(\text{tr}\left(\boldsymbol{\Omega}\left(-\frac{1}{2}\left(\boldsymbol{\Lambda}^{1/2}\boldsymbol{S}_0\boldsymbol{\Lambda}^{1/2} - \frac{c_0^2}{n_0}\boldsymbol{\Lambda}^{1/2}\boldsymbol{S}_0\boldsymbol{w}_0\boldsymbol{w}_0^\top\boldsymbol{S}_0\boldsymbol{\Lambda}^{1/2}\right.\right.\right.\right.$$

$$\left.\left.\left.\left.+\frac{c_0^2}{n_0}\boldsymbol{S}_0\boldsymbol{w}_0\boldsymbol{w}_0^\top\boldsymbol{S}_0 + \boldsymbol{T}\right)\right)\right)\right)$$

$$= \det(\boldsymbol{\Omega})^{(n_0+n)/2}\exp\left(-\frac{1}{2}\text{tr}\left(\boldsymbol{\Omega}\left(\boldsymbol{S}_c - n_c^{-1}\boldsymbol{S}_c\boldsymbol{w}_c\boldsymbol{w}_c^\top\boldsymbol{S}_c\right)\right)\right)$$

$$\times \exp\left(-\frac{n_c}{2}(\boldsymbol{v} - n_c^{-1}\boldsymbol{\Omega}\boldsymbol{S}_c\boldsymbol{w}_c)^\top\boldsymbol{\Omega}^{-1}(\boldsymbol{v} - n_c^{-1}\boldsymbol{\Omega}\boldsymbol{S}_c\boldsymbol{w}_c)\right), \tag{67}$$

and, hence,

$$\boldsymbol{v}|\boldsymbol{\Omega}, \delta, \boldsymbol{x}_{1:n} \sim \mathcal{N}_p\left(\boldsymbol{\Omega}\boldsymbol{a}_c, n_c^{-1}\boldsymbol{\Omega}\right),$$

$$\boldsymbol{\Omega}|\delta, \boldsymbol{x}_{1:n} \sim \mathcal{W}_p\left(n + n_0 + p + 2, \boldsymbol{V}_c\right),$$

with n_c, $\boldsymbol{S}_c$, $\boldsymbol{w}_c$, $\boldsymbol{a}_c$, and $\boldsymbol{V}_c$ defined in the statement of the theorem. The rest of the proof follows from the proof of Theorem 1. □

Remark 1 Another possibility to incorporate the spike and slab prior to the optimal portfolio choice problem is based on the prior given by

$$\boldsymbol{v}|\boldsymbol{\Omega}, \boldsymbol{\Delta} \sim \mathcal{N}_p\left(c_0 n_0^{-1}\boldsymbol{\Omega}\boldsymbol{S}_0\boldsymbol{w}_0, n_0^{-1}\boldsymbol{\Delta}^{1/2}\boldsymbol{\Omega}\boldsymbol{\Delta}^{1/2}\right), \tag{68}$$

$$\boldsymbol{\Omega} \sim \mathcal{W}_p\left(n_0 + p + 2, \boldsymbol{S}_0^{-1} + \frac{\boldsymbol{w}_0\boldsymbol{w}_0^\top}{c_0^{-2}n_0 - \boldsymbol{w}_0^\top\boldsymbol{S}_0\boldsymbol{w}_0}\right), \tag{69}$$

$$\boldsymbol{\Delta} = \text{diag}(d_{0,1}\delta_1 + (1-\delta_1), \ldots, d_{0,p}\delta_p + (1-\delta_p)),$$

$$\delta_i|v_i \sim Be(v_i) \quad \text{and} \quad v_i \sim Unif(0,1), \tag{70}$$

where the symbol $Be(v)$ denotes the Bernoulli distribution with the probability of success v and the symbol $Unif(0, 1)$ stands for the uniform distribution on the interval $[0, 1]$. Finally, $\boldsymbol{\Omega}$ is independent of $\boldsymbol{\Delta}$, and (v_i, δ_i), $i = 1, \ldots, p$, are mutually independent. The quantities $d_{0,i}$, $i = 1, \ldots, p$, are some small positive numbers.

The results of Theorems 1 and 2 provide a simple and efficient way of how the samples from the posterior distribution of the tangency portfolio weights can be drawn. Moreover, in contrast to the MCMC methods, they lead to the exact simulation procedures which can be implemented in practice by generating random objects from the well-known univariate distributions. This will considerably reduce the computation time when the proposed stochastic representations are used in practice.

Drawn samples from the posterior distribution of a linear combination of the tangency portfolio weights can be used for constructing Bayesian inference procedures for each weight of the tangency portfolio separately. This is achieved by computing the sample mean, the sample median, and concentration sets which provide a good proxy for Bayesian point estimators and credible intervals.

Another important application of the results presented in Theorems 1 and 2 leads to the simplification in the computation of posterior mean and posterior variances. Namely, the application of (55) leads to

$$\mathbb{E}\left(\boldsymbol{m}^\top \boldsymbol{\nu}|\delta, \boldsymbol{x}_{1:n}\right) = (n_c(\delta) + p + 2)\frac{\boldsymbol{m}^\top \boldsymbol{w}_c(\delta)}{n_c(\delta) - \boldsymbol{w}_c(\delta)^\top \boldsymbol{S}_c(\delta)\boldsymbol{w}_c(\delta)},$$

and

$$\mathbb{V}ar\left(\boldsymbol{m}^\top \boldsymbol{\nu}|\delta, \boldsymbol{x}_{1:n}\right) = (n_c(\delta) + p + 2)$$
$$\times\left(2\frac{(\boldsymbol{m}^\top \boldsymbol{w}_c(\delta))^2}{(n_c(\delta) - \boldsymbol{w}_c(\delta)^\top \boldsymbol{S}_c(\delta)\boldsymbol{w}_c(\delta))^2} + \frac{\boldsymbol{m}^\top \boldsymbol{S}_c(\delta)^{-1}\boldsymbol{m}}{n_c(\delta) - \boldsymbol{w}_c(\delta)^\top \boldsymbol{S}_c(\delta)\boldsymbol{w}_c(\delta)}\right),$$

where $n_c(\delta)$, $\boldsymbol{w}_c(\delta)$, and $\boldsymbol{S}_c(\delta)$ are given in the statement of Theorem 1. Since $\boldsymbol{m}$ is an arbitrary vector, we get that

$$\hat{\boldsymbol{\nu}}(\delta) = \mathbb{E}\left(\boldsymbol{\nu}|\delta, \boldsymbol{x}_{1:n}\right) = (n_c(\delta) + p + 2)\frac{\boldsymbol{w}_c(\delta)}{n_c(\delta) - \boldsymbol{w}_c(\delta)^\top \boldsymbol{S}_c(\delta)\boldsymbol{w}_c(\delta)},$$

and

$$\boldsymbol{V}_{\hat{\boldsymbol{\nu}}}(\delta) = \mathbb{V}ar\left(\boldsymbol{\nu}|\delta, \boldsymbol{x}_{1:n}\right) = (n_c(\delta) + p + 2)$$
$$\times\left(2\frac{\boldsymbol{w}_c(\delta)\boldsymbol{w}_c(\delta)^\top}{(n_c(\delta) - \boldsymbol{w}_c(\delta)^\top \boldsymbol{S}_c(\delta)\boldsymbol{w}_c(\delta))^2}\right.$$
$$\left. + \frac{1}{n_c(\delta) - \boldsymbol{w}_c(\delta)^\top \boldsymbol{S}_c(\delta)\boldsymbol{w}_c(\delta)}\boldsymbol{S}_c(\delta)^{-1}\right).$$

Using that

$$\mathbb{V}ar\left(\boldsymbol{\nu}|\boldsymbol{x}_{1:n}\right) = \mathbb{E}\left(\mathbb{V}ar\left(\boldsymbol{\nu}|\delta, \boldsymbol{x}_{1:n}\right)\right) + \mathbb{V}ar\left(\mathbb{E}\left(\boldsymbol{\nu}|\delta, \boldsymbol{x}_{1:n}\right)\right),$$

we get the Bayesian estimate of the tangency portfolio weights expressed as

$$\hat{\boldsymbol{w}}_{TP;L} = \frac{1}{\gamma}\frac{1}{B}\sum_{b=1}^{B}\hat{\boldsymbol{\nu}}(\delta^{(b)}) \tag{71}$$

together with the posterior covariance matrix given by

$$\boldsymbol{V}_{\hat{\boldsymbol{w}}_{TP;L}} = \frac{1}{\gamma^2}\left(\frac{1}{B}\sum_{b=1}^{B}\boldsymbol{V}_{\hat{\boldsymbol{v}}}(\delta^{(b)}) + \frac{1}{B-1}\sum_{b=1}^{B}(\hat{\boldsymbol{v}}(\delta^{(b)}) - \gamma\hat{\boldsymbol{w}}_{TP;L})(\hat{\boldsymbol{v}}(\delta^{(b)}) - \gamma\hat{\boldsymbol{w}}_{TP;L})^\top\right), \tag{72}$$

where $\delta^{(b)}$, $b = 1, \ldots, B$, are generated independently from the standard exponential distribution.

Similarly, the Bayesian estimate of the tangency portfolio weights under the spike and slab prior is obtained, and it is expressed as

$$\hat{\boldsymbol{w}}_{TP;S} = \frac{1}{\gamma}\frac{1}{B}\sum_{b=1}^{B}\hat{\boldsymbol{v}}(\boldsymbol{\Lambda}^{(b)}) \tag{73}$$

together with the posterior covariance matrix given by

$$\boldsymbol{V}_{\hat{\boldsymbol{w}}_{TP;S}} = \frac{1}{\gamma^2}\left(\frac{1}{B}\sum_{b=1}^{B}\boldsymbol{V}_{\hat{\boldsymbol{v}}}(\boldsymbol{\Lambda}^{(b)}) + \frac{1}{B-1}\sum_{b=1}^{B}(\hat{\boldsymbol{v}}(\boldsymbol{\Lambda}^{(b)}) - \gamma\hat{\boldsymbol{w}}_{TP;S}) \times(\hat{\boldsymbol{v}}(\boldsymbol{\Lambda}^{(b)}) - \gamma\hat{\boldsymbol{w}}_{TP;S})^\top\right), \tag{74}$$

where

$$\begin{aligned}
\hat{\boldsymbol{v}}(\boldsymbol{\Lambda}) &= \mathbb{E}\,(\boldsymbol{v}|\boldsymbol{\Lambda}, \boldsymbol{x}_{1:n}) = (n_c + p + 2)\frac{\boldsymbol{w}_c(\boldsymbol{\Lambda})}{n_c - \boldsymbol{w}_c(\boldsymbol{\Lambda})^\top \boldsymbol{S}_c(\boldsymbol{\Lambda})\boldsymbol{w}_c(\boldsymbol{\Lambda})},\\
\boldsymbol{V}_{\hat{\boldsymbol{v}}}(\boldsymbol{\Lambda}) &= \mathbb{V}ar\,(\boldsymbol{v}|\boldsymbol{\Lambda}, \boldsymbol{x}_{1:n}) = (n_c + p + 2)\\
&\quad\times\left(2\frac{\boldsymbol{w}_c(\boldsymbol{\Lambda})\boldsymbol{w}_c(\boldsymbol{\Lambda})^\top}{(n_c - \boldsymbol{w}_c(\boldsymbol{\Lambda})^\top \boldsymbol{S}_c(\boldsymbol{\Lambda})\boldsymbol{w}_c(\boldsymbol{\Lambda}))^2}\right.\\
&\quad\left.+\frac{1}{n_c - \boldsymbol{w}_c(\boldsymbol{\Lambda})^\top \boldsymbol{S}_c(\boldsymbol{\Lambda})\boldsymbol{w}_c(\boldsymbol{\Lambda})}\boldsymbol{S}_c(\boldsymbol{\Lambda})^{-1}\right),
\end{aligned}$$

and n_c, $\boldsymbol{w}_c(\boldsymbol{\Lambda})$, and $\boldsymbol{S}_c(\boldsymbol{\Lambda})$ are given in the statement of Theorem 2. The realizations of $\boldsymbol{\Lambda}^{(b)}$ for $b = 1, \ldots, B$ are obtained independently from the distribution of $\boldsymbol{\Lambda}$ specified in (54).

4 Summary

Bayesian portfolio selection attracts both researchers and practitioners of the financial sector today. While some approaches concentrate on the derivation of the posterior distribution of optimal portfolio weights, other methods deal with out-of-sample forecasting and portfolio selection by employing the posterior predictive distribution of the portfolio return. In most of the cases, the resulting Bayesian inference procedures are obtained by using the MCMC, while exact sampling approaches have recently been introduced in Bodnar et al. (2017), Bauder et al. (2021), Bodnar et al. (2022c), and Bodnar et al. (2024b), which employ stochastic representations derived from both the posterior distribution and the posterior predictive distribution.

In this chapter, we contribute to the existent literature by proposing new priors for Bayesian portfolio selection which are based on the Laplace distribution and on the spike and slab approach. Under both priors, stochastic representation for a linear combination of the tangency portfolio weights is derived which can be used to efficiently sample from the posterior distribution of the tangency portfolio weights. Moreover, the derived stochastic representations are used to simplify the computation of the posterior mean of the portfolio weights and their posterior covariance matrix. For example, in the case of the Laplace prior, it can be computed for any level of precision by drawing samples from the standard univariate exponential distribution.

Acknowledgments The authors would like to thank Professor Stepan Mazur, Professor Pär Österholm, and anonymous reviewers for their constructive comments that improved the quality of this work. Olha Bodnar acknowledges valuable support from the internal grant (Rörlig resurs) of Örebro University.

References

Aguilar, O. & West, M. (2000). Bayesian dynamic factor models and portfolio allocation. *Journal of Business & Economic Statistics, 18*(3), 338–357.

Alfelt, G. & Mazur, S. (2022). On the mean and variance of the estimated tangency portfolio weights for small samples. *Modern Stochastics: Theory and Applications, 9*(4), 453–482.

Avramov, D. & Zhou, G. (2010). Bayesian portfolio analysis. *Annual Review of Financial Economics, 2*(1), 25–47.

Bai, Z. D. & Silverstein, J. W. (2010). *Spectral Analysis of Large Dimensional Random Matrices*. Springer.

Baker, S. R., Bloom, N., & Davis, S. J. (2016). Measuring economic policy uncertainty. *The Quarterly Journal of Economics, 131*(4), 1593–1636.

Bauder, D., Bodnar, R., Bodnar, T., & Schmid, W. (2019). Bayesian estimation of the efficient frontier. *Scandinavian Journal of Statistics, 46*(3), 802–830.

Bauder, D., Bodnar, T., Mazur, S., & Okhrin, Y. (2018). Bayesian inference for the tangent portfolio. *International Journal of Theoretical and Applied Finance, 21*(08), 1850054.

Bauder, D., Bodnar, T., Parolya, N., & Schmid, W. (2020). Bayesian inference of the multi-period optimal portfolio for an exponential utility. *Journal of Multivariate Analysis, 175*, 104544.
Bauder, D., Bodnar, T., Parolya, N., & Schmid, W. (2021). Bayesian mean–variance analysis: Optimal portfolio selection under parameter uncertainty. *Quantitative Finance, 21*(2), 221–242.
Bawa, V. S., Brown, S. J., & Klein, R. W. (1979). *Estimation risk and optimal portfolio choice*. North-Holland.
Berger, J. O. & Pericchi, L. R. (1996). The intrinsic Bayes factor for model selection and prediction. *Journal of the American Statistical Association, 91*(433), 109–122.
Berger, J. O. & Pericchi, L. R. (2001). Objective Bayesian methods for model selection: Introduction and comparison. In P. Lahiri (Ed.), *Model selection*. Lecture Notes-Monograph Series (Vol. 38, pp. 135–207). Institute of Mathematical Statistics.
Bernardo, J. M. & Smith, A. F. (2000). *Bayesian theory* (Vol. 405). Wiley.
Black, F. & Litterman, R. (1992). Global portfolio optimization. *Financial Analysts Journal, 48*(5), 28–43.
Bodnar, O. (2009). Sequential surveillance of the tangency portfolio weights. *International Journal of Theoretical and Applied Finance, 12*(06), 797–810.
Bodnar, O. & Bodnar, T. (2010). On the unbiased estimator of the efficient frontier. *International Journal of Theoretical and Applied Finance, 13*(07), 1065–1073.
Bodnar, O., Bodnar, T., & Niklasson, V. (2024a). Constructing Bayesian tangency portfolios under short-selling restrictions. *Finance Research Letters, 62*, 105065.
Bodnar, O., Bodnar, T., & Niklasson, V. (2024b). Incorporating different sources of information for Bayesian optimal portfolio selection. *Journal of Business & Economic Statistics, 43*(2), 365–377.
Bodnar, O., Bodnar, T., & Parolya, N. (2022a). Recent advances in shrinkage-based high-dimensional inference. *Journal of Multivariate Analysis, 188*, 104826.
Bodnar, T., Dette, H., Parolya, N., & Thorsén, E. (2022b). Sampling distributions of optimal portfolio weights and characteristics in small and large dimensions. *Random Matrices: Theory and Applications, 11*(01), 2250008.
Bodnar, T., Dmytriv, S., Okhrin, Y., Otryakhin, D., & Parolya, N. (2024c). *HDShOP: High-Dimensional Shrinkage Optimal Portfolios*. R package version 0.1.5.
Bodnar, T., Dmytriv, S., Okhrin, Y., Parolya, N., & Schmid, W. (2021). Statistical inference for the expected utility portfolio in high dimensions. *IEEE Transactions on Signal Processing, 69*, 1–14.
Bodnar, T., Dmytriv, S., Parolya, N., & Schmid, W. (2019a). Tests for the weights of the global minimum variance portfolio in a high-dimensional setting. *IEEE Transactions on Signal Processing, 67*(17), 4479–4493.
Bodnar, T., Gupta, A. K., & Parolya, N. (2014). On the strong convergence of the optimal linear shrinkage estimator for large dimensional covariance matrix. *Journal of Multivariate Analysis, 132*, 215–228.
Bodnar, T., Gupta, A. K., & Parolya, N. (2016a). Direct shrinkage estimation of large dimensional precision matrix. *Journal of Multivariate Analysis, 146*, 223–236.
Bodnar, T., Lindholm, M., Niklasson, V., & Thorsén, E. (2022c). Bayesian portfolio selection using VaR and CVaR. *Applied Mathematics and Computation, 427*, 127120.
Bodnar, T., Mazur, S., & Okhrin, Y. (2017). Bayesian estimation of the global minimum variance portfolio. *European Journal of Operational Research, 256*(1), 292–307.
Bodnar, T., Mazur, S., & Podgórski, K. (2016b). Singular inverse Wishart distribution with application to portfolio theory. *Journal of Multivariate Analysis, 143*, 314–326.
Bodnar, T., Mazur, S., Podgórski, K., & Tyrcha, J. (2019b). Tangency portfolio weights for singular covariance matrix in small and large dimensions: Estimation and test theory. *Journal of Statistical Planning and Inference, 201*, 40–57.
Bodnar, T., Niklasson, V., & Thorsén, E. (2024d). Volatility sensitive Bayesian estimation of portfolio VaR and CVaR. *Journal of Risk, 26*(4), 1–29.

Bodnar, T., Okhrin, O., & Parolya, N. (2019c). Optimal shrinkage estimator for high-dimensional mean vector. *Journal of Multivariate Analysis, 170*, 63–79.

Bodnar, T., Okhrin, Y., & Parolya, N. (2023a). Optimal shrinkage-based portfolio selection in high dimensions. *Journal of Business & Economic Statistics, 41*(1), 140–156.

Bodnar, T. & Parolya, N. (2024). Reviving pseudo-inverses: Asymptotic properties of large dimensional Moore-Penrose and ridge-type inverses with applications. *arXiv preprint arXiv:2403.15792*.

Bodnar, T., Parolya, N., & Schmid, W. (2018). Estimation of the global minimum variance portfolio in high dimensions. *European Journal of Operational Research, 266*(1), 371–390.

Bodnar, T., Parolya, N., & Thorsén, E. (2023b). Dynamic shrinkage estimation of the high-dimensional minimum-variance portfolio. *IEEE Transactions on Signal Processing, 71*, 1334–1349.

Bodnar, T., Parolya, N., & Thorsén, E. (2024e). Two is better than one: Regularized shrinkage of large minimum variance portfolios. *Journal of Machine Learning Research, 25*, 1–32.

Bodnar, T. & Schmid, W. (2009). Econometrical analysis of the sample efficient frontier. *European Journal of Finance, 15*, 317–335.

Britten-Jones, M. (1999). The sampling error in estimates of mean-variance efficient portfolio weights. *The Journal of Finance, 54*(2), 655–671.

DasGupta, A. (2008). *Asymptotic theory of statistics and probability*. Springer.

DeMiguel, V., Garlappi, L., & Uppal, R. (2009). Optimal versus naive diversification: How inefficient is the 1/N portfolio strategy? *The Review of Financial Studies, 22*, 1915–1953.

Drin, S., Mazur, S., & Muhinyuza, S. (2024). A test on the location of tangency portfolio for small sample size and singular covariance matrix. *Modern Stochastics: Theory and Applications, 12*, 1–17.

Frost, P. A. & Savarino, J. E. (1986). An empirical Bayes approach to efficient portfolio selection. *Journal of Financial and Quantitative Analysis, 21*(3), 293–305.

Gibbons, M. R., Ross, S. A., & Shanken, J. (1989). A test of the efficiency of a given portfolio. *Econometrica, 57*, 1121–1152.

Greyserman, A., Jones, D. H., & Strawderman, W. E. (2006). Portfolio selection using hierarchical Bayesian analysis and MCMC methods. *Journal of Banking & Finance, 30*(2), 669–678.

Gupta, A. K., Varga, T., & Bodnar, T. (2013). *Elliptically contoured models in statistics and portfolio theory*. Springer.

Held, L. & Bové, D. S. (2014). *Applied statistical inference: Likelihood and Bayes*. Springer Science & Business Media.

Ingersoll, J. (1987). *Theory of financial decision making*. Rowman & Littlefield.

Ishwaran, H. & Rao, J. S. (2005). Spike and slab variable selection: Frequentist and Bayesian strategies. *The Annals of Statistics, 33*, 730–773.

Javed, F., Mazur, S., & Thorsén, E. (2024). Tangency portfolio weights under a skew-normal model in small and large dimensions. *Journal of the Operational Research Society, 75*(7), 1395–1406.

Jobson, J. D. & Korkie, B. (1980). Estimation for Markowitz efficient portfolios. *Journal of the American Statistical Association, 75*(371), 544–554.

Jorion, P. (1986). Bayes-stein estimation for portfolio analysis. *Journal of Financial and Quantitative Analysis, 21*(3), 279–292.

Kan, R. & Smith, D. R. (2008). The distribution of the sample minimum-variance frontier. *Management Science, 54*(7), 1364–1380.

Kan, R. & Zhou, G. (2007). Optimal portfolio choice with parameter uncertainty. *Journal of Financial and Quantitative Analysis, 42*(3), 621–656.

Karlsson, S., Mazur, S., & Muhinyuza, S. (2021). Statistical inference for the tangency portfolio in high dimension. *Statistics, 55*(3), 532–560.

Kass, R. E. & Raftery, A. E. (1995). Bayes factors. *Journal of the American Statistical Association, 90*(430), 773–795.

Lassance, N., Vanderveken, R., & Vrins, F. (2023). On the combination of naive and mean-variance portfolio strategies. *Journal of Business & Economic Statistics, 42*(3), 875–889.

Ledoit, O. & Wolf, M. (2017). Nonlinear shrinkage of the covariance matrix for portfolio selection: Markowitz meets goldilocks. *The Review of Financial Studies, 30*(12), 4349–4388.
Ledoit, O. & Wolf, M. (2020). Analytical nonlinear shrinkage of large-dimensional covariance matrices. *The Annals of Statistics, 48*(5), 3043–3065.
Li, Q. & Lin, N. (2010). The Bayesian elastic net. *Bayesian Analysis, 5*(1), 151–170.
Markowitz, H. (1952). Portfolio selection. *The Journal of Finance, 7*, 77–91.
Merton, R. C. (1972). An analytic derivation of the efficient portfolio frontier. *Journal of Financial and Quantitative Analysis, 7*(4), 1851–1872.
Muhinyuza, S., Bodnar, T., & Lindholm, M. (2020). A test on the location of the tangency portfolio on the set of feasible portfolios. *Applied Mathematics and Computation, 386*, 125519.
Muirhead, R. J. (1982). *Aspects of multivariate statistical theory*. Wiley.
Okhrin, Y. & Schmid, W. (2006). Distributional properties of portfolio weights. *Journal of Econometrics, 134*(1), 235–256.
Park, T. & Casella, G. (2008). The Bayesian lasso. *Journal of the American Statistical Association, 103*(482), 681–686.
Rachev, S. T., Hsu, J. S., Bagasheva, B. S., & Fabozzi, F. J. (2008). *Bayesian methods in finance*. Wiley.
Shapiro, A., Zhou, E., & Lin, Y. (2023). Bayesian distributionally robust optimization. *SIAM Journal on Optimization, 33*(2), 1279–1304.
Stambaugh, R. F. (1997). Analyzing investments whose histories differ in length. *Journal of Financial Economics, 45*(3), 285–331.
Stein, C. (1956). Inadmissibility of the usual estimator for the mean of a multivariate normal distribution. In *Proceedings of the Third Berkeley Symposium on Mathematical Statistics and Probability, Volume 1: Contributions to the Theory of Statistics* (pp. 197–206). University of California Press.
Sundberg, R. (2019). *Statistical modelling by exponential families* (Vol. 12). Cambridge University Press.
Tu, J. & Zhou, G. (2010). Incorporating economic objectives into Bayesian priors: Portfolio choice under parameter uncertainty. *Journal of Financial and Quantitative Analysis, 45*(4), 959–986.
Winkler, R. L. (1973). Bayesian models for forecasting future security prices. *Journal of Financial and Quantitative Analysis, 8*, 387–405.
Winkler, R. L. & Barry, C. B. (1975). A Bayesian model for portfolio selection and revision. *The Journal of Finance, 30*(1), 179–192.
Zellner, A. & Chetty, V. K. (1965). Prediction and decision problems in regression models from the Bayesian point of view. *Journal of the American Statistical Association, 60*(310), 608–616.

Predictive Decision Synthesis for Portfolios: Betting on Better Models

Emily Tallman and Mike West

Abstract We discuss and develop Bayesian dynamic modeling and predictive decision synthesis for portfolio analysis. The context involves model uncertainty with a set of candidate models for financial time series with main foci in sequential learning, forecasting, and recursive decisions for portfolio reinvestments. The foundational perspective of Bayesian predictive decision synthesis (BPDS) defines novel, operational analysis and resulting predictive and decision outcomes. A detailed case study of BPDS in financial forecasting of international exchange rate time series and portfolio rebalancing, with resulting BPDS-based decision outcomes compared to traditional Bayesian analysis, exemplifies and highlights the practical advances achievable under the expanded, subjective Bayesian approach that BPDS defines.

1 Introduction

Portfolio optimization is a major area of application and success of Bayesian decision analysis, going back at least to Markowitz (1952) and arguably earlier (de Finetti, 1940; Markowitz, 2006; Rubinstein, 2006). More recent Bayesian literature has emphasized financial time series modeling and forecasting to link with portfolio decision analysis (e.g., Quintana & West, 1987; West & Harrison, 1989a, chap. 15; West & Harrison, 1997, chap. 16; Aguilar & West, 2000; Karlsson, 2013; Zhao et al., 2016; Gruber & West, 2016, 2017; Prado et al., 2021, chap. 10), along with continuing concern for customizing utility specifications to guide Bayesian portfolio selection (e.g., Irie & West, 2019, and the references therein).

E. Tallman
Department of Statistical Science, Duke University, Durham, NC, USA
e-mail: emily.tallman@duke.edu

M. West (✉)
The Arts & Sciences Distinguished Professor Emeritus of Statistics & Decision Sciences, Department of Statistical Science, Duke University, Durham, NC, USA
e-mail: mike.west@duke.edu

S. Mazur, P. Österholm (eds.), *Recent Developments in Bayesian Econometrics and Their Applications*, https://doi.org/10.1007/978-3-032-00110-8_10

Here we explicitly address portfolio decision goals when multiple forecasting models are to be explored and combined. Accounting for model uncertainty in pure forecasting is routine (e.g., West & Harrison, 1997; Jacobson & Karlsson, 2004; Andersson & Karlsson, 2008; Cheng et al., 2012; Aye et al., 2015; Wang et al., 2016; Steel, 2020; Lavine et al., 2021; Wang et al., 2023; Bernaciak & Griffin, 2024), but decision goals are not generally incorporated in these statistical approaches. In portfolio analysis, the main goal of forecasting asset returns is selecting desirable portfolios; we argue that this should be a key focus in combining relevant forecasting models. Recent Bayesian "goal-focused" methods aim to incorporate specific prediction goals beyond those of traditional Bayesian model averaging (e.g., Eklund & Karlsson, 2007; Pettenuzzo & Ravazzolo, 2016; Lavine et al., 2021; Loaiza-Maya et al., 2021; West, 2020; McAlinn, 2021; Bernaciak & Griffin, 2024). However, this literature rarely addresses explicit decision goals. A related literature concerns combining portfolio rules (e.g., Kan & Zhou, 2007; Demiguel et al., 2009; Tu & Zhou, 2011; Füss et al., 2024), though without explicit consideration of forecasting models.

Bayesian Predictive Decision Synthesis (BPDS), introduced in Tallman and West (2023), closes these gaps. BPDS is a fully Bayesian, theoretically founded framework to evaluate, compare, and combine sets of candidate models based on both their anticipated and historical decision outcomes and pure predictive performance and validity. We develop BPDS here in the setting of financial forecasting for portfolio decisions. BPDS is an expansion of Bayesian Predictive Synthesis (BPS) (McAlinn & West, 2019; McAlinn et al., 2020; McAlinn, 2021; Johnson & West, 2023), with explicit incorporation of the decision context and goals. BPDS aims to more highly weight models and outcomes that are better performing in both recommended (Bayesian decision-theoretic) optimal decisions and more traditional statistical forecasting accuracy metrics. The context of sequential portfolio decision analysis is a canonical setting for the development and exploitation of BPDS for improved predictive decision-making in the central practical context of model uncertainty.

Sune Karlsson has contributed to Bayesian time series analysis and forecasting over many years. Most relevant here are questions of model/prior specification and model uncertainty in Bayesian vector autoregressive and time-varying autoregressive models, with additional connections in addressing model uncertainty and combination for prediction and decisions (e.g., Jacobson & Karlsson, 2004; Eklund & Karlsson, 2007; Andersson & Karlsson, 2008; Karlsson, 2013). In these areas and allied technical contexts, this paper links to some of the core areas of Karlsson's interests and contributions to Bayesian analysis in time series and financial econometrics.

2 Bayesian Predictive Decision Synthesis

2.1 Model Uncertainty and Portfolio Decisions

Quantities of interest and relevant notation are defined as follows, indexed by time t and implicitly representing the process of sequential analysis as time progresses. This is a standard setting of Bayesian model uncertainty analysis, indexed by time. We use the standard notation that $\mathcal{D}_{t-1}$ represents all information and data accrued up to and including time $t-1$, the point at which a decision is to be made to target some outcome at time t. Relevant quantities include:

- A forecast outcome q-vector $\mathbf{y}_t$. Examples are the time t daily returns on q assets, representing the percent change in price from time $t-1$ to time t, or a set of returns over a multi-period path ahead from times t to $t+h$ for some portfolio horizon h.
- An explicit decision problem targeting the outcome at time t, in this case selecting a set of portfolio weightings $\mathbf{x}_t$ on the q assets.
- A set of J models $\mathcal{M}_j$, $j = 1{:}J$, such as a set of dynamic regression models for the asset return vector with different predictors across models, or other distinct model forms.
- Initial model probabilities $\pi_{tj} = Pr(\mathcal{M}_j|\mathcal{D}_{t-1})$, such as from traditional Bayesian model averaging (BMA) based on prior data and information, and/or from historical performance in goal-focused forecasting, such as Bayesian adaptive variable selection (AVS: Lavine et al., 2021, a special case of BPS)—or related methods (e.g., Loaiza-Maya et al., 2021).
- Predictive densities provided by each model $\mathcal{M}_j$, $p_{tj}(\mathbf{y}_t|\mathcal{M}_j, \mathcal{D}_{t-1})$, for example, a one-step predictive for that day's FX returns or a multistep predictive for daily returns over the next h days.
- Model $\mathcal{M}_j$ applies decision analysis with its utility function $u_{tj}(\mathbf{y}_t, \mathbf{x}_t)$ to evaluate the model-specific optimal portfolio $\mathbf{x}_{tj}$ for time t. This would be the optimal portfolio decision if the decision-maker restricted the analysis only to include $\mathcal{M}_j$ and its associated utility function.

Faced with this model uncertainty question, the decision-maker aims to synthesize information across models to forecast $\mathbf{y}_t$ and decide on a final portfolio weight vector $\mathbf{x}_t$ under whatever overall utility function is chosen. This is then repeated sequentially over time.

2.2 BPDS

In the above setting, BPDS has the following ingredients. This follows and extends the main methodological developments from the BPDS theory in Tallman and West (2023).

2.2.1 Initial Mixture and Baseline Model

BPDS requires a "baseline" predictive density $p_0(\mathbf{y}_t|\mathcal{M}_0, \mathcal{D}_{t-1})$ based on a notional additional model $\mathcal{M}_0$ and a corresponding baseline optimal decision $\mathbf{x}_{t0}$. The baseline serves as an over-dispersed "safe haven" model whose predictions more heavily favor regions of the $\mathbf{y}_t$ outcome space than are less well supported by the initial mixture of the J models. The initial probabilities π_{tj} are extended and renormalized to include a nonzero π_{t0}—an initially "small" probability on $\mathcal{M}_0$, representing the concern that "all (initial J) models are wrong," i.e., the issue of model set incompleteness.

We thus have the density $p_t(\mathbf{y}_t, \mathcal{M}_j|\mathcal{D}_{t-1}) = \pi_{tj} p_{tj}(\mathbf{y}_t|\mathcal{M}_j, \mathcal{D}_{t-1})$ defined over models $j = 0{:}J$ and outcomes $\mathbf{y}_t$ jointly. The margin for $\mathbf{y}_t$, simply $p_t(\mathbf{y}_t|\mathcal{D}_{t-1}) = \sum_{j=0:J} \pi_{tj} p_{tj}(\mathbf{y}_t|\mathcal{M}_j, \mathcal{D}_{t-1})$, is called the *initial mixture*.

This framework reduces to traditional (sequential, dynamic) Bayesian model uncertainty analysis when $\pi_{t0} = 0$ with the other π_{tj} appropriately based on historical data (West & Harrison, 1997, Sect. 12.2). In that special case, the initial mixture $p_t(\mathbf{y}_t|\mathcal{D}_{t-1})$ is simply that of standard BMA.

2.2.2 BPDS Model Probabilities and Densities

BPDS modifies the initial density $p_t(\mathbf{y}_t, \mathcal{M}_j|\mathcal{D}_{t-1})$ to an updated BPDS density $f_t(\mathbf{y}_t, \mathcal{M}_j|\mathcal{D}_{t-1}) \propto \alpha_{tj}(\mathbf{y}_t) p_t(\mathbf{y}_t, \mathcal{M}_j|\mathcal{D}_{t-1})$, where the $\alpha_{tj}(\mathbf{y}_t)$ are positive, model-dependent and outcome-dependent weights. That is,

$$f(\mathbf{y}_t, \mathcal{M}_j|\mathcal{D}_{t-1}) = k_t \pi_{tj} \alpha_{tj}(\mathbf{y}_t) p_{tj}(\mathbf{y}_t|\mathcal{M}_j, \mathcal{D}_{t-1}), \quad \mathbf{y}_t \in \mathbb{R}^q, \ j = 0{:}J, \tag{1}$$

with k_t normalizing over both outcomes and models, i.e.,

$$k_t^{-1} = \sum_{j=0:J} \pi_{tj} a_{tj} \quad \text{where} \quad a_{tj} = \int_{\mathbf{y}_t \in \mathbb{R}^q} \alpha_{tj}(\mathbf{y}_t) p_j(\mathbf{y}_t|\mathcal{M}_j, \mathcal{D}_{t-1}) d\mathbf{y}_t. \tag{2}$$

The implied margin for $\mathbf{y}_t$ is the *BPDS mixture*

$$f(\mathbf{y}_t|\mathcal{D}_{t-1}) = \sum_{j=0:J} \tilde{\pi}_{tj} f_{tj}(\mathbf{y}_t|\mathcal{M}_j, \mathcal{D}_{t-1}), \tag{3}$$

where, for each model $j = 0{:}J$,

$$f_{tj}(\mathbf{y}_t|\mathcal{M}_j, \mathcal{D}_{t-1}) = a_{tj}^{-1} \alpha_{tj}(\mathbf{y}_t) p_j(\mathbf{y}_t|\mathcal{M}_j, \mathcal{D}_{t-1}) \quad \text{and} \quad \tilde{\pi}_{tj} = k_t \pi_{tj} a_{tj}. \tag{4}$$

The $f_{tj}(\cdot|\cdot)$ are *BPDS model predictive densities*, the $\tilde{\pi}_{tj}$ are *BPDS model probabilities*, and the mixture $f(\mathbf{y}_t|\mathcal{D}_{t-1})$ is the *BPDS predictive density*.

2.2.3 Portfolio BPDS Scores and Relaxed Entropic Tilting

BPDS model weights are generally and uniquely defined as

$$\alpha_{tj}(\mathbf{y}_t) = \exp\{\boldsymbol{\tau}_t'\mathbf{s}_{tj}(\mathbf{y}_t, \mathbf{x}_{tj})\} \tag{5}$$

with ingredients discussed below. The model-specific optimal portfolio vectors $\mathbf{x}_{tj}$ play critical roles in these BPDS model weightings.

First, $\mathbf{s}_{tj}(\mathbf{y}_t, \mathbf{x}_{tj})$ is a vector of *scores* depending on outcome $\mathbf{y}_t$ and the optimal portfolio $\mathbf{x}_{tj}$ under $\mathcal{M}_j$. Here $\mathbf{x}_{tj}$ is generated by optimizing whatever utility function is chosen for the decision analysis under $\mathcal{M}_j$. The elements of $\mathbf{s}_{tj}(\mathbf{y}_t, \mathbf{x}_{tj})$ are realized values of k utility functions at the $\mathcal{M}_j$-optimal portfolio $\mathbf{x}_{tj}$ were $\mathbf{y}_t$ to be the actual outcome. These utility functions are chosen so that higher scores are desirable. This allows for scoring models based on multiple metrics, expanding to multi-criteria decision analysis. Restrictions on score functions depend on the context and models involved. An example of a class of relevant score functions is given in the following section.

Second, $\boldsymbol{\tau}_t$ is a *BPDS tilting vector* that defines relative weights and the directional impact of changes in score elements. Larger values in $\boldsymbol{\tau}_t$ lead to more appreciable BPDS weightings, while $\boldsymbol{\tau}_t = \mathbf{0}$ recovers the prior mixture. The term tilting relates to the Bayesian decision-theoretic foundation and derivation of Eq. (5) which is based on relaxed entropic tilting (RET: see Tallman & West, 2022, 2023, section 3; West, 2023, sect. 2.3).

The specification of $\boldsymbol{\tau}_t$ references expected scores. Taking expectations over $(\mathbf{y}_t, \mathcal{M}_j)$ jointly, the initial model $p_t(\mathbf{y}_t, \mathcal{M}_j|\mathcal{D}_{t-1})$ has *initial expected score* $E_p[\mathbf{s}_{tj}(\mathbf{y}_t, \mathbf{x}_{tj})|\mathcal{D}_{t-1}]$. High scores are desirable. The RET framework aims to achieve scores higher than initial, asking for the BPDS distribution $f_t(\mathbf{y}_t, \mathcal{M}_j|\mathcal{D}_{t-1})$ to have expected score *at least* $E_p[\mathbf{s}_{tj}(\mathbf{y}_t, \mathbf{x}_{tj})|\mathcal{D}_{t-1}] + \boldsymbol{\epsilon}_t$ for some nonnegative k-vector $\boldsymbol{\epsilon}_t$ with at least one positive element. With this constraint, RET chooses $f(\cdot|\cdot)$ to minimize the Küllback-Leibler (KL) divergence of $p(\cdot|\cdot)$ from $f(\cdot|\cdot)$, yielding two key results: (i) $f(\cdot|\cdot)$ is defined in Eq. (1) with weighting function precisely as in Eq. (5), and (ii) $f(\cdot|\cdot)$ achieves *exactly* the expected score improvement bound specified by $\boldsymbol{\epsilon}_t$, i.e., $E_f[\mathbf{s}_{tj}(\mathbf{y}_t, \mathbf{x}_{tj})|\mathcal{D}_{t-1}] = E_p[\mathbf{s}_{tj}(\mathbf{y}_t, \mathbf{x}_{tj})|\mathcal{D}_{t-1}] + \boldsymbol{\epsilon}_t$. The RET theory shows that any feasible choice of $\boldsymbol{\epsilon}_t$ corresponds to a unique $\boldsymbol{\tau}_t$. The decision-maker has the liberty to choose $\boldsymbol{\epsilon}_t$ in each specific portfolio context; various choices are explored in the portfolio examples below.

3 Risk : Return Score Functions for Portfolios

The section temporarily drops the time index t from the notation, for clarity. Thus $\mathbf{y}$ is a vector of asset returns, and $\mathbf{s}_j(\mathbf{y}, \mathbf{x}_j)$ a vector score on $\mathcal{M}_j$ using model-specific portfolio vector $\mathbf{x}_j$ were $\mathbf{y}$ to be the observed outcome.

3.1 Utility Theory, Bivariate Scores, and Risk Tolerance

Consider $k = 2$ score dimensions and the class of vector scores

$$\mathbf{s}_j(\mathbf{y}, \mathbf{x}_j) = (e_j, -e_j^2/2)',$$

where $e_j = r_j - r^*$ with *portfolio return* $r_j = \mathbf{x}_j'\mathbf{y}$ and a chosen *target portfolio return* r^*.

These bivariate scores are motivated by the classic, bounded-above risk-averse exponential utility function for return r given by $U(r) = -\exp(-r/d)$ for some risk tolerance level parameter $d > 0$. A second-order Taylor series approximation around $r = r^*$ gives $U(r) \approx U_q(r)$ with quadratic function $U_q(r) = U(r^*)\{1 - e/d + e^2/(2d^2)\}$ and $e = r - r^*$. This is a very accurate approximation across ranges of r, r^* that correspond to realistic models for return prediction in portfolio analyses, especially for daily returns. Up to constants, $U_q(r)$ essentially reduces to the Markowitz function $e - e^2/(2d)$, maximized at $r = r^* + d$. Were we to consider BPDS based on only a univariate score, then $s_j(\mathbf{y}, \mathbf{x}_j) = e_j - e_j^2/(2d)$ is faithful to this natural utility function and theoretical considerations. This has the contextual interpretation of d as a chosen level of risk tolerance. The implied BPDS weight function would then be $\alpha_j(\mathbf{y}, \mathbf{x}_j) = \exp\{\tau(e_j - e_j^2/(2d))\}$.

In contrast, choosing the suggested and more general bivariate score $\mathbf{s}_j(\mathbf{y}, \mathbf{x}_j) = (e_j, -e_j^2/2)'$ allows flexibility in addressing risk tolerance. This leads to $\alpha_j(\mathbf{y}, \mathbf{x}_j) = \exp\{\tau_1 e_j - \tau_2 e_j^2/2\}$ which agrees with the univariate case above if/when $\tau = \tau_1$ and $d = \tau_1/\tau_2$; this defines a maximizing target return $r^* + d$ at this level of risk tolerance. Hence the bivariate score allows more flexibility in the role of the return/risk elements and their impact on the BPDS analysis while overlaying and replicating the classical exponential utility-based example as a special case. Operationally, it also yields the direct interpretation of τ_1/τ_2 as the implied level of risk tolerance.

3.2 Bivariate Scores and Tilting Vectors

In this section, we step aside from the mixture context to focus on generating portfolio-relevant insights from RET applied to a single model. This aids in understanding the roles of implied tilting vectors in the portfolio context and the relation to risk tolerance, among other details.

For this, consider predictions from any initial model $\mathbf{y} \sim p(\mathbf{y})$ such that the implied return $r = \mathbf{x}'\mathbf{y} \sim p(r)$ has mean $E_p(r) = r^* + f$, where f can be positive or negative, and variance $V_p(r) = q$. We highlight aspects of the implied distribution of the bivariate score $\mathbf{s}(\mathbf{y}, \mathbf{x}) = (e, -e^2/2)'$ involving excess return $e = r - r^*$ and practical implications for BPDS analysis.

3.2.1 Distributions of Scores

Whatever the distribution of r and the score vector under $p(\mathbf{y})$ may be, knowing only $E_p(r) = r^* + f$ and $V_p(r) = q$ implies $E_p[\mathbf{s}(\mathbf{y}, \mathbf{x})] = (f, -(f^2 + q)/2)'$. To explore the distribution of $\mathbf{s}(\mathbf{y}, \mathbf{x})$ further, useful insights arise in assuming a normal approximation to $p(r)$.

With daily returns on FX and stock indices, f and $q^{1/2}$ are typically of the order of 0.01 or less; this is relevant in exploring score uncertainty via the variance matrix $\mathbf{V}_p = V_p[\mathbf{s}(\mathbf{y}, \mathbf{x})]$. Were $p(r)$ to be normal, then it is easy to show that

$$V_p(-e^2/2) = (f^2 + q/2)V_p(e).$$

Hence, e is more uncertain than $-e^2/2$ if $f^2 + q/2 < 1$. With practical levels of small % returns, this is very likely, and $V_p(-e^2/2)$ is typically much smaller than $V_p(e)$. That is, we expect to be much more uncertain about the first score element than the second.

While the two score elements are deterministically related, their joint distribution is not wholly degenerate since knowing $-e^2/2$ does not precisely determine e.

In fact, the covariance is easily shown to be $C(e, -e^2/2) = -fq$. The implied score correlation is then $\rho \equiv Cor(e, -e^2/2) = -f_0(f_0^2 + 0.5)^{-1/2}$, where $f_0 = f/\sqrt{q}$. Note that (i) ρ has the sign of $-f$ and depends only on f_0, the expected excess return in standardized $N(0, 1)$ units, (ii) ρ is a decreasing function of f_0 and $|\rho|$ is an increasing function of f_0, (iii) over the range $f_0 = -1$ to $f_0 = 1$, the correlation decreases from roughly 0.8 to -0.8, (iv) for any given q, the absolute value $|\rho|$ is an increasing function of $|f|$, and (v) for any given f, the value $|\rho|$ is a decreasing function of q.

3.2.2 RET Distribution and Tilting Vector in the Normal Example

Suppose that $p(\mathbf{y})$ is normal. For a given tilting vector $\boldsymbol{\tau} = (\tau_1, \tau_2)'$ and any $\mathbf{x}$, the resulting RET distribution $f(\mathbf{y}) \propto \exp\{\boldsymbol{\tau}'\mathbf{s}(\mathbf{y}, \mathbf{x})\}p(\mathbf{y})$ is then also normal with easily computed moments. The implied $f(r)$ on the return $r = \mathbf{x}'\mathbf{y}$ alone is then normal with moments that are easily shown to be

$$E_f(r) = r^* + (f + q\tau_1)/(1 + q\tau_2) \quad \text{and} \quad V_f(r) = q/(1 + q\tau_2).$$

Hence the expected score $\mathbf{m} = (m_1, m_2)'$ under $f(\cdot)$ has elements

$$\begin{aligned} m_1 &= E_f(r - r^*) = (f + q\tau_1)/(1 + q\tau_2), \\ m_2 &= -E_f[(r - r^*)^2]/2 = -\{q + (1 + q\tau_2)m_1^2\}/\{2(1 + q\tau_2)\}. \end{aligned} \tag{6}$$

These equations are easily solved for $\boldsymbol{\tau}$ as a function of $\mathbf{m}$ to give *analytic* evaluation of the tilting vector. This can be shown to give

$$\tau_1 = \frac{-m_1 q - m_1^2 f - 2m_2}{q(m_1^2 + 2m_2 f)} \quad \text{and} \quad \tau_2 = \frac{-m_1^2 - 2m_2 - q}{q(m_1^2 + 2m_2)}.$$

With $m_1 > f$ and $\tau_1 > 0$ it follows that $\tau_2 > 0$ as is desirable. Then Eq. (6) leads to $\tau_1 > \tau_2 f$ or $d > f$, where $d = \tau_1/\tau_2$ has, as noted earlier, interpretation in terms of risk tolerance. We also easily have that

$$d = \tau_1/\tau_2 = (m_1 q + m_1^2 f + 2m_2 f)/(m_1^2 + 2m_2 + q). \tag{7}$$

Some examples below choose r^* so that $E_p(r) = r^*$, i.e., $f = 0$ in the above. In that case, Eq. (7) reduces to the simple form $d = m_1 q/(m_1^2 + 2m_2 + q)$. This special case leads to $\rho = 0$, i.e., the two score dimensions are uncorrelated.

The above expressions for d relate to our earlier comment about the bivariate score allowing the analysis to more flexibly reflect actual attitudes to risk through the direct specification of expected target scores $(m_1, m_2)'$, any choice of which directly determines risk tolerance. Some examples follow.

3.2.3 Examples: Tilting Vectors and Risk Tolerance

Numerical examples highlight the relationships between target expected scores and the implied tilting vector $\boldsymbol{\tau}$ and associated risk tolerance $d = \tau_1/\tau_2$.

Specifying target expected scores $\mathbf{m}$ under $f(\cdot|\cdot)$ can be simply addressed through utilizing percent improvements over the initial expectations under $p(\cdot|\cdot)$. Figures 1, 2, and 3 display values of τ_1, τ_2, and $d = \tau_1/\tau_2$ for a few practical values of $m = r^* = E_p(r)$ and $q = V_p(r)$ and against different levels of percent target improvements in each of the score dimensions; the percent levels are denoted by m_1^* and m_2^*, where $100m_1^*$ is the percent improvement in expected return over m and $100m_2^*$ is the percent improvement in $-E_p[(r - r^*)^2]/2$. The general lack of correlation between the two score dimensions is due to the lack of prior score correlation. Then, the risk tolerance-level τ_1/τ_2 may become large for small values of m_2^* and large values of m_1^*, indicating that these values should be carefully set to properly constrain risk.

4 Case Study: Data, Models, and Portfolio Construction

4.1 Setting and Data

The study involves an extension of the example applying BPDS to portfolio analysis in Tallman and West (2023), with more recent data, adapted models, and an in-depth investigation of the BPDS hyperparameters used.

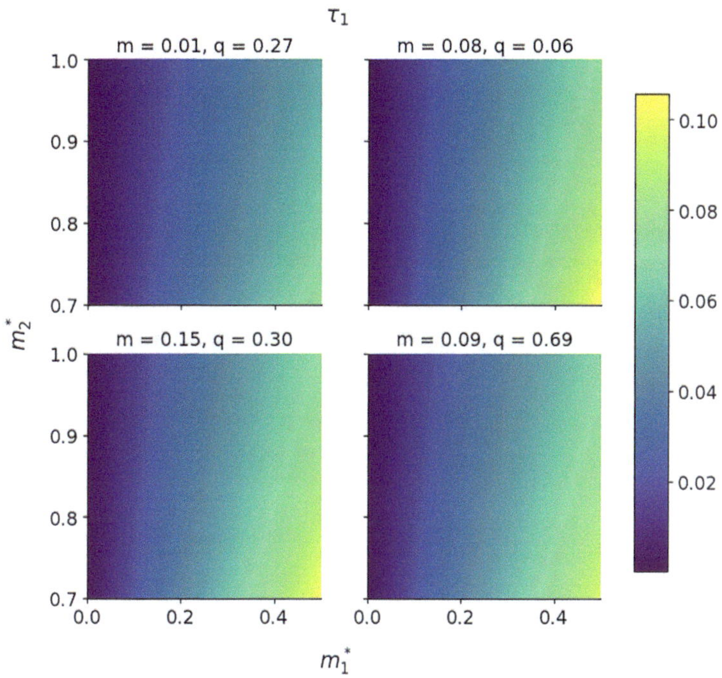

Fig. 1 τ_1 across ranges of score improvement percentages m_1^* and m_2^*, and $m = E(r), q = V(r)$

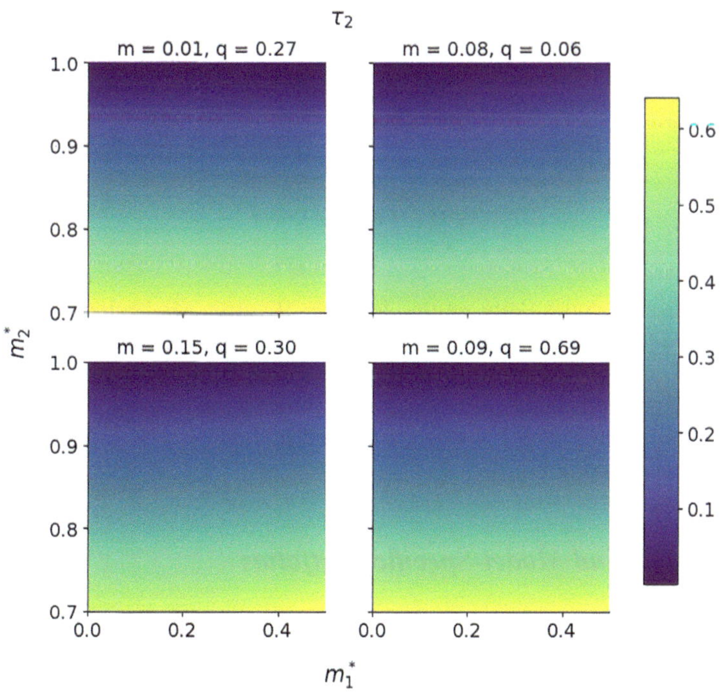

Fig. 2 τ_2 across ranges of score improvement percentages m_1^* and m_2^*, and $m = E(r), q = V(r)$

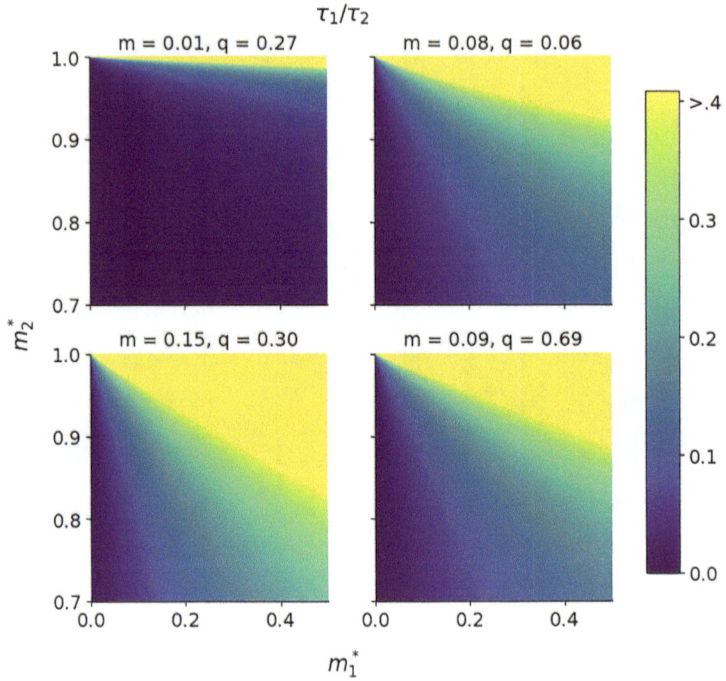

Fig. 3 τ_1/τ_2 across ranges of score improvement percentages m_1^* and m_2^*, and $m = E(r), q = V(r)$

Table 1 FX series in the portfolio case study

Ticker	Currency	Ticker	Currency
AUD	Australian Dollar	JPY	Japanese Yen
EUR	Euro	NOK	Norwegian Krone
NZD	New Zealand Dollar	ZAR	South African Rand
GBP	UK Pound Sterling	CHF	Swiss Franc
CAD	Canadian Dollar		

The dataset includes daily returns for $q = 9$ currencies beginning in January 2001 and ending in December 2021. The motivation for this asset set is to ensure model performance is not anchored in the performance of the market indices used in Tallman and West (2023), with the goal of providing additional value beyond the market return. The $q = 9$ FX assets are noted in Table 1 with cumulative returns over the time period in Fig. 4.

4.2 Models and Model-Specific Portfolios

The $\mathcal{M}_j$ are time-varying vector autoregressive (TV-VAR) models applied to log FX prices. Details of these standard models are in Tallman and West (2023, Appendix) with full theory in Prado et al. (2021, sect. 10.8). On day $t - 1$ each

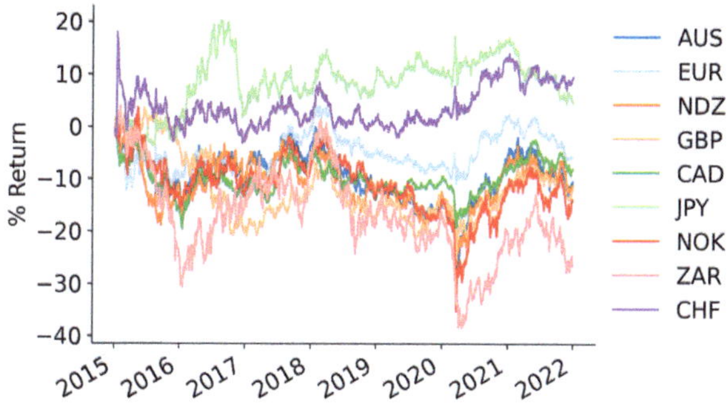

Fig. 4 Cumulative FX returns for the $q = 9$ currencies

model predicts the vector $\log(\mathbf{p}_t)$, where $\mathbf{p}_t$ is the vector of asset prices on day t. These transform to returns via $\mathbf{y}_t = \mathbf{p}_t/\mathbf{p}_{t-1} - 1$.

The data up to the beginning of 2015 is used to fit a large set of models. Data over 2015–2018 inclusive is then used to select a subset of "better-performing models." The final period 2019–2021 inclusive then defines a holdout test period. The larger initial set contains models of AR order $p \in \{1, 2, 3\}$, reflecting potential $1-3$ day momentum effects in FX prices. In each model, each univariate series y_{ti} is predicted by all values of $\mathbf{y}_{t-p:t-1}$ with differing coefficients for each asset. These coefficients evolve over time as multivariate random walks; each model uses a state evolution discount factor $\delta = 0.9995$ to govern the degree of these changes. The model set is further expanded with a range of possible discount factors β governing the evolution of the time-varying volatility matrix in each model (Aguilar & West, 2000; Irie & West, 2019; Prado et al., 2021, chaps. 9 & 10). Discount factor values $\beta \in \{0.94, 0.98, 0.995\}$ define differing degrees of change in levels of volatility of each of the assets over time, as well as—critically for portfolio analysis—in the interdependencies among the assets as represented by time-varying covariances, with smaller values allowing for more adaptation. This gives nine different model parameter (β, p) combinations.

In each TV-VAR model $\mathcal{M}_j$, standard Bayesian forward-filtering analysis applies, and forecasting for daily portfolio rebalancing uses Monte Carlo samples of the predictive distributions for log-prices mapped to the mean vector and covariance matrix on the percent return scale. Model-specific decisions $\mathbf{x}_{tj}$ use standard Markowitz mean-variance optimization: These minimize predicted portfolio variances among sum-to-one portfolios with daily expected returns constrained to specified target levels r^* (e.g., Prado et al., 2021, sect. 10.4.7). The initial set of $J = 27$ BPDS model/decision pairs is completed by the use of three different benchmark targets $r^* \in \{0.05, 0.10, 0.15\}$ coupled with each of the nine TV-VAR models. The actual target return for each model $\mathcal{M}_j$ on any day t is $r^*_{tj} = \min\{\max(\mathbf{f}_{tj}, 10^{-6}), r^*\}$, where $\mathbf{f}_{tj}$ is the mean vector of predicted returns. This adaptive return target helps down-weight desired returns when the predictions favor small values; this helps to prevent large portfolio positions that would result from using the benchmark target. It also ensures that a portfolio that will lose money is never targeted in expectation. On each day, the resulting optimized portfolio weight vector $\mathbf{x}_{tj}$ is thus a vector of asset weights that has two constraints: They sum to one, i.e., $\mathbf{x}_{tj}'\mathbf{1} = 1$, and they have the specified target $E_{p_j}(\mathbf{x}_{tj}'\mathbf{y}_t|\mathcal{M}_j, \mathbf{D}_{t-1}) = r_{tj}^*$ for the expected return that day.

4.3 General Comments on TV-VAR Modeling of FX Series

While the choice of the $\mathcal{M}_j$ is not the main theme of this chapter, we comment further on our use of TV-VAR models for log FX prices. In contrast to a common approach in mathematical finance that emphasizes models of empirical returns (e.g., Tsay, 2010, and the references therein), models for prices directly can sometimes better exploit market fluctuations to generate improved short-term predictive value and decision outcomes. Daily FX prices often evidence short-term "momentum" effects, drifting up or down over a few days, while being relatively stable over other periods and then sometimes more erratic. This can be modeled in returns if desired but is sometimes more sensitively represented and interpretable on the price scale. Informed interventions are encouraged in Bayesian forecasting (e.g., Harrison & West, 1985; West & Harrison, 1986; West & Harrison, 1989b; West & Harrison, 1997, chap. 11), and then interpretation of model parameters is particularly key. Further, differencing to use returns can obscure small changes that can be of practical import in resulting portfolio decisions. This is exacerbated by the fact that differencing also induces artificial lag-1 dependencies that can additionally obscure real changes in cross-series volatility and trends.

TV-VAR models for log or raw prices embed simpler models for returns as special cases. Consider a TV-VAR model for log prices where the intercept is close to 0, the TV-VAR lag-1 coefficient is close to 1, and the TV-VAR coefficients at higher lags are also close to 0. This defines a model that is approximately that of zero mean returns, where the returns are (as usual) defined as first differences of log prices. In the limit with an exact 0 or 1 coefficient, this canonical theoretical model is recovered exactly. The TV-VAR modeling strategy thus allows an empirical evaluation of the relevance of the traditional return-focused model. Standard Bayesian model comparison based on model marginal likelihoods (products of one-step ahead predictive p.d.f.s evaluated at the outcomes e.g., West & Harrison, 1997, sect. 12.2) generally favor TV-VAR lags 1 or 2 on daily FX data, indicating statistical support for the more general, encompassing TV-VAR model.

Typical patterns of estimated TV-VAR coefficients for any daily FX series j are as follows. First, the lag-1 autoregressive coefficient is close to 1 and highly significant, while the lag-2 coefficient is small, though over some periods quite significant. Time trajectories of these coefficients of lagged series j on itself over 1 and 2 days tend to be fairly stable except in really volatile market periods. Second, the lag-1 and lag-2 coefficients of the model for series j on the other FX series are typically inferred to be small for most series, with some being clearly statistically significant alone and with strong *negative dependence* between the two lags. This latter point is key: In periods of stability of FX prices over a few days, these lagged predictors have little or no effect with contributions from 1- and 2-day lags canceling each other; however, in periods of either a short-term directional drift in prices or a more erratic, volatile period, the model can more adequately predict relative to a model that differences away the subtle 1 and 2 day effects. Even though inferred lag-2 coefficients (and most cross-series lag-1 coefficients) are small, their combined predictive impact can be meaningful. Finally, in periods of more major changes in financial markets, the lag-1 and lag-2 coefficients representing cross-series relationships can markedly respond.

The FX case study involves a time period with changes in both volatilities of individual series and in cross-series relationships—the latter both contemporaneous and lagged. Simple models of returns that do not admit the potential for both drifts over time and more abrupt changes in volatility and cross-series parameters are naturally disadvantaged in terms of predictive ability and hence in resulting portfolio decisions. Then, coupled with statistical support for models with 1- or 2-day lags on a purely predictive basis is the empirical performance of resulting portfolios, discussed below.

4.4 Selection of Models for BPDS Analysis

From the initial set of 27 models, a greedy strategy defines a subset of "good" models from analysis over the initial time period 2015–2018. This is to reduce cross-model dependencies in the resulting smaller set and to focus on some of the potentially more useful predictive and decision models. This is done sequentially, reducing from the full model set at each step. At each step, (i) consider the set of remaining models that have historical daily returns whose empirical correlation with those of each of the models already selected is lower than a defined *correlation bar* taken as 0.95, (ii) among these, identify the model with the highest realized 1-day portfolio Sharpe ratio (SR—computed as realized daily mean return divided by realized daily return standard deviation over the historical data, multiplied by $\sqrt{252}$ to transform to the annual scale), and (iii) add this model to the selected set until there are no models left with a positive Sharpe ratio or small enough correlation.

As the models share the same mathematical foundation, there is a high level of cross-model dependence in the initial set, so a rather high correlation bar is used to yield several models in the selected set. The bar at 0.95 led to $J = 7$ selected models with variability in the defining model order and volatility matrix discount factor parameters. This suffices for our study; the goal is not necessarily to create an optimal model set that will lead to the highest returns, but rather a well-performing model set for demonstrating aspects and benefits of BPDS in portfolio decisions.

One relevant detail to note is that the returns for January 15, 2015 were removed from the analysis. That was the day the Swiss Franc (CHF) was decoupled from the Euro, which resulted in an immediate gain of nearly 30% relative to the US dollar. This led to extreme returns across all models and greatly down-weighted the impact of performance throughout the rest of the test period. To avoid biasing model

Table 2 Annualized values of realized Sharpe ratios (SRs) and maximum daily return correlation ρ with other selected models within the set for the $J = 7$ models selected based on data over 2015–2018

β	r^*	p	SR	ρ	β	r^*	p	SR	ρ
0.995	0.15	1	0.41	0.89	0.98	0.15	2	0.20	0.95
0.97	0.15	1	0.40	0.89	0.995	0.05	3	0.11	0.95
0.995	0.1	2	0.25	0.95	0.97	0.1	3	0.10	0.92
0.995	0.15	3	0.22	0.95					

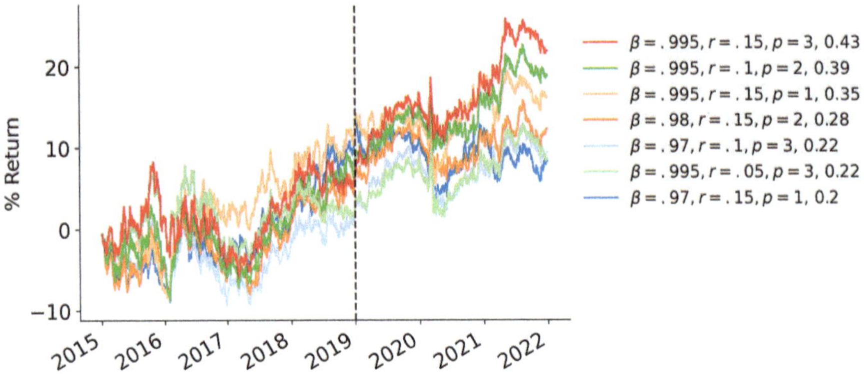

Fig. 5 Cumulative portfolio returns over 2015–2018 for the $J = 7$ selected models. The empirical SR for each is displayed after the label in the legend indicating the model-specific parameters $(\beta, r = r^*, p)$

selection on this single-day event, this day is ignored in the analysis. The selected models, their Sharpe ratios, and maximum return correlations with other selected models over 2015–2018 are shown in Table 2. Cumulative returns and Sharpe ratios from portfolios under each of these models over the entire period are shown in Fig. 5.

5 Initial BPDS Analysis

5.1 Portfolio Setting: Forecast Horizon and Portfolio Targets

We first revisit examples in Tallman and West (2023) with our new data and expanded model set. The score function is $\mathbf{s}_{tj}(\mathbf{y}_t, \mathbf{x}_{tj}) = [\mathbf{x}_{tj}'\mathbf{y}_t, -(\mathbf{x}_{tj}'\mathbf{y}_t - r_t^*)^2/2]'$, with further relevant details discussed in Sect. 2 above. The r_t^* in the first dimension is dropped for simplicity and ease of specification for the target score improvement in terms of a percentage. As r^* is just a constant with respect to $\mathbf{y}_t$ and $\mathbf{x}_{tj}$, this does not change any previous results and has no effect on the resulting analysis. Note that, unlike the settings in Tallman and West (2023), r_t^* now depends on t, selected such that $r_t^* = \sum_{j=0:J} \pi_{tj} r_{tj}^*$, a dynamic target based on the specified targets in the underlying models. This does not affect the mechanics of the BPDS methodology; it simply ensures that BPDS is more adaptive to exploit the fact that the model targets are now changing over time. Several BPDS specifications are evaluated to help understand how BPDS is applied and to explore robustness. One main variant explored is the portfolio forecast horizon. With this motivation, results from evaluating both $h = 1$ and $h = 5$ day ahead portfolios are included. The 1-day portfolios are the portfolios introduced previously, calculated using the predicted next-day returns. The 5-day portfolios are made by fitting each TV-VAR model to daily returns, as in the 1-day case, but then forecasting asset returns 5 days into the future. This gives predictions $p_{tj}(\mathbf{y}_t|\mathcal{M}_j, \mathcal{D}_{t-1})$, where now $\mathbf{y}_t$ is redefined as $\mathbf{y}_t = \mathbf{p}_{t+4}/\mathbf{p}_{t-1} - 1$. The moments of this distribution are then used to define 5-day ahead portfolios with the same sum-to-one constraint along with the revised target expected return in $\mathcal{M}_j$ as $\min\{\max(\mathbf{f}_{tj}^{(5)}, 10^{-6}), r_{tj}^*\}$, where $\mathbf{f}_{tj}^{(5)}$ is the mean forecast vector 5-days ahead. Though calculated using a 5-day portfolio, these portfolios will still be updated each day; a given portfolio will only be held for a single day before rebalancing. The interest in 5-day-ahead analyses is partly contextual in that such portfolios can have better risk profiles and smaller movements in portfolio weights. Also, models might be more easily differentiated, as single-day forecasts tend to be more similar across models, while the longer horizon can show more distinctions.

Further explorations consider ranges of chosen percent target score improvements, with BPDS target expected scores set as the element-wise product vector $\mathbf{m}_t = \bar{\mathbf{m}}_t * \mathbf{m}^*$, where $\bar{\mathbf{m}}_t = E_p[\mathbf{s}_{tj}(\mathbf{y}_t, \mathbf{x}_{tj})|\mathcal{D}_{t-1}]$ and $\mathbf{m}^* = (m_1^*, m_2^*)'$ for all combinations of $m_1^* \in \{1.05, 1.3\}$ and $m_2^* \in \{0.8, 0.90, 0.95, 0.99\}$. The initial probabilities are set using discounted BMA probabilities (Zhao et al., 2016); that is, $\pi_{tj} \propto \pi_{t-1,j}^{\alpha} p_j(\mathbf{y}_{t-1}|\mathcal{D}_{t-2})$ at each t. The discount factor $\alpha = 0.8$ provides discounting of historical model weights between days and hence—relative to standard BMA which has $\alpha = 1$—analysis avoids the eventual degeneration of model probabilities. The BPDS portfolio is calculated using Markowitz optimization, just as in each of the models, but now with a target return of $r_t^* + d_t$ with risk tolerance $d_t = \tau_{t1}/\tau_{t2}$. This is the value of r that maximizes $\boldsymbol{\tau}_t'\mathbf{s}_t(r) = \tau_{t1}r - \tau_{t2}/2(r - r_t^*)^2$, the weighted sum of the score function utilizing BPDS weighting vector $\boldsymbol{\tau}_t$. We additionally restrict the tilting vector such that $d_t < r_t^*$ so that the portfolio target can be at most doubled; this is in order to obviate large values earlier seen in Fig. 3 and is enforced in the computations to evaluate $\boldsymbol{\tau}_t$ at each time point.

5.2 Some Summary Results

Comparative results on cumulative returns and Sharpe ratios are in Fig. 6. Additional comparisons are with the discounted BMA analysis ($\pi_{t0} = 0$, $\boldsymbol{\tau}_t = \mathbf{0}$ for all t, discount factor $\alpha = 0.8$, and target return r_t^*), simply called "BMA." We use this version as a comparison since our initial mixture is rooted in the discounted BMA weights, so any changes in BPDS relative to the BMA comparison will be because

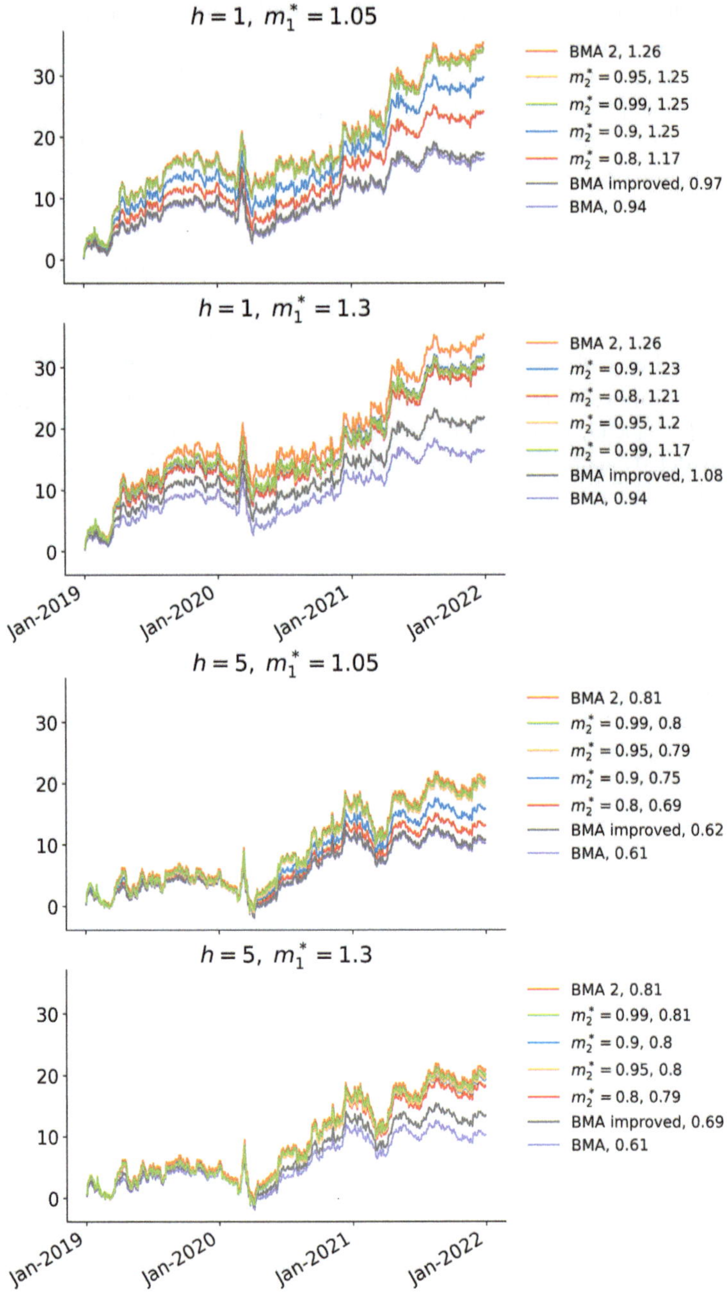

Fig. 6 Cumulative returns under BPDS with discounted BMA initial probabilities and return target $r_t^* + \tau_{t1}/\tau_{t2}$ for a variety of score targets and subject to $\tau_{t1}/\tau_{t2} < r_t^*$. This analysis covers the secondary testing period 2019–2021 with portfolio horizon $h = 1$ (upper) and $h = 5$ (lower). Annualized values of realized Sharpe ratios are displayed after the model labels in the legends

of the BPDS procedure and not the discount factor. Then, in an effort to advantage BMA we added extensions using target returns $(1 + m_1^*)r_t^*$ and $2r_t^*$, labeled as "improved BMA" and "BMA 2," respectively. These connect to the fact that m_1^* is the desired percent return score improvement in BPDS, and $2r_t^*$ is the upper bound on $r_t^* + \tau_{t1}/\tau_{t2}$ which defines the target return for BPDS.

BPDS is able to achieve superior cumulative returns and Sharpe ratios compared to BMA and "improved BMA," across all settings of $\mathbf{m}_t$. The 5-day portfolios improve over BMA comparisons but generally see relatively lackluster results compared to the 1-day portfolios. This is mainly due to the use of initial BMA-based weights. BMA focuses wholly on one-step ahead predictive accuracy, so the initial probabilities used here in BPDS generally favor models based on predictive performance at this short-term horizon. This represents disconnect in BPDS when focused on 5-day portfolios. A relevant extension would be weighting models using 5-day forecast accuracy/likelihood weightings. Then, there is no obvious pattern of improvement across different values of m_{t1}^* and m_{t2}^*, partly due to the restriction placed on $d_t = \tau_{t1}/\tau_{t2}$ that emphasizes control on the risk:reward behavior regardless of the target setting and drives most of the results.

BPDS outcomes are generally close to those of BMA 2 but do not empirically improve upon the latter in this first set of comparisons. This indicates that most of the performance gains come from the target increase and that the tilting and variation in the target increase are not as helpful. However, note that the methodology for setting the score improvement here is rather simplistic, using a grid of percent improvements. In particular, this does not consider the underlying distribution of the score function which implies, for example, that a 10% decrease in variance could be a fairly large jump, whereas a 5% increase in the target return itself is a fairly small improvement. This observation is key to the methodological extension in the next section that explicitly involves consideration of aspects of the predictive distributions for scores in setting the BPDS targets and yields substantial advances in BPDS performance and dominance over the BMA extensions.

6 Practical Portfolio BPDS: Structured Score Targets

6.1 Score Standardization and Eigenscore Targets

Recent use of BPDS in economic applications in Chernis et al. (2024) introduced target expected score specifications that reflect the importance of relative scales and dependencies among the elements of each $\mathbf{s}_{tj}(\mathbf{y}_t, \mathbf{x}_{tj})$ under the initial distribution $p_t(\mathbf{y}_t, \mathcal{M}_j|\mathcal{D}_{t-1})$ at each time $t-1$. We have discussed aspects of this above in Sect. 3.2.1 and revisit it here. A key point is to exploit aspects of the distribution of scores to guide choices of relevant targets that are not wildly inconsistent with the distribution. This aids in avoiding overly aggressive or inconsistent targets such as those highlighted in the naive score targets used in the previous section.

At each time $t-1$, write $\bar{\mathbf{m}}_{tp}$ and $\mathbf{V}_{tp}$ for the mean vector and variance matrix of $\mathbf{s}_{tj}(\mathbf{y}_t, \mathbf{x}_{tj}) = [\mathbf{x}_{tj}'\mathbf{y}_t, -(\mathbf{x}_{tj}'\mathbf{y}_t - r_t^*)^2/2]'$ under $p_t(\mathbf{y}_t, \mathcal{M}_j|\mathcal{D}_{t-1})$. With eigendecomposition $\mathbf{V}_{tp} = \mathbf{E}_t\mathbf{D}_t^2\mathbf{E}_t'$ where $\mathbf{D}_t = \text{diag}(\delta_{t1}, \delta_{t2})$ with positive elements, the *eigenscore* $\mathbf{D}^{-1}\mathbf{E}_t'[\mathbf{s}_{tj}(\mathbf{y}_t, \mathbf{x}_{tj}) - \bar{\mathbf{m}}_{tp}]$ is a *standardized score vector* with zero mean vector and identity variance matrix. Specified shifts in each dimension of the standardized score vector are now on the same scale and so directly comparable, while the orthogonality implies that there will be limited interaction between them in the ensuing BPDS analysis. Hence it is natural and intuitive to define BPDS analysis based on the specification of *standardized expected scores* that then map back to the original score scale. That is, with a specified target vector $\mathbf{m}_{t\epsilon} = (m_{t\epsilon 1}, m_{t\epsilon 2})'$ on the standardized scale, the implied target score on the original scale is $\mathbf{m}_t = \bar{\mathbf{m}}_t + \mathbf{E}_t\mathbf{D}_t\mathbf{m}_{t\epsilon}$.

We know from Sect. 3.2.1 that $\mathbf{V}_{tp}$ is typically close to diagonal (and exactly diagonal in some cases) with a much larger variance on the first score element than the second. That is, $\mathbf{E}_t \approx \mathbf{I}$ or exactly $\mathbf{I}$, and $\delta_{t1} \gg \delta_{t2}$. Hence $\mathbf{m}_t \approx$ (or exactly $=$) $\bar{\mathbf{m}}_t + \mathbf{D}_t\mathbf{m}_{t\epsilon} = (m_{t\epsilon 1}\delta_{t1}, m_{t\epsilon 2}\delta_{t2})'$. So, with comparable values of the two standardized score targets, the fact that $\delta_{t1} \gg \delta_{t2}$ implies a much greater—and undesirable—increase of the target expected return (first element of $\mathbf{m}_t$) than the risk (second element of $\mathbf{m}_t$). Further insights come from theoretical results in Tallman and West (2023, sect. 3.4) that define approximate expressions for $\boldsymbol{\tau}_t$ when specified target scores are "small deviations" from $\mathbf{m}_{tp}$. In the current setting that theory leads to $\boldsymbol{\tau}_t \approx \mathbf{D}_t^{-1}\mathbf{m}_{t\epsilon} = (m_{t\epsilon 1}/\delta_{t1}, m_{t\epsilon 2}/\delta_{t2})'$. This shows that, for given $\mathbf{m}_{t\epsilon}$ and knowing $\delta_{t1} \gg \delta_{t2}$, there is much greater shrinkage toward zero of τ_{t1} than of τ_{t2}. This implies that a much larger improvement is needed in the return dimension to achieve a similar level of tilting as in the risk dimension.

An initial use of the standardization concept in this setting explores $\mathbf{m}_{t\epsilon} = \epsilon\mathbf{1}$ for some small $\epsilon > 0$ defining equal increments in each score dimension. To assess relevant values of ϵ, we begin with a focus on the target excess return itself. Set $m_{t1} = \bar{m}_{t1} + \phi|\bar{m}_{t1}|$ for some ϕ such that $0 < \phi \ll 1$, so that 100ϕ represents the percent target increase over the initial expected return. Equating this to $m_{t1} = \bar{m}_{t1} + \epsilon_t\delta_{t1}$ yields the (small) $\epsilon_t = \phi\bar{m}_{t,1}/\delta_{t1}$. The implication for the target for the risk element of the score vector is that $\bar{m}_{t2} + \phi\bar{m}_{t1}\delta_{t2}/\delta_{t1}$. This target score strategy is now used in FX reanalysis.

6.2 FX Portfolio Analysis Revisited

6.2.1 Example Analysis with Eigenscores

The analysis of Sect. 5 is repeated, now setting BPDS score targets with the eigenscore strategy. We compare results under a range of percent improvements 100ϕ on the expected return. Note that in this setting there is the result that the implied risk tolerance level $d_t = \tau_{t1}/\tau_{t2} \approx \delta_{t2}/\delta_{t1}$. As a result, the BPDS target return improvement will generally be constant across ϕ, and any improvements for two different values of ϕ will be the result of the tilting and not due to different increases in the BPDS target return.

The analysis is repeated for all $\phi \in \{0.01, 0.025, 0.05, 0.1, 0.15, 0.2\}$. Cumulative returns and Sharpe ratios are shown in Fig. 7. Results are shown from analysis using the approximation $\boldsymbol{\tau}_t \approx (m_{t\epsilon 1}/\delta_{t1}, m_{t\epsilon 2}/\delta_{t2})'$ and from exact calculations; this aids in empirical investigation of the accuracy of the approximation. Resulting differences in $d_t = \tau_{t1}/\tau_{t2}$ and $\boldsymbol{\tau}_t$ between the approximate and exact analyses are in Figs. 8 and 9.

Portfolio results under all BPDS specifications far exceed those under BMA and the improved BMA approaches, using either the exact or approximate calculations of $\boldsymbol{\tau}_t$. The 1-day portfolios have superior results in terms of both returns and Sharpe ratios, again mainly due to the lack of connection between the initial weights and the BPDS score in the 5-day case. There are some small differences between the exact and approximate versions for larger values of ϕ. This is expected as higher ϕ corresponds to higher values of ϵ, and the theoretical approximation is for "small ϵ." The high increases in returns are partly shown in Fig. 8; the return target is drastically increased, leading to a target of a nearly 0.35% daily return. This aggression in terms of seeking high returns leads to the extreme returns seen in Fig. 7. Since the Sharpe ratios are normalized, this BPDS specification can be realistically compared with BMA despite the difference in return scales, and the results demonstrate realistic and marked improvements using BPDS.

6.2.2 Repeat Analysis Using Modified Target Scores

The above results are very positive in terms of BPDS improving decision outcomes. However, the large values of risk tolerance $d_t = \tau_{t1}/\tau_{t2}$ generated overly optimistic return targets. Knowing that $d_t = \delta_{t2}/\delta_{t1}$ and that this ratio cannot be changed by altering ϕ, a modified choice of standardized target score is suggested to address this. This modifies the target to $\epsilon(1, c)'$, where $c > 0$ represents and induces differences in the degrees of tilting in the return and risk dimension of the score. In particular, $c > 1$ increases the role of the risk element and hence can balance an overly optimistic return target. Numerical examples below explore this, with c chosen based on the initial test data analysis; its value is determined so that the average over the test time period of generated values of d_t is equal to d^*, a chosen average threshold on the risk tolerance measure.

The examples here are based on choices $d^* = 0.05h$ and $d^* = 0.1h$, linked to the $h = 1$ and $h = 5$ day horizons for portfolio targets. This implies that, on average over the initial test dataset, the return target increase will be less than either 0.05 or 0.1 for the 1-day portfolio. This will lead to a more practically reasonable target compared to the aggressive 0.35 daily target in the analysis of the previous section, an analysis in which the tolerance was unconstrained. This leads to $c = 7.4$ and $c = 3.7$, respectively, when $h = 1$, and $c = 3.25$ and $c = 1.625$ when $h = 5$. Portfolio outcomes shown here are based on analyses using the exact calculation of

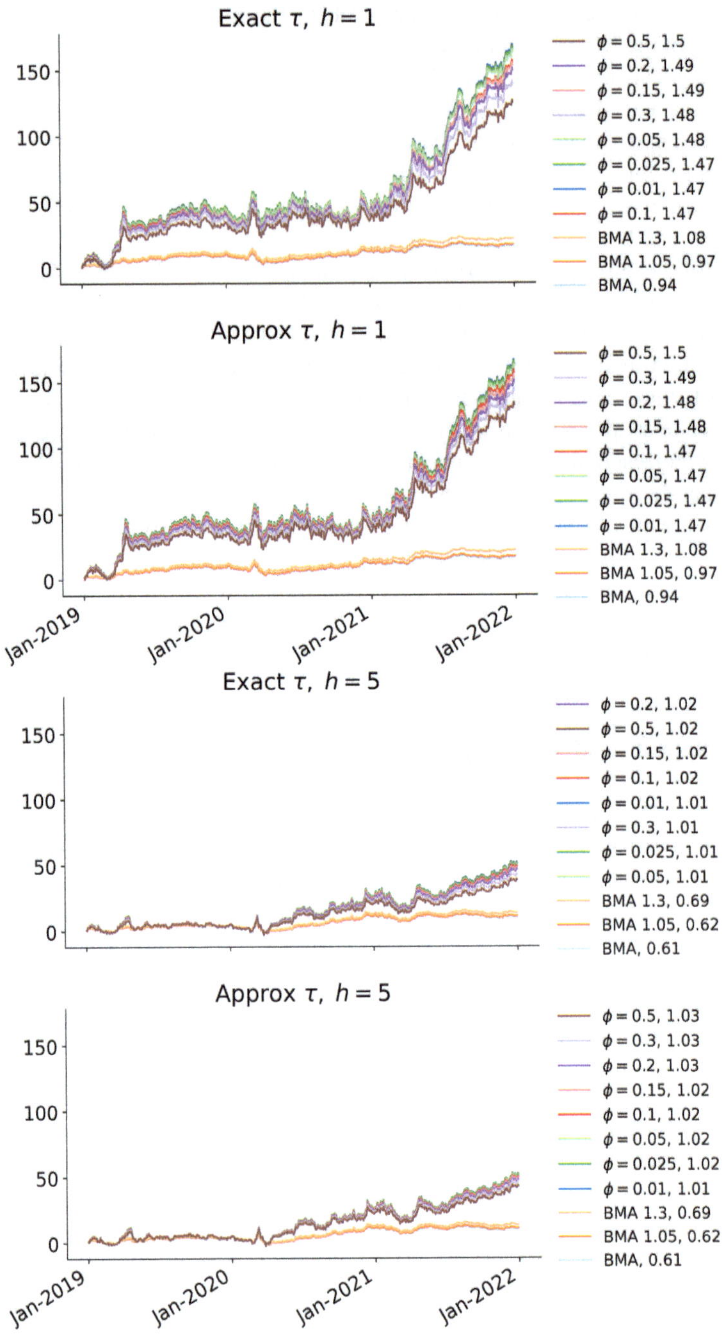

Fig. 7 Cumulative returns in BPDS eigenscore analysis for different values of target expected return defined by ϕ. Results are from the period 2019–2021, for 1-day (upper) and 5-day (lower) portfolios, and with portfolio return target $r_t^* + \tau_{t1}/\tau_{t2}$. Annualized values of realized Sharpe ratios are displayed after each label in each of the figure legends

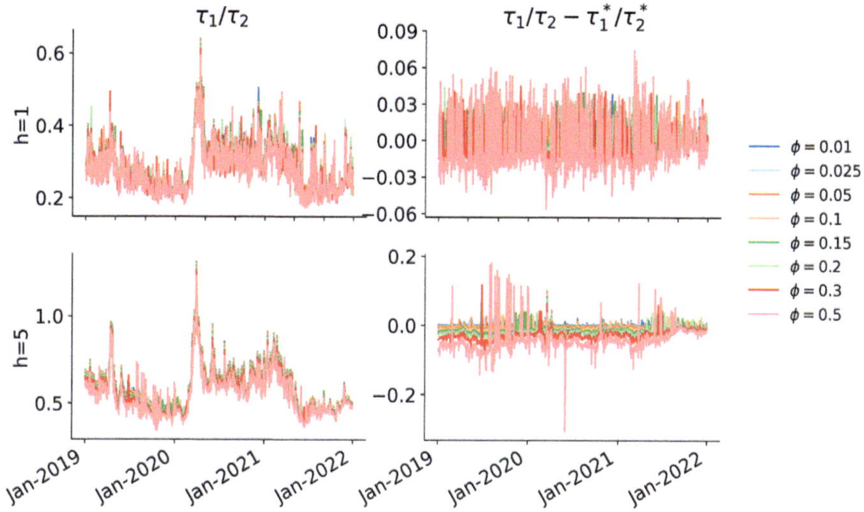

Fig. 8 Return target increase τ_{t1}/τ_{t2} (left) and the difference between the exact τ_{t1}/τ_{t2} and the approximation τ_{t1}^*/τ_{t2}^* (right) using eigenscore BPDS for a variety of ϕ

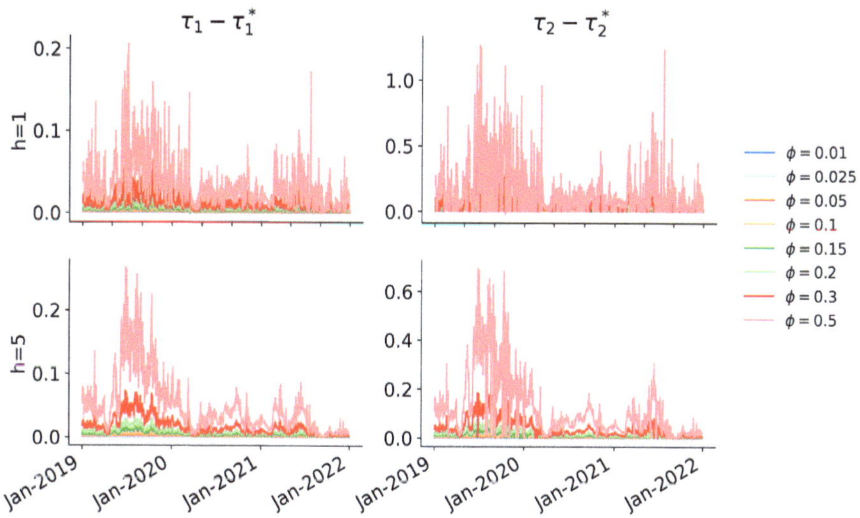

Fig. 9 Difference between the exact $\boldsymbol{\tau}_t$ and the approximation $\boldsymbol{\tau}_t^*$ using eigenscore BPDS for various ϕ

$\boldsymbol{\tau}_t$; the approximate values are now used only as starting points for the optimization to evaluate $\boldsymbol{\tau}_t$. Realized cumulative returns and Sharpe ratios are shown in Fig. 10.

Realized cumulative returns are now on a much more practically reasonable scale, while still maintaining a nice risk-adjusted return. All BPDS specifications for the 1-day portfolios are seen to substantially improve over BMA relative to the initial exploration in Sect. 5. Note also that the empirical cumulative returns under the 5-day strategy are more comparable with those under the 1-day strategy than in the earlier, unconstrained analysis. That said, there is still room for improvement in the multi-day strategies as they are here still using initial model probabilities based on variants of BMA-based weights. As noted earlier, the latter are wholly based on one-step forecast accuracy. This is a main question in terms of future development of more appropriate initial model weightings, among which will be to adapt prior goal-focused scoring approaches to refine the specification of the initial π_{tj} (Lavine et al., 2021; Loaiza-Maya et al., 2021).

There is some—though relatively limited—variation in empirical returns as ϕ is varied, again due almost wholly to differences in resulting tilting. Small values of ϕ are empirically preferable in terms of Sharpe ratios, contrasting with the results from the naive analysis of the previous section. This again demonstrates the importance and value of "small" changes in BPDS target score relative to initial predictions.

Exploration of realized characteristics of portfolio decisions is naturally tied to the time period chosen and displayed. The cumulative return trajectories and Sharpe ratios shown in Fig. 10 are for the full 3-year test period. While not a main point in terms of developing and illustrating the BPDS methodology, it is of contextual interest to look at outcomes over different time periods. Our choice of 2019–2021 inclusive is contextually interesting due to the major differences in economic and market behavior over these years, and some insights are generated by looking at year-specific results. As a snapshot, Table 3 lists realized Sharpe ratios for each specific year under the 1-day portfolio strategy. The main points to note are (i) the year-to-year differences, with 2020 being a standout in terms of relatively poor outcomes across models, and (ii) the strong consistency in Sharpe ratios within each year as ϕ varies to define different BPDS targets.

As already emphasized, the empirical returns and the overall Sharpe ratio under BPDS as shown in Fig. 10 are substantially superior under any of the BPDS variants relative to BMA. The Sharpe ratios per year under BMA are relatively competitive with the exception of 2021, likely because BPDS is rooted in BMA through the initial weights, except for the diffuse baseline. In 2019, BPDS aligned with BMA in terms of performance. In 2020, BPDS struggled to improve over the lackluster BMA because of the minimal information contained in the models and the BMA initial weights. In 2021, BPDS has greatly increased performance, potentially as information stabilized post-pandemic. Regardless, BPDS is able to achieve at least the performance of the initial weighting scheme and also demonstrates the potential to vastly outperform it. Note that these realized Sharpe ratios represent just one aspect of "performance." Additional metrics highlighting differences between the strategies—including purely predictive comparisons in the full space of returns defined by models for $\mathbf{y}_t$—will be of interest in further comparisons.

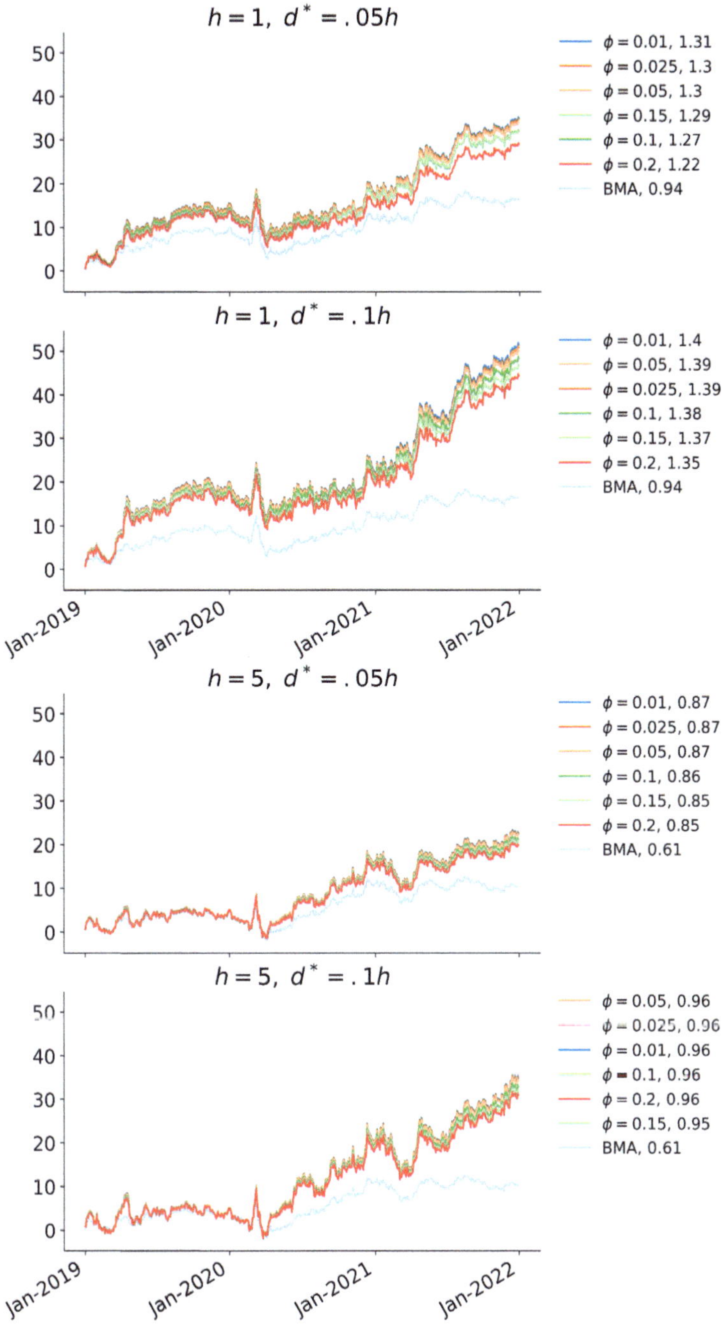

Fig. 10 Cumulative returns using standardized target improvement and return/risk upper constraint d^*. Results are from the period 2019–2021, for 1-day (upper) and 5-day (lower) portfolios, and with portfolio return target $r_t^* + d_t$ with $d_t = \tau_{t1}/\tau_{t2}$ constrained to be less than d^*. Annualized values of realized Sharpe ratios are displayed after ϕ in each legend

Table 3 Annual values of realized Sharpe ratios using standardized target improvement and return/risk constraint d^* for $h = 1$

	$d^* = 0.05$			$d^* = 0.10$		
	2019	2020	2021	2019	2020	2021
$\phi = 0.01$	2.21	0.21	1.98	2.14	0.16	2.34
$\phi = 0.025$	2.21	0.21	1.97	2.14	0.15	2.33
$\phi = 0.05$	2.19	0.22	1.97	2.14	0.16	2.33
$\phi = 0.1$	2.21	0.21	1.89	2.14	0.18	2.29
$\phi = 0.15$	2.20	0.26	1.90	2.12	0.16	2.28
$\phi = 0.2$	2.21	0.13	1.84	2.16	0.14	2.24
BMA	2.17	0.23	0.88	2.17	0.23	0.88

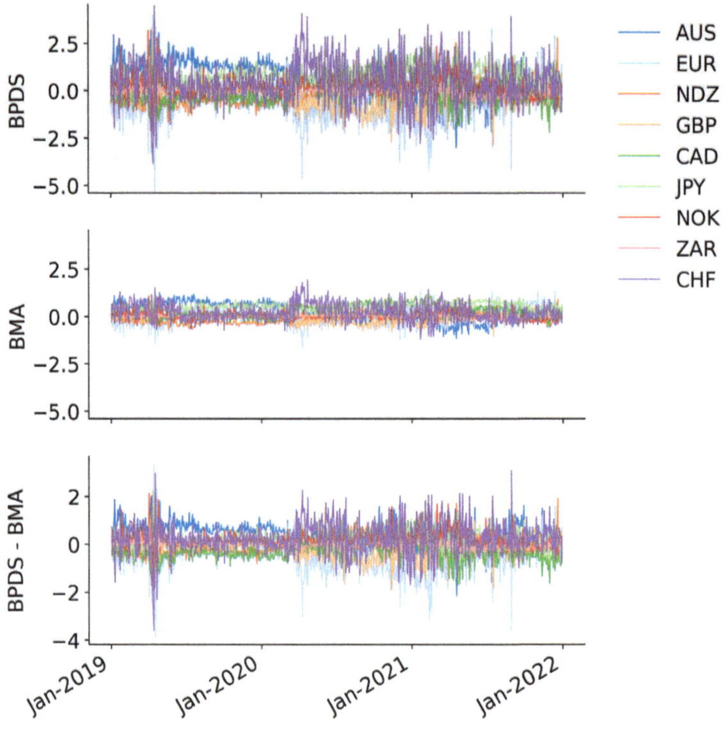

Fig. 11 Trajectories of optimal portfolio weights from constrained BPDS analysis with $h = 1, \phi = 0.01$ and $d^* = -0.1$ (upper) and BMA analysis (middle), together with the differences between weights (lower)

Some of the above discussion is emphasized by viewing how the optimal portfolio weights vary over time, especially in comparing BPDS results with those under BMA. Figure 11 explores this for the constrained BPDS analysis using $h = 1, \phi = 0.01$ and $d^* = 0.1$, i.e., the analysis that is best performing in terms of cumulative returns and Sharpe ratio as seen in Fig. 10. Relative to BMA, the portfolio weights under BPDS are more adaptive and variable over time, consistent with the increased return target and allocation of changing probabilities across models. While the overall patterns over time are similar across series apart from this higher level of variation under BPDS, there are notable differences that partly underlie the improved portfolio performance. BPDS analysis is more active in shorting the Euro and favoring the "safe-haven" Swiss Franc in early 2019, taking similar positions in the early Covid pandemic. In that latter period the BPDS analysis also more heavily shorted CAD relative to BMA's small positive position, likely in part to allow larger positive bets on the Swiss Franc. There are other notable periods of larger differences, some of which may be interpretable in context but the main point being the greater adaptability and responsiveness of BPDS to recent performance, important events, and local changes in FX market conditions.

7 Summary Comments

BPDS is a foundational framework that enables the integration of expected and historical decision outcomes in predictive model uncertainty settings. As exemplified here, BPDS can potentially improve realized decision outcomes and thus serve as an important tool for portfolio managers for model combination, calibration, and evaluation. The case study developed in this chapter demonstrates the potential for BPDS to improve the trade-off between risk and reward in portfolio analysis. It also provides additional methodology for understanding relevant score functions and setting target scores in the portfolio setting. The results further confirm and expand on the positive results originally reported in Tallman and West (2023). The focused development of customized BPDS target scores and the practical results demonstrate the importance of incorporating information about the underlying score distribution and the contextual interpretation of the elements of the score vector. The examples show the potential for BPDS to improve on current Bayesian model averaging methods in portfolio analysis, with the prospects for more adaptive portfolios that have better resulting portfolio outcomes.

There are various potential future directions for applying BPDS in finance and allied areas. One is in expanding the set of score functions to include other relevant metrics, such as portfolio churn or skewness/kurtosis of the return distribution, and other features of portfolio "paths" over multiple forecast horizons (e.g., Irie & West, 2019; Tallman, 2024). Further investigation of setting BPDS target scores is also relevant, especially toward improved understanding of how small changes in the target score lead to changes in the resulting decisions. It is also of interest to consider utilizing initial weights that are customized and potentially more relevant than those based on simple, following prior literature on formal Bayesian model weighting based on historical outcomes with respect to specific predictive and decision goals (e.g., Eklund & Karlsson, 2007; Lavine et al., 2021; Bernaciak & Griffin, 2024; Chernis et al., 2024, and the references therein).

Acknowledgments The research reported here was performed while Emily Tallman was a PhD student in Statistical Science at Duke University (2000–2024). Tallman's research was partially supported by the U.S. National Science Foundation through NSF Graduate Research Fellowship Program grant DGE 2139754. Any opinions, findings, and conclusions or recommendations expressed in this material are those of the authors and do not necessarily reflect the views of the National Science Foundation.

The authors are grateful to the editors and two anonymous referees for their positive and constructive comments on the original version of this work.

Example code implementing the portfolio BPDS analysis is available and directly accessible at github.com/emtall/BPDS-portfolio or via stat.duke.edu/~mw/softwareetc.html. The original version of the work (April 2024) is at https://arxiv.org/abs/2405.01598.

References

Aguilar, O. & West, M. (2000). Bayesian dynamic factor models and portfolio allocation. *Journal of Business and Economic Statistics, 18*, 338–357.

Andersson, M. K. & Karlsson, S. (2008). Bayesian forecast combination for VAR models. In S. Chib, W. Griffiths, G. Koop, & D. Terrell (Eds.), *Bayesian Econometrics: Advances in Econometrics* (Vol. 23, pp 501–524). Emerald Group Publishing Limited.

Aye, G., Gupta, R., Hammoudeh, S., & Kim, W. J. (2015). Forecasting the price of gold using dynamic model averaging. *International Review of Financial Analysis, 41*, 257–266.

Bernaciak, D. & Griffin, J. E. (2024). A loss discounting framework for model averaging and selection in time series models. *International Journal of Forecasting, 40*, 1721–1733.

Cheng, C., Xu, W., & Wang, J. (2012). A comparison of ensemble methods in financial market prediction. In *2012 Fifth International Joint Conference on Computational Sciences and Optimization* (pp. 755–759). Curran Associates.

Chernis, T., Koop, G., Tallman, E., & West, M. (2024). Decision synthesis in monetary policy. Staff Working Paper 2024–30, Bank of Canada. ArXiv: 2406.03321.

de Finetti, B. (1940). The problem of "full-risk insurances". *Journal of Investment Management, 4*, 19–34. 2006 Translation by Luca Barone.

Demiguel, V., Garlappi, L., & Uppal, R. (2009). Optimal versus naive diversification: How inefficient is the $1/n$ portfolio strategy? *Review of Financial Studies, 22*, 1915–1953.

Eklund, J. & Karlsson, S. (2007). Forecast combination and model averaging using predictive measures. *Econometric Reviews, 26*, 329–363.

Füss, R, Glück, T, Koeppel, C., & Miebs, F. (2024). An averaging framework for minimum-variance portfolios: Optimal rules for combining portfolio weights. *Swiss Finance Institute Research Paper 24–10*. Swiss Finance Institute.

Gruber, L. F. & West, M. (2016). GPU-accelerated Bayesian learning and forecasting in simultaneous graphical dynamic linear models. *Bayesian Analysis, 11*, 125–149.

Gruber, L. F. & West, M. (2017). Bayesian forecasting and scalable multivariate volatility analysis using simultaneous graphical dynamic linear models. *Econometrics and Statistics, 3*, 3–22.

Harrison, P. J. & West, M. (1985). Aspects of Bayesian forecasting. *Journal of the Operational Research Society, 36*, 1155.

Irie, K. & West, M. (2019). Bayesian emulation for multi-step optimization in decision problems. *Bayesian Analysis, 14*, 137–160.

Jacobson, T. & Karlsson, S. (2004). Finding good predictors for inflation: A Bayesian model averaging approach. *Journal of Forecasting, 23*, 479–496.

Johnson, M. C. & West, M. (2023). Bayesian predictive synthesis with outcome-dependent pools. Submitted for publication *ArXiv:1803.01984*.

Kan, R. & Zhou, G. (2007). Optimal portfolio choice with parameter uncertainty. *Journal of Financial and Quantitative Analysis, 42*, 621–656.

Karlsson, S. (2013). Forecasting with Bayesian vector autoregression. In G. Elliott & A. Timmermann (Eds.), *Handbook of Economic Forecasting* (Vol. 2, pp 791–897). Elsevier.

Lavine, I., Lindon, M., & West, M. (2021). Adaptive variable selection for sequential prediction in multivariate dynamic models. *Bayesian Analysis, 16*, 1059–1083.
Loaiza-Maya, R., Martin, G. M., & Frazier, D. T. (2021). Focused Bayesian prediction. *Journal of Applied Econometrics, 36*, 517–543.
Markowitz, H. (1952). Portfolio selection. *The Journal of Finance, 7*, 77–91.
Markowitz, H. (2006). De Finetti scoops Markowitz. *Journal of Investment Management, 4*, 5–18.
McAlinn, K. (2021). Mixed-frequency Bayesian predictive synthesis for economic nowcasting. *Journal of the Royal Statistical Society (Ser C), 70*, 1143–1163.
McAlinn, K., Aastveit, K. A., Nakajima, J., & West, M. (2020). Multivariate Bayesian predictive synthesis in macroeconomic forecasting. *Journal of the American Statistical Association, 115*, 1092–1110.
McAlinn, K. & West, M. (2019). Dynamic Bayesian predictive synthesis in time series forecasting. *Journal of Econometrics, 210*, 155–169.
Pettenuzzo, D. & Ravazzolo, F. (2016). Optimal portfolio choice under decision-based model combinations. *Journal of Applied Econometrics, 31*, 1312–1332.
Prado, R., Ferreira, M. A. R., & West, M. (2021). *Time Series: Modeling, Computation and Inference* (2nd ed.). Chapman & Hall/CRC Press.
Quintana, J. M. & West, M. (1987). An analysis of international exchange rates using multivariate DLMs. *The Statistician (Now: Journal of the Royal Statistical Society, Ser D), 36*, 275–281.
Rubinstein, M. (2006). Bruno de Finetti and mean-variance portfolio selection. *Journal of Investment Management, 4*, 3–4.
Steel, M. F. J. (2020). Model averaging and its use in economics. *Journal of Economic Literature, 58*, 644–719.
Tallman, E. (2024). *Bayesian Predictive Decision Synthesis: Methodology and Applications*. PhD Thesis, Duke University.
Tallman, E. & West, M. (2022). *On entropic tilting and predictive conditioning*. Technical Report, Department of Statistical Science, Duke University. arXiv2207.10013
Tallman, E. & West, M. (2023). Bayesian predictive decision synthesis. *Journal of the Royal Statistical Society (Ser B), 86*, 340–363.
Tsay, R. S. (2010). *Analysis of Financial Time Series*. Wiley Series in Probability and Statistics (3rd ed.). Wiley.
Tu, J. & Zhou, G. (2011). Markowitz meets Talmud: A combination of sophisticated and naive diversification strategies. *Journal of Financial Economics, 99*, 204–215.
Wang, X., Hyndman, R. J., Li, F., & Kang, Y. (2023). Forecast combinations: An over 50-year review. *International Journal of Forecasting, 39*, 1518–1547.
Wang, Y., Ma, F., Wei, Y., & Wu, C. (2016) Forecasting realized volatility in a changing world: A dynamic model averaging approach. *Journal of Banking and Finance, 64*, 136–149.
West, M. (2020). Bayesian forecasting of multivariate time series: Scalability, structure uncertainty and decisions (with discussion). *Annals of the Institute of Statistical Mathematics, 72*, 1–44.
West, M. (2023). Perspectives on constrained forecasting. *Bayesian Analysis, 19*, 1013–1039.
West, M. & Harrison, P. J. (1986). Monitoring and adaptation in Bayesian forecasting models. *Journal of the American Statistical Association, 81*, 741–750.
West, M. & Harrison, P. J. (1989a). *Bayesian Forecasting and Dynamic Models* (1st ed.). Springer.
West, M. & Harrison, P. J. (1989b). Subjective intervention in formal models. *Journal of Forecasting, 8*, 33–53.
West, M. & Harrison, P. J. (1997). *Bayesian Forecasting and Dynamic Models* (2nd ed.). Springer.
Zhao, Z. Y., Xie, M., & West, M. (2016). Dynamic dependence networks: Financial time series forecasting & portfolio decisions (with discussion). *Applied Stochastic Models in Business and Industry, 32*, 311–339.

MIX
Papier aus verantwortungsvollen Quellen
Paper from responsible sources
FSC® C105338

If you have any concerns about our products,
you can contact us on
ProductSafety@springernature.com

In case Publisher is established outside the EU,
the EU authorized representative is:
Springer Nature Customer Service Center GmbH
Europaplatz 3, 69115 Heidelberg, Germany

Printed by Libri Plureos GmbH
in Hamburg, Germany